IEE ELECTROMAGNETIC WAVES SERIES 42

Series Editors: Professor P. J. B. Clarricoats
Professor J. R. Wait
Professor E. V. Jull

Spectral theory and excitation of open structures

Other volumes in this series:

Spectral theory and excitation of open structures

Victor P. Shestopalov
and
Youri V. Shestopalov

The Institution of Electrical Engineers

Published by: The Institution of Electrical Engineers, London,
United Kingdom

© 1996: The Institution of Electrical Engineers

The Institution of Electrical Engineers,
Michael Faraday House,
Six Hills Way, Stevenage,
Herts. SG1 2AY, United Kingdom

British Library Cataloguing in Publication Data

A CIP catalogue record for this book
is available from the British Library

ISBN 0 85296 876 0

Printed in England by Galliard (Printers) Ltd., Norfolk

Contents

Preface

The physical principles of open electrodynamic structures: open resonators, open waveguide resonators, open waveguides and open diffraction gratings are different from those of closed ones because of radiation loss, edges, multi-connected cross-sections, and the necessity to take into account the behaviour of the electromagnetic field at infinity. That is why the spectrum of eigenoscillations is not real, additional demands to the energy relations in different space domains appear, and the statements of spectral problems change. The latter form a special class of non-self-adjoint boundary eigenvalue problems of mathematical physics with a spectral parameter entering in a nonlinear way. Such problems require working out new methods of solution, which have not been elaborated until recently.

This monograph deals with foundations of rigorous spectral theory for open electrodynamic structures constructed by the authors and their co-workers on the basis of the method of operator-valued functions of one or several complex variables. This has been applied for the first time by Shestopalov (1979, 1980a, 1980b) in the studies of the problems of normal waves in transmission lines and developed for the problems of fundamental frequencies in two-dimensional circular open resonators (Poyedinchuk, 1983a, 1983b, Koshparenok *et al.*, 1985, Shestopalov, 1987) open waveguide resonators (Poyedinchuk *et al.*, 1986, Pochanina *et al.*, 1989a, 1989b, Pochanina *et al.*, 1991) and diffraction gratings (Sirenko and Shestopalov, 1985, 1986, 1987a, 1987b, 1987c, Sirenko *et al.*, 1985a, 1985b, 1988, Shestopalov and Sirenko, 1989). The main advantage of this method is connected with the possibility of uniting the homogeneous (spectral) and inhomogeneous (excitation) problems arising in mathematical models of electrodynamics because it allows the simultaneous determination of the domains of spectra localisation (where the excitation problems have no solutions) and the resolvent set of operators (where these problems are uniquely solvable), i.e. to prove the existence and uniqueness of solutions to boundary-value problems.

Various statements of spectral problems for Helmholtz and Maxwell's equations are known together with different "spectral theories". Here we will treat the spectral theory not as the end in itself but as a tool to study the phenomena modelled by the boundary-value problems for these equations. Physical processes are observed at real frequencies and continuation to the complex domain is a mathematical method. The data obtained at complex frequencies gives essential information about "real" solutions at real frequencies. Here it is important to construct analytical continuations of solutions and operators of the problems (considered as a function of the spectral parameter) to an appropriate complex manifold of the spectral parameter variation (and to apply in this way the "correct" conditions at infinity). Then one can naturally use the idea that a holomorphic function is actually determined by its singularities. The character of the latter gives a lot of essential data concerning the properties of solutions and the spectra distributions.

The above conclusions may be considered as an informal scheme for creating the spectral theory. We have tried to realise this programme with comprehensive mathematical completeness.

From the viewpoint of applications one should note that the spectral problems considered within the frame of this theory are reduced to determination of characteristic numbers of operator-valued functions which admits effective analytical and numerical treatment on the basis of the mathematically justified notions of the generalised dispersion equation and implicit operator-valued functions. The complex "roots" of these equations (and the corresponding eigenfunctions) enable one to describe the resonance properties of open structures. Now one can state that the dispersion equations which have been widely used before in the complex domain acquire a distinct physical meaning.

The monograph consists of four parts. Part 1 presents the spectral theory and the methods for solutions to excitation problems for open and shielded cylindrical resonators and waveguides with dielectric inclusions, in particular with noncompact boundaries of the transversal cross-section together with the mathematical background based on the original methods of the spectral theory of operator-valued functions. Part 2 deals with development of the theory for open waveguide resonators. Part 3 considers spectral problems for diffraction gratings and establishes appropriateness connecting free oscillations with the stationary scattering and diffraction.

The authors are grateful to Dr. E. V. Chernokozhin, Professor A. S. Ilyinsky, Dr. A. E. Poyedinchuk, Dr. Y. K. Sirenko and Dr. Y. G. Smirnov who made valuable contributions to the research.

List of notations

E, H	electric and magnetic intensities
$\exp(-j\omega t)$	time dependence
ω	cyclic frequency
$j = \sqrt{-1}$	
$k = 2\pi\lambda$	wave number
λ	wavelength
$\epsilon,\ \mu$	permittivity and permeability of substances
$\epsilon_0,\ \mu_0$	permittivity and permeability of vacuum
n	interior unit normal
$x_1,\ x_2,\ x_3$	Cartesian co-ordinates
$\hat{i}_1,\ \hat{i}_2,\ \hat{i}_3$	unit vectors in x_1, x_2 and x_3 directions
a^*	complex conjugate of a
I	identity operator
A^{-1}	operator inverse to A
δ_{mn}	Kronecker delta
Δ	Laplacian

Introduction

In this monograph we have tried to give from a single point of view the analysis of physical processes in open cylindrical resonators, waveguides and diffraction gratings on the basis of a specially constructed spectral theory.

It is known that studies of open electrodynamic structures may be carried out in terms of the stationary scattering or stationary diffraction. The latter is based on the solution to Maxwell's equations with the time dependence $\exp(-j\omega t)$ and its main purpose is to determine the spatial field structure. The stationary scattering deals with the time evolution of fields and requires the statement of compactly supported initial conditions. Here it is important to determine the parts of the energy of an initial excitation going off to infinity and remaining in a bounded region when $t \to \infty$.

The problems of the stationary scattering and diffraction are connected by means of Laplace's transform calculated with respect to the frequency parameter $\omega\sqrt{\epsilon_0\mu_0}$. The solutions may be obtained as linear combinations of quasi-stationary states (free oscillations and waves) whose spatial field structure does not vary in time and which are described with the help of nontrivial solutions to homogeneous boundary-value problems for reduced Maxwell's equations. Consequently, the connection between the models of scattering and diffraction may be effectively realised if to begin with the solution to spectral problems for Maxwell's equations considered with respect to various appropriate spectral parameters.

In order to illustrate this scheme let us consider the initial boundary-value problem for the system of Maxwell's equations connected with determination of free oscillations of an open resonator placed in the free space

$$\frac{\partial \mathbf{u}}{\partial t} = A\mathbf{u}, \quad t > 0, \qquad \mathbf{u}|_{t=0} = \mathbf{u}_0$$

where

$$A = \begin{bmatrix} 0 & c \cdot \text{curl} \\ -c \cdot \text{curl} & 0 \end{bmatrix} \qquad \mathbf{u} = \begin{bmatrix} \mathbf{E} \\ \mathbf{H} \end{bmatrix}$$

is Maxwell's operator acting in a Banach (Hilbert) space H with a dense domain of definition

$$D(A) = \{\mathbf{u} : A\mathbf{u} \in H, \ \mathbf{u} = \begin{bmatrix} \text{div} \, \mathbf{E} \\ \text{div} \, \mathbf{H} \end{bmatrix} = 0$$

$\mathbf{u}$ satisfies appropriate boundary-value conditions$\}$

$\mathbf{E}$ and $\mathbf{H}$ are vectors of electric and magnetic intensities. The choice of H is governed by the demands of a model, in particular, by the conditions at infinity based on the general energy requirement

$$P_0 = \int (|\mathbf{E}_0|^2 + |\mathbf{H}_0|^2) \, d \, v < \infty, \qquad \mathbf{u}_0 = \begin{bmatrix} \mathbf{E}_0 \\ \mathbf{H}_0 \end{bmatrix}$$

It is natural to assume that free oscillations occurring in the absence of external sources due to the initial energy should attenuate in time.

The initial boundary-value problem for the system of Maxwell's equations is equivalent to

$$p\hat{\mathbf{u}} = A\hat{\mathbf{u}} + \mathbf{u}_0$$

where $\hat{\mathbf{u}}$ is Laplace's transform of $\mathbf{u}$ (calculated under the assumption of its existence) and

$$\mathbf{u}(t) = \frac{1}{2\pi j} \int_{a-j\infty}^{a+j\infty} \exp(pt)\mathbf{u}_0 \, d \, p$$

where the line $\text{Re} \, p = a > s_0 > 0$ does not contain the spectrum of A. The latter describes the time evolution of free oscillations. In order to "calculate" this magnitude one should know the spectrum of A, i.e. the properties of an analytical continuation of the operator-valued function $(pI - A)^{-1}$ considered with respect to the (complex) parameter p.

One of effective ways to solve this problem is in passing to the equation

$$-j\lambda \mathbf{v} = A\mathbf{v}, \quad \mathbf{u}(t) = \mathbf{v} \exp(-j\lambda t)$$

where $\mathbf{v}$ satisfies definite conditions at infinity and may be interpreted as resonator's eigenoscillation corresponding to the fundamental frequency λ. It is natural to assume that $\text{Im}\,\lambda \leq 0$, i.e. the oscillation process attenuates in time. This method is used in the classical problems of the stationary diffraction theory where the fields and sources vary in time as $\exp(\pm j\omega t)$ (ω is a given excitation frequency).

Operators stating the conditions at infinity are nonlinear functions of λ, e.g. in the case of the Reichardt-Sveshnikov radiation conditions for the problem of free oscillations in open cylindrical resonators with compact boundaries of the transversal cross-section it is required that

$$\mathbf{v}(\mathbf{r}) = \sum_{m=-\infty}^{+\infty} a_m H_m^{(1)}(\lambda r) \exp(jm\phi) \quad \text{for sufficiently large} \quad r > 0$$

$[\mathbf{r} = (r, \phi)]$ where (analytical continuation of) Hankel's functions $H_m^{(1)}(z)$ and the operator of the spectral problem are considered on an appropriate Riemann surface. For the considered problems of free oscillations in open cylindrical resonators with compact and noncompact boundaries Λ may be chosen as the infinite-sheet Riemann surface of the analytical continuation of the fundamental solution $H_0^{(1)}(\lambda|\mathbf{r} - \mathbf{q}|)$ coinciding with that of the function $\ln \lambda$.

This approach may be justified within the frames of the following assumptions. Let the spectrum of A consist of eigenvalues forming a discrete set of points with finite multiplicities and $\|pI - A\| \to 0$ uniformly with respect to $\arg\lambda$ over a certain system of circles $C_n : |\lambda| = R_n$ $(n = 1, 2, \ldots)$, $R_n \to \infty$. Then one can expect to obtain an asymptotic expansion

$$\mathbf{u}(t) \sim \sum_{p_k} c_k \mathbf{w}_k \exp(p_k t), \quad t \to \infty$$

where $\mathbf{w}_k$ are the eigenfunction of A corresponding to the eigenvalues ip_k and c_k is uniquely determined by $\mathbf{u}_0$. Hence the time evolution of a free oscillation may be described by eigenfunctions and eigenvalues of Maxwell's operator if one can effectively determine its (point) spectrum.

Spectral properties of Maxwell's operator are studied within the frame of the conventional scheme presented by Krasnushkin (1947) and justified by Samarsky and Tikhonov (1947) for empty shielded cylindrical waveguides (tubes). It is assumed that under certain conditions spatial domains bounded by closed or open cylindrical surfaces acquire the properties of (cylindrical) waveguides and (or) resonators, i.e. they may support (free) oscillations and (or) running (normal) waves (eigenwaves). As has been indicated above, propagation and oscillations of electromagnetic fields in an isotropic medium are described by the system of Maxwell's equations

$$\begin{cases} \text{curl } \mathbf{E} = -j\omega\mu\mathbf{H} \\ \text{curl } \mathbf{H} = j\omega\epsilon\mathbf{E} \end{cases}$$

where ϵ and μ are dielectric permeability and permittivity of the medium which do not vary along the longitudinal co-ordinate x_3. To consider cylindrical resonators and waveguides (adding to the system of Maxwell's equations the boundary-value conditions for $\mathbf{E}$ and $\mathbf{H}$ on perfectly conducting surfaces, on the boundaries between different media and the conditions at infinity in open domains) and to look for solutions of Maxwell's equations in the form

$$\mathbf{E}, \mathbf{H} = \mathbf{E}, \mathbf{H}(\mathbf{r}) \exp(j\gamma x_3)$$

we get the problem of normal waves propagating along a cylindrical guiding surface and corresponding to certain complex values of the longitudinal wavenumber γ, which is regarded as a spectral parameter. For $\gamma = 0$ we have the specific problem of determining the solutions which do not depend on x_3 and correspond to certain complex values of the spectral parameter ω which are called fundamental frequencies.

A great number of papers and works are devoted to the investigation of fundamental frequencies and normal waves in different types of resonators and waveguides [a detailed list of references is presented, for example, by Ilyinsky and Shestopalov (1989)]. We shall not make a review of this literature because it is a very specialised and complicated matter, turning readers' attention only to several recent works by Shestopalov (1991, 1993), Ilyinsky and Ena (1991), Smirnov (1993) and Lozhechko and Shestopalov (1994c, 1995) developing the method of operator-valued functions for solving nonlinear eigenvalue problems in electrodynamics which is applied in this monograph. One has to acknowledge the studies of the method of boundary integral equations with strongly elliptic Fredholm operators for different types of boundary-value problems for the Helmholtz equation in Sobolev spaces [see von Petersdorff (1989) and the references in this paper].

Most of the results which have been obtained until recently are related to shielded waveguides (transmission lines), open waveguides with compact boundaries or with noncompact boundaries having a given (co-ordinate) shape. In this monograph we will study rather general families of both shielded waveguides and open structures with noncompact boundaries of cross-sections which have not been treated before generalising the classical results of Samarsky and Tikhonov (1947), Tikhonov and Samarsky (1977) (see also Ilyinsky *et al.*, 1993) concerning the existence of normal waves in empty tubes and based on the methods of the spectral theory of self-adjoint operators. In fact, among the main features of the boundary eigenvalue problems arising in the theory of guided waves in open structures with inhomogeneous dielectric filling are the following: it is not possible to carry out correct mathematical treatment of eigenvalue problems using standard approaches of the spectral theory of differential operators because the corresponding boundary-value problems are non-self-adjoint (and for open structures they are stated in unbounded domains with irregular boundaries containing "edges") and their boundary "transmission" operators have rather complicated character. In addition to this, radiation conditions at infinity, which model radiation losses, depend on a spectral parameter. The above conclusions were the main reasons to develop the spectral theory of open structures presented in this monograph.

The essence of this theory is that the spectral problems for a wide class of two-dimensional open resonators, open waveguides and gratings are equivalently reduced by the methods of the Riemann-Hilbert problem (Shestopalov, 1970), boundary integral equations and operator pencils to the problems on characteristic numbers for canonical Fredholm operator-valued functions $I - B(\lambda)$ [in particular, by separation of variables in local co-ordinates for open resonators with circular cross-sections and equivalent regularisation carried out by Poyedinchuk (1984)], which in several cases turn out to be finite-meromorphic with respect to the spectral parameter. An equivalent reduction of spectral problems to Fredholm equations opens the way for elaborating effective analytical and numerical methods for the determination of the point spectrum of Maxwell's operator. Such methods are presented in this book. It is established that the calculation of the spectrum of fundamental frequencies and normal waves can be carried out by mathematically justified algorithms for which the estimation of the convergence rate is obtained.

Various excitation (diffraction) problems for open resonators and gratings with different types of inclusions (together with the determination of Green's functions for the corresponding boundary-value problems) have been considered. Let us note only some of them. The excitation problems for open circular resonators are reduced to the second kind uniquely solvable Hilbert-Schmidt system of coupled functional equations with trigonometric kernels. For numerical computations the truncation method is applied and an asymptotic estimate for the accuracy of approximate solutions is obtained. Excitation problems for a wide family of open resonators with noncompact boundaries are reduced in the generalised statement to (abstract) Fredholm operator pencils and it is proved that the uniqueness may be violated only in the lower half-plane of the complex frequency parameter (on the first sheet of the Riemann surface).

The spectral theory of open electrodynamic structures (determination of their free oscillations) is constructed in the monograph by the method of analytical continuation of the classical diffraction problems (Reichardt, 1960, Wineberg, 1972, Sanchez-Palencia, 1980) to the complex domain of the frequency spectral parameter (or of the spectral parameter corresponding to the longitudinal wavenumber in the problems of normal waves which are not directly related to any "classical" excitation problems). The most important problem here is the replacement of the Sommerfeld radiation conditions by the radiation conditions at infinity which could be applied for complex values of a spectral parameter varying in a complex manifold. The classical diffraction problem is stated at real frequencies as a boundary-value problem with the Sommerfeld radiation conditions and then its analytical continuation to the Riemann surface Λ is constructed with the help of the Reichardt-Sveshnikov radiation conditions at infinity. Such a "continued" problem turns out to be uniquely solvable in the whole Riemann surface except for not more than a countable set of (spectral) points $\lambda = \lambda_n$ (situated out of the closed upper half-plane of the first sheet of Λ) which are the finite-order poles of the analytical continuation of the resolvent of the operator corresponding to the initial boundary-value problem.

There is of principle methodological difference in the solution of spectral and diffraction (excitation) problems. Let us illustrate this difference taking as an example a diffraction problem whose solution admits an integral representation

$$u(\mathbf{r}) = \int_L z(\mathbf{q})\Phi(\mathbf{r},\mathbf{q};\omega)\,\mathrm{d}\,l_{\mathbf{q}}$$

where $\Phi(\mathbf{r},\mathbf{q};\omega)$ is a given (kernel) function, $z(\mathbf{y})$ is the sought for function. Other types of solution's representations may be obtained if to expand the function sought in terms of appropriate (basis) functions. The boundary-value conditions yield the equation for u

$$\gamma z(\mathbf{r}) + \int_L z(\mathbf{q})\Phi_1(\mathbf{r},\mathbf{q};\omega)\,\mathrm{d}\,l_{\mathbf{q}} = F(\mathbf{r})$$

where $\Phi_1(\mathbf{r},\mathbf{q};\omega)$ and γ are given. If one can solve this integral (functional) equation and prove its equivalence to the (uniquely solvable) diffraction problem, the (unique) solution u to the latter is obtained in the form of the potential presented above.

In the case of a spectral (homogeneous) problem the notion of uniqueness becomes invalid and the equivalence between the homogeneous operator equation (with $F \equiv 0$) and the initial spectral problem plays a crucial role. The former, in accordance with the idea of the method, has no independent role and is considered as a tool for the solution to the spectral problem. One could recall in this way the "classical" dispersion equations of the theory of planar dielectric waveguides what makes clear why we call such homogeneous equations arising in the spectral theory the generalised dispersion equations. In terms of the spectral theory the equivalence means that the set of the "roots" of generalised dispersion equations (or characteristic numbers of the corresponding operator-valued functions) coincides with the set of eigenvalues of the initial spectral problem. The proof of equivalence is often neglected (i.e. the operator equations are considered without establishing a connection with initial eigenvalue problems) what yields incorrect results.

It is natural to consider the operators of generalised dispersion equations as functions of several complex variables where one is chosen as a spectral parameter and the rest which describe the shape and dielectric filling of a resonator or a waveguide may be called nonspectral parameters. In a number of cases the spectral problems are reduced to determination of zeros of infinite determinants which may be written in the form $F(w,\omega) = \det[I - A(w,\omega)]$ where ω and w denote a spectral and (one) nonspectral parameter. The equation $F(w,\omega) = 0$ defines implicit functions $\omega(w)$ which may be called generalised dispersion curves (or relations). The study of such generalised dispersion relations enables us to describe a lot of important physical phenomena. It has been shown, in particular, that the smoothness of spectral characteristics as functions of non-spectral parameter(s) may be violated in the "interaction" zone of free oscillations corresponding to different fundamental frequencies, i.e. when they approach each other in the complex space. When introducing the notion of interaction we assume that it is the change in qualitative and quantitative composition of partial components of the eigenoscillation under the conditions close to the existence of other eigenoscillations.

Let us describe some specific classes of open structures treated in the monograph.

Shielded waveguides in which every volume domain violating the regularity of a waveguide channel is considered as an open waveguide resonator are studied as a special class of open waveguides. The diffraction losses in such open waveguide resonators are caused by radiation not to the free space but to semi-infinite regular waveguides. Therefore, quasi-stationary oscillations may be considered here as eigenoscillations and the corresponding frequencies as their fundamental frequencies. The latter are determined by the singularities of the analytical continuation of Green's functions (which is the kernel of the resolvent of the diffraction problem) and, therefore, one can call them waveguide's spectrum. Its properties essentially depend on radiation conditions determining diffraction communication between the volumes of a resonator and surrounding space based on the physical principle of the absence of waves coming at certain frequencies from semi-infinite waveguide channels.

The main theoretical results, which are formulated for open waveguide resonators loaded with two identical parallel-plane waveguides, may be generalised for waveguides with arbitrary number of radiation channels, as well as for rectangular, circular and other types of waveguides.

The spectral theory of open waveguide resonators have been developed by Kirilenko *et al.* (1974, 1979), Yashina (1978), Kirilenko (1978, 1980), Rud' (1984, 1985), Kirilenko and Rud' (1980), Kirilenko and Masalov (1982), Rud' *et al.* (1986), Poyedinchuk *et al.* (1986), Shestopalov (1987) and Sirenko *et al.* (1987).

Diffraction gratings are considered in the monograph as a special class of open structures with noncompact (periodic) boundaries on the basis of the study of corresponding boundary eigenvalue problems for homogeneous Maxwell's equations with respect to the complex frequency spectral parameter in the complete statement, i.e. with appropriate radiation conditions which take into account periodicity and admit analytical continuation to a complex domain (to the Riemann surface of the analytical continuation of the specially introduced canonical Green's function).

The uniqueness theorems for excitation problems have been proved and the foundations of the spectral theory of diffraction gratings have been laid by Sirenko and Shestopalov (1985) Sirenko *et al.* (1985) and Sirenko (1986). The general scheme of investigating the spectral problems and the basic results presented in these works (reduction to the problems on characteristic numbers for canonical Fredholm operator-valued functions, spectra discreteness for the free oscillations and general description of the domains of its localisation on the Riemann surface) are similar to those obtained for open resonators and waveguide resonators.

Owing to the results of solution to the spectral problems, various resonance and anomalous phenomena have been explained which are observed in the diffraction problems when a spectral parameter is close to its resonant values. The dynamic theory of gratings has been constructed (see the monograph by Sirenko and Shestopalov, 1989).

Let us indicate the special importance of the results obtained within the frame of the spectral theory for the studies of holographic gratings, in particular, with continuous and piece-wise constant refraction indices. Spacious calculation and theoretical material is presented in the papers by Sirenko (1978, 1983a, 1983b), Sirenko and Shestopalov (1982, 1987), Yatsik (1988) and Sirenko *et al.* (1990).

The study of open structures with noncompact nonperiodic boundaries of the cross-section occupies a special place in the monograph. We consider (in Chapters 2 and 3) several broad families of open cylindrical resonators and waveguides formed by perfectly conducting plane screens with arbitrary finite perturbations in the form of arbitrary cavities filled with piece-wise homogeneous medium, (layered) dielectric "lugs" situated on the screen, dielectric pivots and strips placed over the screen. They have distinct resonance properties caused by the presence of cavities which enable one to effectively control their scattering characteristics. An image slot line formed by a plane screen with longitudinal slot and a rectangular waveguide beneath may serve as a "canonical" example. Section 1 is specially devoted to the detailed description of such structures. The only general demand here is that the cross-sections of the considered open structures may differ from the half-plane only inside the finite ball and all dielectric inclusions lie inside this ball.

A mathematical model for the study of free oscillations, normal waves, scattering (inverse scattering) and diffraction of electromagnetic waves has been constructed for such open structures by Lozhechko and Shestopalov (1994a, 1994b, 1994c, 1995) on the basis of the special form of the Reichardt-Sveshnikov radiation conditions at infinity proposed by Ilyinsky and Shestopalov (1989). The discreteness of the free oscillations and normal wave spectra has been proved. Investigations of these open structures essentially complement both the "mathematical" and "physical" parts of the monograph. In fact, the results obtained within the frame of this model (in particular, the existence and uniqueness theorems for several types of excitation problems) are based on considerations of the problems in the generalised statement and the technique and methods applied essentially differ from the approaches using boundary integral equations or infinite matrix operators.

Asymptotic methods developed in Chapter 1 on the basis of the results obtained by Chernokozhin and Shestopalov (1995) enable one to obtain the explicit formulae for determination of fundamental frequencies with guaranteed accuracy in the form of asymptotic series. Such exact and reliable analytical results are of principal importance for practical calculations of normal waves, fundamental frequencies and other characteristics of guiding systems and resonators. They allow one to guarantee the accuracy and to decrease computation time in some cases up to more than two orders (from minutes to seconds). These results naturally form the basis of several applied programs for numerical analysis of transmission lines and other types of guiding, scattering and resonance systems.

Part 1 Foundations of spectral theory

The approximate approach which prevailed during the last two decades in the theory of open resonators was based on the assumption that the dimensions of mirrors and distance between them are much less than the wavelength. The corresponding heuristic and asymptotic methods (of geometric and physical optics, the beam method, the methods of parabolic equation and model problems etc.) allow description of only a part of the fundamental frequencies spectrum related to "bouncing ball" or "whispering gallery" oscillations. It was quite sufficient for the needs of coherent optics where laser devices were applied. Since the above assumption does not hold in the millimetre and submillimetre wavelength ranges, such asymptotic methods turn out to be unacceptable. The situation becomes sharper when dielectric inclusions are placed in resonator's volume and especially when their dimensions are commensurable with the wavelength. Hence the creation of rigorous mathematical models which do not require *a priori* assumptions on the dimensions of open resonators and dielectric inclusions becomes the barest necessity.

Determination of the resonance characteristics of wave fields having differing nature (electromagnetic, acoustic) as well as the modelling of open (closed) resonators, waveguides, transmission lines, multi-layered dielectric waveguides etc. are the classical branches of modern mathematical physics which have been developed since the beginning of this century. Nevertheless, there are still many gaps both in the theory and availability of reliable methods for effective solution to various applied problems. In fact, from the mathematical viewpoint, the spectral characteristics of wave fields are directly connected with the eigenvalues of certain (non-self-adjoint) boundary-value problems.

Questions inevitably arise concerning the existence of these eigenvalues and their properties as functions of various parameters, as well as effective ways of eigenvalues calculation. These problems remain unsolved for a majority of open structures.

The basic idea of the approach developed recently in a number of books and articles [see, e.g. the review in the book by Ilyinsky and Shestopalov (1989)] is to collect spectral data of guiding systems and resonators by solving such eigenvalue problems (on normal waves and fundamental frequencies) which turn out to be nonlinear with respect to the spectral parameter $z = \gamma$ or $z = \omega$, where ω is a frequency of electromagnetic field and γ is a longitudinal wavenumber. They are reduced to the problems of characteristic numbers (or generalised dispersion equations) determining $z = z(\mathbf{a})$ implicitly as a function of the vector $\mathbf{a}$ of nonspectral parameters characterising the considered structure [one may call $z(\mathbf{a})$ generalised dispersion curves]. In particular, one of the $\bar{a}$ components (a_0) may serve as a small parameter (e.g. a_0 may be chosen as the diameter of a boundary irregular perturbation) and the spectral data is obtained as asymptotic series with respect to a_0 (see Section 1.9). A sufficient volume of spectral data enables one to reconstruct all the most important characteristics of the structure (device).

The main goal of our studies is thus to elaborate mathematical models of the wave and oscillation processes on the basis of operator methods for the analysis of "nonlinear" eigenvalue problems, which have been originally designed as the most convenient and universal tool for solving the multi-parameter spectral problems arising in the models of wave processes.

In the first Part we develop rigorous mathematical methods for solving the boundary-value problems arising in the mathematical models of the problems of free and forced oscillations in open cylindrical resonators and eigenwaves in open waveguides. The essence of the models is in reduction to the problems of characteristic numbers in the form of a generalised dispersion equation for the multi-parameter operator-valued function (see Section 1.4).

Chapter 1 deals with the studies of the spectral and excitation problems for open resonators. This Chapter also contains the majority of mathematical methods of spectral and potential theory originally developed by the authors which are applied in the monograph.

Consideration of the spectral and excitation problems in the generalised statement and equivalent reduction to (abstract) operator pencils allow us to extend the results (in Chapter 2) to a wide family of open resonators with noncompact boundaries of the cross-section. A lot of numerical data concerning various resonance effects in different types of open resonators is also presented in this Chapter.

Chapter 3 considers eigenwaves in open waveguides. The universal methods presented are applicable to all known types of transmission lines and dielectric waveguides. Special attention is paid to plane slotted structures. The completeness of the system of normal waves of an arbitrary shielded waveguide with arbitrary inclusions (in the form of dielectric pivots and strips) together with the proof of existence and general description of the distribution of the eigenwave spectrum generalise the classical results of Samarsky and Tikhonov (1947) obtained for empty shielded waveguides.

Chapter 1

Spectral theory and excitation of open resonators

Let us clarify first of all the state of the problem, presenting a brief survey of the known approximate models for open resonators with dielectric inclusions.

Approximate Fredholm integral equations have been obtained by Fox and Lee (1961) with the help of the Kirchhof method for resonators with rectangular, circular and parabolic (perfectly-conducting) mirrors. This work has generated a series of papers. Boyd and Gordon (1961) have solved the Fredholm equation analytically in spherical co-ordinates. Boyd and Kogelnik (1962) have classified open resonators with respect to their diffraction losses (they have stated that there are domains of high and low losses for different distances between the mirrors). Streifer and Gamo (1966) have reduced approximate integral equations for non-confocal resonators to infinite-matrix equations (unfortunately, they have not presented an explicit form of basis functions). Buldirev and Fradkin (1964) have given a mathematically consistent deduction of the boundary integral equations and verified the domains of the variation of resonator's parameters where the model Boyd and Kogelnik (1962) may be approximately applied. These conditions have an asymptotic character and do not allow qualitative estimation of the inaccuracy of approximate models. Let us note that the equations obtained by Buldirev and Fradkin (1964) are also approximate and cannot be used for the description of spectral characteristics of open resonators in the resonant wavelength range.

The methods of geometric optics have been applied by Bikov (1965, 1966,

1968) for resonators with elliptic mirrors. Unlike the approach proposed by Buldirev and Fradkin (1964), the corresponding "obvious" models neglect diffraction losses and enable one to calculate only approximate values of the real parts of the "bouncing ball" eigenfrequencies.

Approximate representations for the real parts of fundamental frequencies have been obtained by Weinstein (1966) with the help the parabolic equation method for confocal, concentric and strip resonators.

Influence of various inhomogeneities (shifts and warps of mirrors etc.) on the spectral characteristics of open resonators have been investigated with the use of different approximate approaches by Averbakh *et al.* (1965), Vlasov and Talanov (1965), Weinstein (1966), Gloge (1966), Lazutkin (1968), Boitzov and Fradkin (1968), Voitovich and Nefedov (1970), Epishin (1978) and others. In several cases accurate results have been obtained by the methods of perturbations and parabolic equation. The typical difficulty here is determination of eigenvalues of complex matrices whose absolute values are close (Nefedov, 1982). But application of the perturbation technique for asymptotic integral equations cannot be mathematically justified by known methods [see, e.g. the book by Kato (1972)] since the corresponding operators are non-self-adjoint and eigenfunctions do not form a basis (Weinstein, 1966).

Further development of asymptotic models for open resonators was carried out mainly within the frames of the same approaches. Slavyanov (1973) proposed a general method for constructing the short-wave asymptotics of the solutions to approximate integral equations which enabled him to explicitly determine the influence of the aberration of mirrors on eigenoscillations. A matrix representation of the operator defined by approximate integral equations has been constructed by Popov (1974, 1977). Floquet's solution of a parabolic equation connected with the eigenfunctions of a "stable" open resonator with infinite mirrors has been obtained by Babich and Buldirev (1972). Such basis-solutions of a parabolic equation are concentrated for an "unstable" resonator in the vicinity of its axis and calculations of diffraction losses are reduced to the search for the roots of infinite determinants. The theory of "mode characteristics" proposed by Harrington and Mautz (1971) and developed by Haikonen (1975) allows one to approximate results without attracting the method of integral equations. Cullen *et al* (1972) and Erickson (1977) applied a variational approach (see also Nikolsky, 1967) which enabled one to make more precise some asymptotic formulae for the real parts of fundamental frequencies. Lower and upper estimates for diffraction losses of confocal open resonators have been verified by Popov (1974) on the basis of the spectral theory of normal non-self-adjoint operators. Approximate formulae for the complex "bouncing ball" fundamental frequencies have been obtained by Lazutkin (1981) but they require specifications in order to be applicable for the calculation of diffraction losses.

There are also various approximate models of open resonators with inclusions, for example, in the form of a dielectric prism (constructed, in particular, with the use of impedance boundary-value conditions) proposed by Vakhitov (1967), Fong and Shung (1971), Nefedov (1979) and others. Determination of the fundamental frequencies in open resonators with dielectric films (Zakharov and Troitsky, 1970), plates (Petrushin and Tretyakova, 1983, Androsov *et al.*, 1983), cylinders (Rajevsky, 1970) and lenses (Boitzov, 1968) have been approximately reduced to different types of dispersion equations.

Spectral characteristics of resonators with inclusions have been investigated by Belostotsky *et al.* (1983) on the basis of the boundary integral equation method developed by Vasiljev and Materikova (1965) by means of the Fock integral equation (Fock, 1970) under the assumption that the size of inclusions are commensurable with the wavelength but much less than that of the mirrors. Approximate analytical results have been obtained for the inclusions in the form of a dielectric ball and circular cylinder.

There are weaknesses in all known approximate models of open resonators with dielectric inclusions.

It is known (Weinstein, 1951) that shielded volume resonators with perfectly conducting walls have the discrete spectrum of fundamental frequencies. It is not possible to prove similar results for open resonators within the frames of the approximate models discussed above, nor to estimate the boundaries of applicability of approximate results and to apply the corresponding approaches in the resonant wavelength range (see Arnold, 1972).

Creation of the new methods for solving non-self-adjoint boundary-value problems in open domain modelling the oscillation processes in open resonators becomes a vital necessity. Foundations of such models have been laid by Feld and Sukharevsky (1969), Shestopalov (1971, 1975, 1983), Sologub (1971), Zakharov and Pimenov (1982) and Ilyinsky and Shestopalov

(1989). Borzenkov and Sologub (1975), Melezhik (1980) and Koshparenok *et al.* (1980) have applied the method of the Riemann-Hilbert problem (Shestopalov, 1971) for the studies of open resonators in the form of perfectly conducting slotted cylinders and strips. Vinogradov *et al.* (1981) and Tuchkin and Shestopalov (1982) developed the semi-inversion technique for some three-dimensional resonators and arbitrary unclosed cylinders. Sikhinin (1981) investigated qualitative problems of the theory of free oscillations in open acoustic resonators formed by a finite number of arbitrary unclosed cylinders. The spectrum discreteness has been proved there by the reduction to Fredholm integral equations with logarithmic kernels.

Chapter 1 deals with foundations of rigorous mathematical models of excitation and oscillation processes in open resonators. The models created are based on the general concepts of analytical continuation to the complex domain of the frequency spectral parameter, equivalent reduction to operator equations where operators are considered as functions of one or several complex variables, and application of the methods of spectral theory of operator-valued functions.

Special attention is paid in the first Sections to the open resonators whose transversal cross-sections are formed by a finite number of closed and unclosed circular arcs with circular dielectric inclusions. The spectral problem for such resonators (i. e. determination of their fundamental frequencies) is reduced with the help of the method of the Riemann-Hilbert problem (Shestopalov, 1971) by variables separation in the local co-ordinates to the problem of characteristic numbers for a canonical Fredholm finite-meromorphic operator-valued function (of the complex frequency spectral parameter varying in the Riemann surface Λ of the analytical continuation of the fundamental solution to the two-dimensional Helmholtz equation) with an infinite-matrix operator having a trigonometric kernel. An equivalent reduction to a Fredholm operator yields the discreteness of the free oscillations spectra (which coincides with the set of eigenvalues of the initial boundary-value problem, i. e. with its point spectrum), as well as the Fredholm property for the excitation problems proved in the same manner. The uniqueness and existence theorems for these problems are then naturally obtained as a result of determining the domains in Λ (subsets of the resolvent set) which do not contain eigenwaves.

Effective numerical methods and algorithms for calculating the characteristic numbers of matrix operator-valued functions have been constructed and justified on the basis of the truncation technique and the theory of "eigenconvergence" proposed by Vainikko and Karma (1974).

The approach applied for circular cylindrical resonators is based on the possibility of a solution's expansion in terms of appropriate combinations of functions forming complete systems in the functional (Hilbert) space naturally connected with edge conditions. A completely different technique using solution representations in the form of (Green's) generalised potentials is applied in Sections 1.7 and 1.8 for the so-called slotted structures with plane transversal geometry. In this way the theory of potentials for a special class of two-dimensional domains has been developed (Section 1.5) together with the elements of the spectral theory of integral operator-valued functions with meromorphic kernels. The results obtained (in particular, for the problems with the Reichardt-Sveshnikov radiation conditions at infinity) enable us to prove the discreteness of the free oscillations spectra and the Fredholm property of excitation problems for a broad family of open slotted resonators.

Asymptotic methods elaborated in Section 1.9 essentially differ from those applied before (see the survey presented above) since they enable one both to prove the existence of and to obtain approximate formulae for complex fundamental frequencies with guaranteed accuracy.

A special place is occupied by the theory of the Morse critical points (Section 1.10) which is used as a tool to explain and justify various resonant phenomena on the basis of the theory of smooth mappings given by the functions of several complex variables.

1.1 Cross-sections of open resonators: canonical structures

Here we will describe the families of two-dimensional (plane) domains forming the cross-sections of cylindrical resonators and waveguides considered in Chapters 1 – 3.

Basic features of the approach presented are the treatment of open cylindrical structures whose cross-sections may be formed by domains with noncompact boundaries, in particular, by combinations of open and closed domains, with application of the Sveshnikov-Riechardt partial radiation conditions which generalise the classical ones used for cylindrical scatterers with compact boundaries. Such problems have been considered before for local bodies and periodical structures (Sveshnikov, 1950, Sveshnikov and Ilyinsky, 1968 and Shestopalov and Sirenko, 1989). On the other hand, the well developed approaches (the methods of integral equations, partial domains etc.) cannot be applied for solving the problems considered due to the complicated character of (non-compact) boundaries.

The structures considered model perfectly conducting surfaces with arbitrary finite perturbations including (layered) dielectric "lugs" and strips placed over the screen, and, in particular, most of the known types of transmission lines and dielectric waveguides. They have distinct resonance properties caused by the presence of cavities which enable one to effectively control their scattering characteristics. Image slot line, formed by a plane screen with a longitudinal slot and a rectangular waveguide beneath it, may serve as an example of such a scatterer.

We will consider resonators and waveguides filled with piece-wise homogeneous medium and formed by perfectly conducting cylindrical screens. The two-dimensional domains forming their cross-sections by the plane $x_3 = 0$ (in the Cartesian co-ordinate system x_1, x_2, x_3) will be conditionally divided into several families. We will describe in what follows the boundaries of these cross-section domains.

The family CB of open domains with compact boundaries

Sub-family CB^e: the boundary is formed by a finite number of closed and (or) unclosed non-intersecting finite piece-wise smooth curves ("strips")

$$S_i \ (i = 1, 2, \ldots, N_S) \quad S_l \cap S_m = \emptyset \ (l \neq m)$$

$$S = \bigcup_{i=1}^{N_S} S_i \subset B_O^R = \{\mathbf{r} = (x_1, x_2) : |\mathbf{r}| = \sqrt{x_1^2 + x_2^2} < R\}$$

Sub-family CB^ϵ: the domain of the family CB^ϵ contains a finite number of "dielectric inclusions" $\Omega_\epsilon = \bigcup_{i=1}^{N_\epsilon} \Omega_{\epsilon_i} \subset B_O^R$, $\Omega_{\epsilon_l} \cap \Omega_{\epsilon_m} = \emptyset, l \neq m$ such that $\epsilon(\mathbf{r}) = \epsilon_0 = \text{const}, \mathbf{r} \notin \Omega_\epsilon, \epsilon(\mathbf{r}) = \epsilon_i(\mathbf{r}), \mathbf{r} \in \Omega_{\epsilon_i}, \mu(\mathbf{r}) = \mu_0 = \text{const}, \mathbf{r} \notin \Omega_\epsilon,$ $\mu(\mathbf{r}) = \mu_i(\mathbf{r}), \mathbf{r} \in \Omega_{\epsilon_i}$, and it is assumed that either $D = \partial\Omega_\epsilon \cap S_m = \emptyset$ for

$m = 1, 2, \ldots, N_{eS} \leq N_S$, or $\partial\Omega_\epsilon \cap S_m = S_m$ for $m = 1, 2, \ldots, N_{\epsilon S}$, $N_{\epsilon S} = N_S - N_{eS}$.

We will assume in what follows that the permittivity and permeability of the medium, $\epsilon(\mathbf{r})$ and $\mu(\mathbf{r})$, are continuous functions inside dielectric inclusions Ω_ϵ with only possible gap lines on their boundaries Γ_ϵ [e.g. $\epsilon(\mathbf{r})$ and $\mu(\mathbf{r})$ may be piece-wise constant functions] and resonator itself is placed in a homogeneous medium with complex permittivity and permeability, ϵ_0, μ_0 (Im $\epsilon_0 \geq 0$, Im $\mu_0 \geq 0$).

It is assumed that the boundaries of the domains from the family CB^ϵ consist of two components, S and Γ_ϵ.

The important particular case here is

Sub-family CB_c^ϵ of the domains with compact circular boundaries when the strips (S) coincide with circular arcs

$$S_i \subset B_{O_i}^{a_i} = \{\mathbf{r} : |\mathbf{r}-\mathbf{r}_i| = \sqrt{(x_1 - x_{1i})^2 + (x_2 - x_{2i})^2} < a_i\}, \quad m = 1, 2, \ldots, N_S$$

and the inclusions (D) are the circles $\Omega_{\epsilon_m} = B_{O_m}^{a_m}$, $m = N_S+1, N_S+2, \ldots, M$, $M = N_S + N_\epsilon$. $O_n = \mathbf{r}_n = (x_{1n}, x_{2n})$ are the origins of the local polar co-ordinate systems (r_n, ϕ_n), $l_{in} = l_{ni}$ is the distance between the origins O_i, O_n, ρ_{in}, ψ_{in} are the co-ordinates of the origin of the nth co-ordinate system with respect to the ith co-ordinate system, $2\theta_i$ is the angle dimension of S_i' which is the supplement of S_i to a complete circle and ψ_i is the orientation angle of the center of S_i' in the ith co-ordinate system $(i = 1, 2, \ldots, M)$. Axes of all the co-ordinate systems are parallel to each other.

CB_c^e denotes the domains whose (compact circular) boundaries consist only of the strips (S) coinciding with circular arcs (without dielectric inclusions).

Families NB of open domains with non-compact boundaries

Sub-family NB_p: the domain of the family NB_p is formed by the half-plane $\Omega^1 = \{\mathbf{r} : x_2 > 0\}$ and several finite domains Ω_i^2, $i = 1, 2, \ldots, N_2$, $\Omega_i^2 \cap \Omega_m^2 = \emptyset, l \neq m$, we put $\Omega^2 = \bigcup_{i=1}^{N_2} \Omega_i^2$ assuming that $\Omega^2 \subset \{\mathbf{r} : x_2 < 0\}$, all the domains are bounded by closed simply-connected piece-wise smooth contours and have different filling: $\epsilon = \epsilon(\mathbf{r}) = \epsilon_n(\mathbf{r})$, $\mathbf{r} \in \Omega^n, n = 1, 2, \epsilon_2(\mathbf{r}) =$

$\epsilon_{2i}(\mathbf{r})$, $\mathbf{r} \in \Omega_i^2$, $\mu = \mu(\mathbf{r}) = \mu_n(\mathbf{r})$, $\mathbf{r} \in \Omega^n$, $n = 1, 2$, $\mu_2(\mathbf{r}) = \mu_{2i}(\mathbf{r})$, $\mathbf{r} \in \Omega_i^2$, Ω^2 and Ω^1 have a common part of the boundaries, an interval $\Gamma = \bigcup_{i=1}^{N_2}(\partial\Omega_i^2 \cap \partial\Omega^1) \in \{\mathbf{r} : x_2 = 0\}$ with several intervals ("slots") Γ_L, $\Gamma_L \subseteq \Gamma$ coinciding with the gap line of the functions $\epsilon(\mathbf{r}), \mu(\mathbf{r})$: $\Gamma_\epsilon = \Gamma_L$.

Sub-family NB_b: the domain of the family NB_p is formed by Ω^1 and Ω^2 where the boundary $\Gamma = \partial\Omega^1$ of the domain Ω^1 differs from the straight line $\{\mathbf{r} : x_2 = 0\}$ on a finite interval, i.e. $\Gamma = \Gamma^R \cup \Gamma^I$, where $\Gamma^R = \{\mathbf{r} : x_2 = 0, |x_1| > a\}$ is the regular part of the boundary formed by two beams, the irregular part $\Gamma^I = \Gamma_M^I \cup \Gamma_S^I$ is an unclosed piece-wise smooth contour, $\Omega^2 = \bigcup_{i=1}^{N_2} \Omega_i^2$, $\Omega^2 \subset R^2 \setminus \bar{\Omega}^1$, the structure of Ω^2 and the dielectric filling are the same as in the family NB_p and a common part of the boundaries of Ω^1, Ω^2 coincides with a part of Γ^I. The gap lines $\Gamma_\epsilon = \Gamma_S^I$ of the functions $\epsilon(\mathbf{r}) = \epsilon_i(\mathbf{r})$, $\mathbf{r} \in \Omega^i$, $\mu(\mathbf{r}) = \mu_i(\mathbf{r})$, $\mathbf{r} \in \Omega^i$, $i = 1, 2$ coincide with a finite number of unclosed contours ("slots") $\Gamma_S^I = \bigcup_{i=1}^{M_S} \Gamma_{Si}^I \subset \Gamma^I$.

Sub-family NB_u: Ω^1 and Ω^2 are unbounded domains with the common boundary $\Gamma = \partial\Omega^1 = \partial\Omega^2$ which differs from the straight line $\{\mathbf{r} : x_2 = 0\}$ on a finite interval ($\Gamma = \Gamma^R \cup \Gamma^I$, $\Gamma^R = \{\mathbf{r} : x_2 = 0, |x_1| > a\}$, $\Gamma^I = \Gamma_M^I \cup \Gamma_S^I$ is an unclosed piece-wise smooth contour) and $\bar{\Omega}^1 \cup \Omega^2 = R^2$. The gap lines Γ_ϵ of the functions $\epsilon(\mathbf{r}) = \epsilon_i(\mathbf{r})$, $\mathbf{r} \in \Omega^i$, $\mu(\mathbf{r}) = \mu_i(\mathbf{r})$, $\mathbf{r} \in \Omega^i$, $i = 1, 2$ coincide with a finite number of unclosed contours ("slots") $\Gamma_S^I = \bigcup_{i=1}^{M_S} \Gamma_{Si}^I \subset \Gamma^I$.

We define the sub-families $NB_b^r \subset NB_b$ and $NB_u^r \subset NB_u$ assuming that the slots (i.e. the multi-component curve Γ_ϵ) are formed by rectilinear intervals $\Gamma_S^I = \bigcup_{i=1}^{M_S} \Gamma_{Si}^I$, $\Gamma_{Si}^I \subset \{\mathbf{r} : x_2 = \text{const}\}$, $i = 1, 2, \ldots, M_S$.

Sub-family NB_c: $\Omega^1 = \Omega$ is assumed to be an arbitrary open domain whose boundary is formed by an unbounded open piece-wise smooth contour $\Gamma = \partial\Omega$ differing from the straight line $\{\mathbf{r} : x_2 = 0\}$ on a finite interval, a multi-connected (perfectly conducting) component in the form of a finite number of closed and (or) unclosed non-intersecting finite piece-wise smooth curves: the "strips" S_i, $i = 1, 2, \ldots, N_1$, $S_l \cap S_m = \emptyset, l \neq m$; $S = \bigcup_{i=1}^{N_1} S_i \subset \Omega^1$; $S \subset \partial\Omega^1$, which have no common points with Γ, and the "apertures" $S_a = \bigcup_{i=1}^{N_a} S_{ai}$ such that each of unclosed finite curves S_{ai} has one common point with Γ. Ω may contain a finite number of finite dielectric inclusions

$\Omega_\epsilon = \bigcup_{i=1}^{N_3} \Omega_{i\epsilon} \subset \Omega$, $\Omega_{l\epsilon} \cap \Omega_{m\epsilon} = \emptyset, l \neq m$, $\epsilon_1(\mathbf{r}) = \epsilon_0 = \text{const}$, $\mathbf{r} \notin \Omega_\epsilon$,
$\epsilon_1(\mathbf{r}) = \epsilon_{1i}(\mathbf{r})$, $\mathbf{r} \in \Omega_i^1$, $\mu_1(\mathbf{r}) = \mu_0 = \text{const}$, $\mathbf{r} \notin \Omega_\epsilon^1$, $\mu_1(\mathbf{r}) = \mu_{1i}(\mathbf{r})$, $\mathbf{r} \in$
Ω_i^1, $\partial\Omega_i^1 \cap S = \emptyset$, $i = 1, 2, \ldots, N_3$ (see Figure 1.1). Here the gap lines Γ_ϵ of
the functions $\epsilon(\mathbf{r})$ and $\mu(\mathbf{r})$ may have a complicated structure, in particular,
$\Gamma_\epsilon \supseteq \partial\Omega_\epsilon$ and the gap lines may contain a finite number of unclosed contours
Γ_i (the "layer" boundaries) connecting a pair of different points on $\partial\Omega$.

The general demand for all the families is that all the inhomogeneities
(dielectric inclusions and irregular parts of the boundaries) lie inside a finite
ball B_O^R for a certain $R > 0$. In other words, unbounded domains forming the
cross-sections of the open structures considered may differ from the half-plane
only inside B_O^R.

It is clear that $NB_p \subset NB_b^r \subset NB_b \subset NB_c$.

Now it is easy to describe the families BD of bounded (closed) domains
which model the cross-sections of closed (shielded) cylindrical resonators or
waveguides.

The families BD_c, BD_b, BD_u: the domains $\Omega_c = \Omega \cap G_b$ for the family
BD_c, $\Omega_b = (\Omega^1 \cap G_b) \cup \Omega^2$ and $\Omega_u = (\Omega^1 \cup \Omega^2) \cap G_b$ for the families BD_b, BD_u,
(see the definitions of the families) where G_b is an arbitrary domain bounded
by a closed piece-wise smooth simply-connected contour $\Gamma_b \subset \partial\Omega_c$ $(\partial\Omega_b, \partial\Omega_u)$
containing all the inhomogeneities. For the family BD_c assume that the
boundary of $\Omega_c^i = \Omega^i \cap G_b$, $i = 1, 2$ where $\Omega = \Omega^1$ belongs to the family NB_c
and $\Omega^2 = R^2 \setminus \bar{\Omega}^1$ may contain, as well as Ω^1, a finite number of the gap
lines of the functions $\epsilon(\mathbf{r}), \mu(\mathbf{r})$ in the form of both the closed and unclosed
contours (Γ_i, bounding the "layers" and connecting a pair of different points
on $\partial\Omega_{bc}$), and a multi-connected (perfectly conducting) component of their
boundary (see the definition of the family NB_c)

We also define the sub-families $BD_b^r \subset BD_b$ and $BD_u^r \subset BD_u$ assuming
that their slots are formed by rectilinear intervals $\Gamma_S^I = \bigcup_{i=1}^{M_S} \Gamma_{Si}^I$, $\Gamma_{Si}^I \subset \{\mathbf{r} :$
$x_2 = \text{const}\}$, $i = 1, 2, \ldots, M_S$.

A canonical structure for the families BD_u^r (BD_b^r) is the rectangular
shielded slot line whose cross-section is formed by two rectangular domains

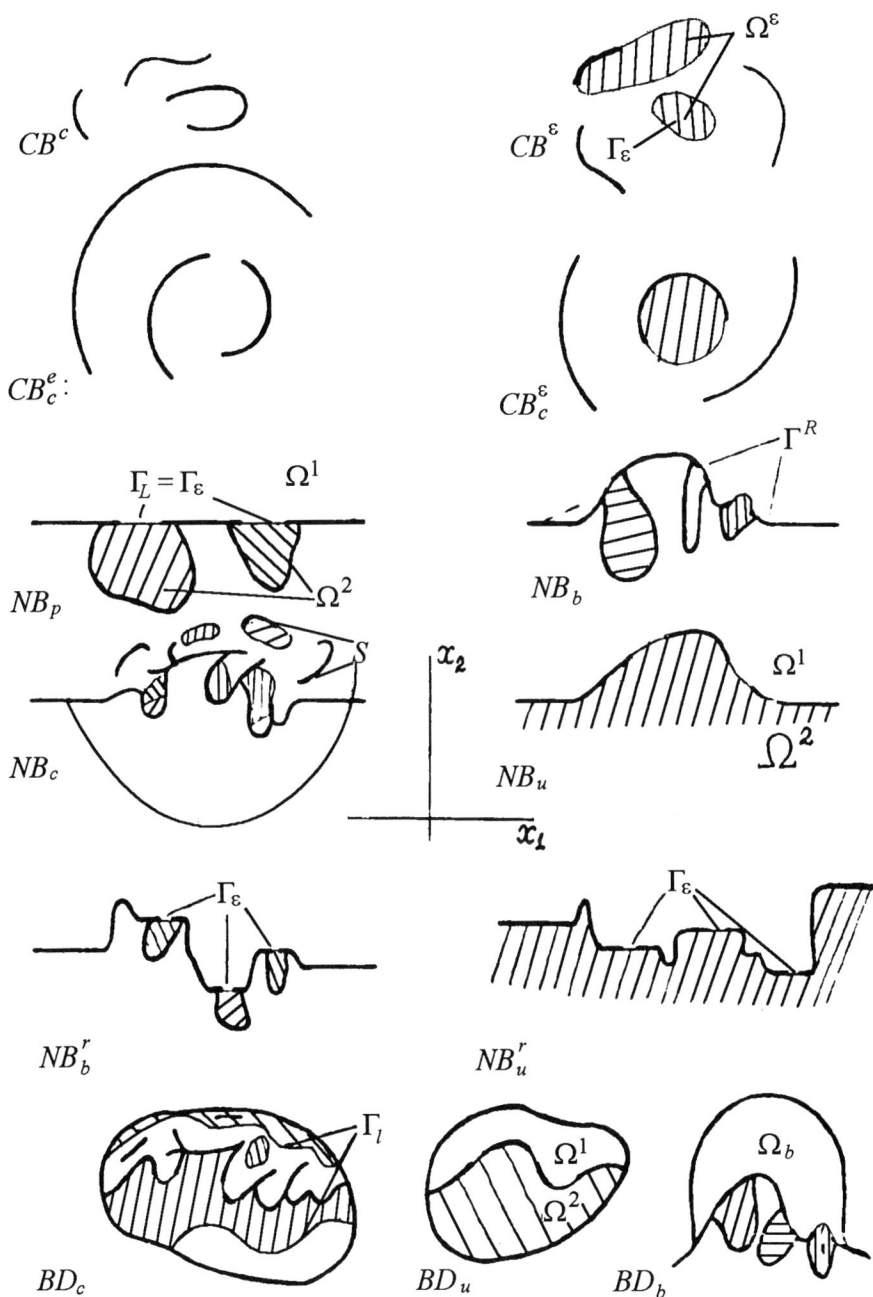

Figure 1.1 *Cross-sections of open resonators*

$\Pi_{ab_1} = \{\mathbf{r} : 0 < x_1 < a, \ 0 < x_2 < b_1\} \subset \Omega^+ = \{\mathbf{r} : x_2 > 0\}$ and $\Pi_{ab_2} = \{\mathbf{r} : 0 < x_1 < a; \ -b_2 < x_2 < 0\} \subset \Omega^- = \{\mathbf{r} : x_2 < 0\}$ with the common part of the boundaries $\Gamma = \partial S_1 \cap \partial S_2 = \{(\mathbf{r}) : x_2 = 0, \ 0 \le x_1 \le a\}$ containing an interval $\Gamma_L = \{\mathbf{r} : x_2 = 0, \ \frac{a}{2} - w = d_1 < x_1 < d_2 = \frac{a}{2} + w\}$ (a slot) with endpoints (edges) $\partial \Gamma_L = \{A_1 = \frac{a}{2} - w, \ A_2 = \frac{a}{2} + w\}$.

A canonical structure for the families NB_b (NB_b^r) is the rectangular image slot line whose cross-section is formed by the half-plane $\Omega^1 = \Omega^+$ and the rectangular domain $\Pi_{ab} = \Omega^2 = \{\mathbf{r} : 0 < x_1 < a; \ -b < x_2 < 0\}$ with an interval (a slot) $\Gamma_L = \{\mathbf{r} : x_2 = 0, \ \frac{a}{2} - w = d_1 < x < d_2 = \frac{a}{2} + w\}$ on the common part of the boundaries of Ω_1, Ω_2.

A canonical structure for the families NB_u (NB_u^r) is the open slot line whose cross-section is formed by the half-planes $\Omega^1 = \Omega^+$ and $\Omega^2 = \Omega^-$ with several intervals (slots) $\Gamma_L = \cup_{i=1}^{N_L} \Gamma_{Li}$, $\Gamma_{Li} = \{\mathbf{r} : x_2 = 0, \ d_i < x_1 < d_{i+1}, \ i = 1, 2, \ldots, N_L\}$ on the common part of the boundaries of Ω^1 and Ω^2.

For the canonical structures we will assume that the dielectric permeability of the medium $\epsilon = \epsilon(\mathbf{r}) = \epsilon_i$, $\mathbf{r} \in \Omega_i$ $(i = 1, 2)$ is a piece-wise constant function.

1.2 Statements of spectral problems

Let us consider open resonators infinite and uniform along the x_3-axis and formed by a finite number N of infinitely thin, perfectly conducting, unclosed circular cylindrical surfaces containing inclusions in the form of M homogeneous isotropic circular dielectric pivots with the axis parallel to that of the resonator's mirrors (the permittivity and permeability of the medium, $\varepsilon(\mathbf{r})$ and $\mu(\mathbf{r})$, depend on spatial co-ordinates). The resonator itself is placed in a homogeneous medium with complex permittivity and permeability, ε_0, μ_0 ($\operatorname{Im} \varepsilon_0 \ge 0$, $\operatorname{Im} \mu_0 \ge 0$).

We will thus assume first that the cross-sections in the $\mathbf{r}$-plane of the open resonators considered belong to the family CB_c^ϵ and then consider resonators of a wider family NB_c (for the definitions see Section 1.1).

Let $\Omega = [R^2 \setminus \cup_{l=1}^2 S_l] \setminus \partial S_3$ and Λ is the Riemann surface of the analytical continuation of the fundamental solution to the two-dimensional Helmholtz equation with respect to the spectral parameter $\kappa = ka$ ($k = \frac{\omega}{c}$, ω is the fre-

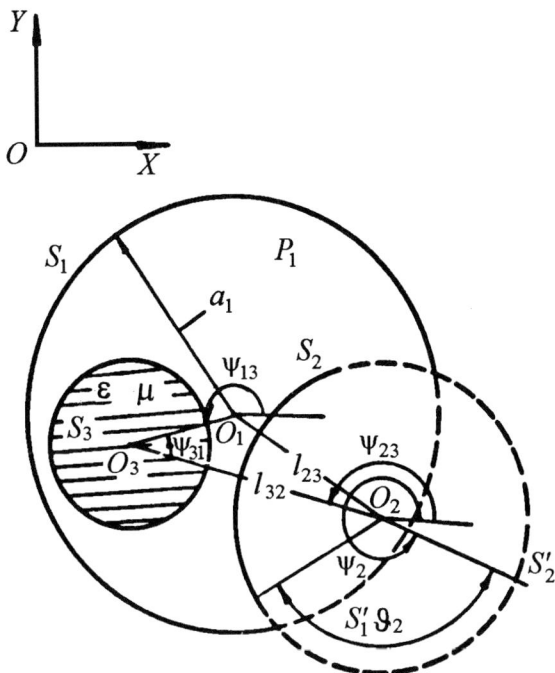

Figure 1.2 *Cross-section of open circular resonators (family CB_c^ϵ)*

quency parameter, c is the light velocity in vacuum and a is the characteristic size of the structure).

We will study the spectrum of eigenoscillations of this electrodynamic structure for two-dimensional oscillations $\left(\frac{\partial}{\partial x_3} = 0\right)$. Taking this into account, together with the presence of inclusions inside the resonator, one can reduce the problem in terms of a homogeneous system of Maxwell's equations to two independent problems for the E and H oscillations.

In other words, one has to determine the values of the spectral parameter $\kappa \in \Lambda$ (eigenvalues) for which there exist non-trivial solution of the homogeneous Helmholtz equation

$$\Delta u(\mathbf{r}) + k^2 \chi(\mathbf{r}) u(\mathbf{r}) = 0, \quad \mathbf{r} \in \Omega = R^2 \setminus (S \cup \Gamma_\epsilon) \tag{1.1}$$

belonging to the functional class

$$M = \{u : \ u \in C^2(\Omega) \cap C^1(\cup_i \bar{\Omega}^{\epsilon_i} \setminus S_\delta)\}$$

where S_δ denotes vicinities of the endpoints of S ("edges" or the "return" points of the boundary $\partial\Omega$ of Ω) and $\Omega^{\epsilon_i} = \{\mathbf{r} : \epsilon(\mathbf{r}) = \epsilon_i = \text{const}\}$, satisfying homogeneous boundary conditions on S

$$u = 0, \quad \text{(E oscillations)} \quad \frac{\partial u}{\partial \mathbf{n}} = 0 \quad \text{(H oscillations)} \tag{1.2}$$

the conjugation conditions on Γ_ϵ

$$u^+ - u^- = 0$$

$$\left[\frac{1}{\mu(\mathbf{r})}\frac{\partial u^+}{\partial \mathbf{n}} - \frac{1}{\mu_0}\frac{\partial u^-}{\partial \mathbf{n}}\right] = 0 \quad \text{(E oscillations)} \tag{1.3}$$

$$\left[\frac{1}{\epsilon(\mathbf{r})}\frac{\partial u^+}{\partial \mathbf{n}} - \frac{1}{\epsilon_0}\frac{\partial u^-}{\partial \mathbf{n}}\right] = 0 \quad \text{(H oscillations)}$$

the Meixner conditions

$$\iint_Q \left(|u|^2 + \left|\frac{\partial u}{\partial x_1}\right|^2 + \left|\frac{\partial u}{\partial x_2}\right|^2\right) d\mathbf{r} < \infty \tag{1.4}$$

for every compactum $Q \subset R^2$ and the radiation conditions introduced by Reichardt (1960): there exists an $R_0 > 0$ such that for all $\mathbf{r} = (r, \phi) : \hat{k}r \geq R_0$, where $\hat{k} = k\sqrt{\epsilon_0\mu_0}$, the following representation holds for the function $u(\mathbf{r})$

$$u(\mathbf{r}) = \sum_{m=-\infty}^{\infty} a_m H_m^{(1)}(\hat{k}r)p_m(\phi) \tag{1.5}$$

where $p_m(\phi) = \exp(jm\phi)$ $(m = 0, \pm1, \pm2, \ldots)$, the series admit double termwise differentiation with respect to $\hat{k}r \geq R_0$ and $\phi \in (0, 2\pi)$; $k = \frac{\kappa}{a}$, $r = \frac{x_1}{\cos\phi} = \frac{x_2}{\sin\phi}$, $\mathbf{n}$ is the unit normal to the boundary of the domain $S \cup \Gamma_\epsilon$; $u^\pm$ and $\frac{\partial u^\pm}{\partial \mathbf{n}}$ are the limiting values of u and $\frac{\partial u}{\partial \mathbf{n}}$ on the boundary, where "+" denotes the values taken from inside the domain and "−" from outside the domain. The Hankel function $H_m^{(1)}(z)$ of the first kind and m-th order is considered here on the Riemann surface Λ,

$$\chi(\mathbf{r}) = \begin{cases} \epsilon(\mathbf{r})\mu(\mathbf{r}), & \mathbf{r} \in \Omega_\epsilon \\ \epsilon_0\mu_0, & \mathbf{r} \in R^2 \setminus (\bar{\Omega}_\epsilon \cup S) \end{cases}$$

$$\gamma(\mathbf{r}) = \begin{cases} \mu(\mathbf{r}), & \text{(E oscillations)} \\ \epsilon(\mathbf{r}), & \text{(H oscillations)} \end{cases}$$

$$\gamma_0 = \begin{cases} \mu_0 & \text{(E oscillations)} \\ \epsilon_0 & \text{(H oscillations)} \end{cases}$$

The form of the radiation condition is connected with the fact that the value of $\kappa = \kappa' + j\kappa''$ is complex. In the case of real κ one can take the Sommerfeld radiation conditions and for complex κ with the positive imaginary part the radiation conditions introduced by Kupradze (1973). The condition given by eqn. 1.5 is the generalisation of the latter which is applied for arbitrary complex κ and may be considered as the continuation of the Sommerfeld condition from the real axis or as the continuation of the Kupradze condition from the upper half-plane of the complex plane κ to the Riemann surface Λ.

The partial radiation conditions have been introduced by Sveshnikov (1950) for the waveguide problems and then modified for various external diffraction problems, for example, on star-type bodies and irregular periodic structures (Ilyinsky and Sveshnikov, 1968). In this way we will also call in what follows the conditions given by eqn. 1.5 the partial Reichardt-Sveshnikov radiation conditions at infinity.

The function $u(\mathbf{r})$ is the longitudinal component of the electromagnetic eigenfield, H_3 in the H-case and E_3 in the E-case. Other components of the E and H eigenfields can be determined from Maxwell's equations:

in the H-case

$$\mathbf{E}(\mathbf{r}) = \frac{j}{k\epsilon}\,\mathrm{curl}[u(\mathbf{r})\hat{i}_3]$$
$$H_{x_3}(\mathbf{r}) = u(\mathbf{r}), \qquad H_{x_1} = H_{x_2} = 0$$

in the E-case

$$\mathbf{H}(\mathbf{r}) = -\frac{j}{k\mu}\,\mathrm{curl}[u(\mathbf{r})\hat{i}_3]$$
$$E_{x_3}(\mathbf{r}) = u(\mathbf{r}), \qquad E_{x_1} = E_{x_2} = 0$$

where, in accordance with the notation introduced in Section 1.1, $\epsilon = \epsilon(\mathbf{r})$ and $\mu = \mu(\mathbf{r})$ are assumed to be piece-wise constant functions.

In order to generalise the statement of the spectral problem given by eqns. 1.1–1.5 for the family NB_c it is sufficient to put in eqn. 1.5

$$p_m(\phi) = 0 \quad (m = -1, -2, \ldots), p_m(\phi) = \cos m\phi \quad \text{in the case of H oscillations}$$

and

$$p_m(\phi) = \sin m\phi \quad (m = 0, 1, 2, \ldots) \quad \text{in the case of E oscillations}$$

Determination of the eigenoscillations spectra for the families CB_c^ϵ and NB_c (and, consequently, for all the sub-families NB_p, NB_b^r, NB_b, NB_c due to the inclusions $NB_p \subset NB_b^r \subset NB_b \subset NB_c$) is thus reduced to the boundary eigenvalue problem given by eqns.1.1–1.5 which is analogous to the corresponding problems for closed (shielded) structures studied, in particular, by Samarsky and Tikhonov (1947), if the outgoing radiation condition given in eqn. 1.5 are replaced by the conditions $u(\mathbf{r}) = 0$, $\mathbf{r} \in G$ in the E-case or $\frac{\partial u(\mathbf{r})}{\partial \mathbf{n}} = 0$, $\mathbf{r} \in G$ in the H-case, where G is a closed surface containing the structure (all the contours S and Γ_ϵ). For such closed domains the corresponding boundary-value problems are self-adjoint, the eigenoscillations spectrum exists and forms a discrete set of real points (if $\operatorname{Im} \epsilon = 0$) with a single accumulation point at infinity (Reed and Simon, 1978). But the condition given by eqn. 1.5 in principle changes the situation and up to now the spectrum existence has not been proved for the considered family of open structures [the results obtained by Arsenjev (1976) are the exceptions and they cannot be directly applied for the considered types of open resonators]. Very few justified methods for the eigenoscillations calculation are available. One has to point out in this way investigations of the asymptotic spectra behaviour, in particular, the "bouncing ball" oscillations described by Weinstein (1966). There are practically no information on the spectra in the "resonance" domain where $|ka| = O(1)$ (a is the characteristic size of the structure).

We will show that the problem given by eqns.1.1–1.5 is equivalent for the families CB_c^ϵ, NB_p, NB_b^r and NB_u^r to the problem on characteristic numbers for a summation-type Fredholm operator-valued function what enables us to prove the discreteness of the eigenoscillations spectrum, to establish the boundaries for the application of asymptotic formulas obtained by Weinstein (1966) and to consider the spectral properties of two-dimensional open resonators with dielectric inclusions when the distances between the mirrors are comparable with the apertures and the mirrors curvature radii (i.e. when $|ka| \sim 1$).

The spectral problem given by eqns.1.1–1.5 is difficult for a direct solution since eigenfunctions must satisfy the partial radiation condition. Hence, we will consider instead an equivalent operator-valued function and find its characteristic numbers [which coincide with (complex) fundamental frequencies of the considered open resonators] by the methods of the spectral theory of operator-valued functions developed for these problems by Ilyinsky and Shestopalov (1989).

Since E and H oscillations are analogous in terms of boundary value problems, we will consider in this Section only the case of H oscillations for the family CB_c^ϵ. The families NB_p, NB_b^r and NB_u^r are considered in Sections 1.7–1.9.

We look for the solution of the boundary-value problem given by eqns. 1.1–1.5 for the family CB_c^ϵ (under the assumption of its existence) as a linear superposition of M Fourier series

$$u(\mathbf{r}) = \begin{cases} \displaystyle\sum_{n=1}^{M}\sum_{m=-\infty}^{\infty} x_m^n G_m(\hat{k}r_n, \hat{a}a_n)\exp(jm\phi_n), & \mathbf{r} \in R^2 \setminus (S \cup \Gamma_\epsilon) \\[4mm] \displaystyle\sum_{m=-\infty}^{\infty} y_m^n \frac{J_m(kr_n\sqrt{\epsilon_n - N\mu_{n-N}})}{J_m(ka_n\sqrt{\epsilon_n - N\mu_{n-N}})}\exp(jm\phi_n), & \mathbf{r} \in \Omega_\epsilon \end{cases}$$

(1.6)

where

$$G_m(\hat{k}r_n, \hat{a}a_n) = \begin{cases} H_m^{(1)}(\hat{k}a_n)J_m(\hat{k}r_n), & r_n \le a_n, n \le N \\ H_m^{(1)}(\hat{k}r_n)J_m'(\hat{k}a_n), & r_n > a_n, n \le N \\ H_m^{(1)}(\hat{k}r_n)J_m(\hat{k}a_n) & n > N \end{cases}$$

Here $H_m^{(1)}(x)$ and $J_m(x)$ are the Hankel functions and the Bessel functions of the order m and $J_m'(x)$ denotes the first derivative of $J_m(x)$. Assume that the sequences x_m^n and y_m^n ($m = \ldots, -1, 0, 1, \ldots,$) belong to the Hilbert space

$$l_2 = \{z_m, m = \ldots, -1, 0, 1, \ldots, : \sum_{m=-\infty}^{\infty} |z_m|^2 < \infty\}$$

(for every $n = 1, 2, \ldots, M$) and the following conditions hold:

(A) the series

$$\sum_{m=-\infty}^{\infty} x_m^n H_m^{(1)}(\hat{k}a_n)J_m(\hat{k}a_n)\exp(jm\phi_n)$$

$$\sum_{m=-\infty}^{\infty} y_m^n \frac{J_m'(ka_n\sqrt{\epsilon_n - N\mu_{n-N}})}{J_m(ka_n\sqrt{\epsilon_n - N\mu_{n-N}})}\exp(jm\phi_n)$$

converge uniformly on Γ_ϵ with respect to ϕ_n;

(B) the series

$$\sum_{m=-\infty}^{\infty} x_m^n H_m^{(1)'}(\hat{k}a_n) J_m(\hat{k}a_n) \exp(jm\phi_n)$$

$$\sum_{m=-\infty}^{\infty} x_m^n H_m^{(1)'}(\hat{k}a_n) J_m'(\hat{k}a_n) \exp(jm\phi_n)$$

converge uniformly on $(\cup_{l=1}^M S_l') \setminus S_\delta$ with respect to ϕ_n, where S_δ are the infinitely small vicinities of the endpoints of S ("edges"), and define absolutely integrable functions of ϕ_n. It is proved subsequently that these conditions imply the fulfillment of the Meixner-type conditions given by eqn. 1.4.

One has to point out that due to the addition theorems, the Sobolev theorem (Sobolev, 1966) and the properties of the cylindrical functions (Watson, 1949) $u(\mathbf{r})$ defined in eqn. 1.6 is an infinitely differentiable function of its variables, satisfies radiation condition given by eqn. 1.5 (for every $k \in C_k = \{k : \arg k < \pi\}$) and eqn. 1.1 (in the classical sense) in $D = R^2 \setminus (S \cup \Gamma_\epsilon)$.

In order to construct the matrix operator-valued function of the problem considered, we will rewrite conjugation conditions given by eqn. 1.3 with the help of eqn. 1.6 and in terms of the unknowns x_p^n and y_p^n $(n = 1, 2, \ldots, N_S)$:

$$x_p^n = \sum_{s=1}^M \sum_{m=-\infty}^{\infty} a_{pm}^{ns} x_m^s$$

$$y_p^n = x_p^n \frac{2 J_p(\hat{k}a_n) g_p^{N_S}}{j\pi \hat{k}a_n [g_n^{N_S} J_p'(\hat{k}a_n) - e_n^{N_S} g_p^{N_S'} J_p(\hat{k}a_n)]}$$

(1.7)

where

$$a_{pm}^{ns}(k) = \frac{R_m(\hat{k}a_s) T_{m-p}(\hat{k}l_{ns})}{H_p^{(1)}(\hat{k}a_n)} \frac{\epsilon_{n-N_S} - \epsilon_0}{\epsilon_{n-N_S} + \epsilon_0} \exp[j(m-p)\psi_{sn}], \quad s \le N_S$$

$$a_{pm}^{ss}(k) = [1 - \Delta_p^n A_{pm}^s(k)]\delta_p^n$$

$$A_{pm}^s(k) = \frac{H_p^{(1)}(\hat{k}a_n) g_p^{N_S} - e_n^{N_S} g_p^{N_S'} H_n^{(1)}(\hat{k}a_n)}{e_n^{N_S} g_p^{N_S'} J_p'(\hat{k}a_n) - g_p^{N_S} J_p'(\hat{k}a_n)}$$

$$a_{pm}^{ns}(k) = \frac{J_m(\hat{k}a_s) H_{m-p}^{(1)}(\hat{k}l_{ns})}{H_p^{(1)}(\hat{k}a_n)} \frac{\epsilon_{n-N_S} - \epsilon_0}{\epsilon_{n-N_S} + \epsilon_0} \exp[j(m-p)\psi_{sn}], \quad n \ne s \quad (1.8)$$

$$\Delta_p^n = \frac{\epsilon_{n-N_S} - \epsilon_0}{\epsilon_{n-N_S} + \epsilon_0} \frac{J_p(\hat{k}a_s)}{H_p^{(1)}(\hat{k}a_n)} \qquad \delta_p^m = \begin{cases} 0, & p \neq m \\ 1, & p = m \end{cases}$$

$$R_m(x) T_{m-p}(y) = \begin{cases} H_m^{(1)'}(x) J_{m-p}(y), & |x| > |y| \\ H_m^{(1)}(y) J_m'(x), & |x| < |y| \end{cases}$$

$$e_n^{N_S} = \sqrt{\frac{\epsilon_0 \mu_{n-N_S}}{\mu_0 \epsilon_{n-N_S}}}$$

$$g_m^{N_S} = J_m(ka_n \sqrt{\epsilon_{n-N}\mu_{n-N}})$$

$$g_m^{N_S'} = J_m'(ka_n \sqrt{\epsilon_{n-N}\mu_{n-N}})$$

Application of boundary-value conditions given by eqn. 1.2 for eqn. 1.6 yields the system of functional equations

$$\sum_{m=-\infty}^{\infty} x_m^n H_m^{(1)'}(ka_m) J_m'(\hat{k}a_m) \exp(jm\phi_n)$$

$$+ \sum_{s \neq n} \sum_{m=-\infty}^{\infty} x_m^s A_m^{sn}(\phi_n) + \sum_{s=1}^{M-N_S} \sum_{m=-\infty}^{\infty} x_m^{N+s} B_m^{N+sn}(\phi_n) = 0 \tag{1.9}$$

where

$$A_m^{sn}(\phi_n) = \hat{R}(ka_s) \sum_p \hat{Q}_p(\hat{k}a_n) T_{m-p}(\hat{k}l_{ns})$$

$$\times \exp[j(m-p)\psi_{sn}] \exp(jp\phi_n)$$

$$B_m^{N+sn}(\phi_n) = J_m(\hat{k}a_{N+s}) \sum_p \hat{R}_p(ka_n) T_{m-p}(\hat{k}l_{ns}) \tag{1.10}$$

$$\times \exp[j(m-p)\psi_{N+sn}] \exp(jp\phi_n)$$

$$\hat{Q}_m = J_m'(x), \quad \hat{T}_m = J_m(x), \quad \hat{R}_m = H_m^{(1)'}(x)$$

$$R_m(x) = \begin{cases} J_m'(x), & a_n < l_{ms} \\ H_m^{(1)'}(x), & a_n > l_{ms} \end{cases} \qquad \hat{Q}_m(x) = \begin{cases} J_m'(x), & a_n < l_{ms} \\ H_m^{(1)'}(x), & a_n > l_{ms} \end{cases}$$

$$\hat{T}_m = \begin{cases} H_m^{(1)}(x), & a_n < l_{ms} \\ J_m(x), & a_n > l_{ms} \end{cases}$$

$S_n \subset P_s$ and P_s denotes the circular arcs which are situated on the boundaries of the dielectric inclusions.

The function $u(\mathbf{r})$ defined by eqn. 1.6 must be continuous on the gap lines of ϵ and μ what yields the system of functional equations

$$\sum_{m=-\infty}^{\infty} x_m^n \exp(jm\phi_n) = 0, \quad |\phi_n - \psi_n| < \theta_n \; (n = 1, 2, \dots, N_S) \tag{1.11}$$

The limiting values of u and $\frac{\partial u}{\partial \mathbf{n}}$ are continuous on $\cup_{i=1}^{N_S} S_i' \setminus S_\delta$ and it follows from eqns. 1.9 and 1.11 that these values coincide on the set $\cup_{i=1}^{N_S} S_i'$.

As mentioned above, the function $u(\mathbf{r})$ given by eqn. 1.6 satisfies the partial radiation conditions. Due to the assumptions (A) and (B) and the addition theorems for the cylindrical functions, eqns. 1.7, 1.9 and 1.11 are equivalent to the boundary-value conditions and the conjugation conditions given by eqns. 1.2 and 1.3. Hence, in accordance with the results of Tikhonov and Samarsky (1978), the existence of a solution to the spectral problem given by eqns. 1.1–1.5 yields eqn. 1.6. On the other hand, if the function $u(\mathbf{r})$ defined by eqn. 1.6 satisfies the assumptions (A) and (B) and the sequences $(x_m^n)_{m=-\infty}^{\infty}$ and $(y_m^n)_{m=-\infty}^{\infty}$ in eqn. 1.6 are the solutions to eqns. 1.7, 1.9 and 1.11, then $u(\mathbf{r})$ is a solution of the considered spectral problem.

Hence the initial spectral problem is equivalent to the infinite system of linear algebraic equations given by eqn. 1.7 and to the system of coupled functional equations 1.9, 1.11 with a trigonometric kernel $\exp(jp\phi)$.

Further construction of the matrix operator-valued functions is connected with the application of the method of the Riemann-Hilbert problem (Shestopalov, 1971). As a result, we obtain the infinite system of linear algebraic equations

$$x_p^n = \sum_{s=1}^{M} \sum_{m=-\infty}^{\infty} a_{pm}^{ns}(k) x_m^s \quad (n = 1, 2, \ldots, M) \tag{1.12}$$

where

$$a_{pm}^{nm}(k) = (-1)^{m+p} \exp[j(m-p)\psi_n] \delta_m(\hat{k}a_n) F_p^m(u_n), \quad u_n = -\cos\theta_n,$$

$$a_{pm}^{ns}(k) = j\pi(\hat{k}a_n)^2 \gamma_p^{sn} \hat{R}_m(\hat{k}a_s)$$
$$\times \sum_q (-1)^q \hat{Q}_q(\hat{k}a_n) \hat{T}_{m-q}(\hat{k}l_{ns}) F_p^q(u_n) \exp[jq(\psi_n - \psi_{sn})]$$
$$n, s \leq N_S, \ s \neq n$$

$$a_{pm}^{ns}(k) = j\pi(\hat{k}a_n)^2 \gamma_p^{sn} J_m(\hat{k}a_s)$$
$$\times \sum_q (-1)^q R_q(\hat{k}a_n) T_{m-q}(\hat{k}l_{ns}) F_p^q(u_n) \exp[jq(\psi_n - \psi_{sn})]$$
$$n \leq N_S, \ s > N_S$$

$$\gamma_p^{sn} = (-1)^p \exp(jm\psi_{sn}) \exp(-jp\psi_n) \quad \delta_m(x) = |m| + j\pi x^2 J_m'(x) H_m^{(1)\prime}(x)$$

$$F_p^m(x) = \begin{cases} -\ln \frac{(1+x)}{2}, & m = p = 0 \\ \frac{1}{m} V_{m-1}^{-1}(x), & m \neq 0, \ p = 0 \\ \frac{1}{p} V_{p-1}^{m-1}(x), & p \neq 0 \end{cases} \qquad (1.13)$$

The properties of $F_p^m(x)$ are given by Shestopalov (1985).

Let us introduce the following notations: $\sigma(\Delta)$ is the spectrum of the initial problem given by eqns. 1.1–1.5 (the set of fundamental frequencies) and $\sigma_H(A)$ is the set of complex numbers $k \in C_k$ for which there exists a non-trivial solution of eqn. 1.12. Consequently, $\sigma(\Delta) \subseteq \sigma_H(A)$.

Let us formulate now the fundamental questions: does the inverse inclusion hold, i.e. is it true that $\sigma(\Delta) = \sigma_H(A)$? What is the structure of $\sigma(\Delta)$, i.e. is this set discrete as in the case of closed waveguides and resonators and has it finite accumulation points? What are the optimal ways of analytical and numerical determination of $\sigma(\Delta)$ and $\sigma_H(A)$? The last problems are connected with a correct replacement of the infinite system given by eqn. 1.12 by a truncated ("cut") one when the "approximate" characteristic numbers are defined as a sequence zeros of the "truncated" determinants. Will these sequences converge to the true characteristic numbers and what will be the rate of convergence? The answers to these questions may be found with the help of the "discrete convergence" theory developed by Vainikko and Karma (1974) and the spectral theory of finite-meromorphic operator-valued functions.

Let us carry out the detailed investigation of eqn. 1.12. Let B be the set of zeros of $H_p^{(1)}(\hat{k}a_n)$,

$$J_p'(ka_n \sqrt{\epsilon_n - N\mu_{n-N}}) J_p(\hat{k}a_n) \sqrt{\frac{\epsilon_0 \mu_{n-N_S}}{\mu_0 \epsilon_{n-N_S}}} - J_p(ka_n \sqrt{\epsilon_n - N\mu_{n-N}})$$

in C_k and $\hat{B} = C_k \setminus B$.

Lemma 1.1: The matrices $\|a_{pm}^{ns}(k)\|_{p,m=-\infty}^{\infty}$ given by eqns. 1.8 and 1.13 define for every $k \in C_k$ the kernel operators in the space l_2.

The proof of this lemma is based on verifying asymptotic estimations for $a_{pm}^{ns}(k)$ in the form of sufficient conditions providing the convergence of the series $\sum_{p,m} |a_{pm}^{ns}(k)|$ for every $k \in C_k$. Hence the matrices $\|a_{pm}^{ns}(k)\|_{p,m=-\infty}^{\infty}$ define bounded operators in l_2 which we will denote by $A^{ns}(k)$. Let $\{\mathbf{f}_p\}_{p=-\infty}^{\infty}$ be an arbitrary basis in l_2. Let us prove that the series $\sum_p (A^{ns}\mathbf{f}_p, \mathbf{f}_p)$ converge absolutely for every $k \in \hat{B}$ which implies the kernel property for the operators $A^{ns}(k)$ (Sobolev, 1966). If $\{e_p\}_{p=-\infty}^{\infty}$ is the canonical basis in l_2 ($e_p = \{\delta_m^p\}_{m=-\infty}^{\infty}$) then $\mathbf{f}_p = \sum_p f_p^n e_n$, $\sum_p |f_p^n|^2 = \sum_n |f_p^n|^2 = 1$ and we get the required property from the chain of inequalities $\sum_p |(A^{ns}\mathbf{f}_p, \mathbf{f}_p)| < \sum_p \sum_{l,m} |a_{lm}^{ns}(k)||f_p^l f_p^m| \le \sum_{l,m} |a_{lm}^{ns}(k)| < \infty$.

We consider the operators $A^{ns}(k) : l_2 \to l_2$ as operator-valued functions of the spectral parameter k defined on $\hat{B} \subset C_k$. Let us prove that the system of these kernel operator-valued functions is equivalent to the problem given by eqns. 1.1–1.5.

Eqn. 1.12 may be represented in the equivalent form

$$\mathbf{x}^n = \sum_{s=1}^{M} A^{ns}(k)\mathbf{x}^s \quad (n = 1, 2, \ldots, M) \tag{1.14}$$

Eqn. 1.14 is the problem on characteristic numbers for a matrix operator-valued function. In order to prove this fact we will introduce the Hilbert space

$$l_2^M = \left\{ \mathbf{x} = \left\| \begin{array}{c} \mathbf{x}^1 \\ \vdots \\ \mathbf{x}^M \end{array} \right\| \quad \mathbf{x}^n = \{x_m^n\}_{m=-\infty}^{\infty} \in l^2 \; n = 1, 2, \ldots \right\} \tag{1.15}$$

related to the inner product $\langle \mathbf{x}, \mathbf{y} \rangle = \sum_{i=1}^{M} (\mathbf{x}^i, \mathbf{y}^i)$. It is easy to see that bounded linear operators in l_2^M are defined by quadratic $M \times M$ matrices whose elements are bounded operators in l_2.

Let B^{ns} be the system of bounded operators in l_2. Then one can construct the operator matrix $B = \|B^{ns}\|_{n,s=1}^{M}$ such that

$$B\mathbf{x} = \left\| \begin{array}{c} \sum_{s=1}^{M} B^{1s}\mathbf{x}^s \\ \vdots \\ \sum_{s=1}^{M} B^{Ms}\mathbf{x}^s \end{array} \right\| \tag{1.16}$$

It is obvious that B is a bounded linear operator in l_2^M. Let us refer to the system of operator-valued functions $A^{ns}(k)$ the matrix operator-valued function $A(k) = \|A^{ns}(k)\|_{n,s=1}^{M}$ acting in l_2^M and defined by eqn. 1.16. Then the infinite system of linear homogeneous algebraic equations given by eqn. 1.12 may be considered as an operator equation in l_2^M

$$[I - A(k)]\mathbf{x} = \mathbf{o} \tag{1.17}$$

where I is the unit operator in l_2^M and $\mathbf{o}$ is the zero element of l_2^M. Let us denote by $\rho(A)$ the set of regular points $k \in \hat{B}$ of $I - A(k)$ where the inverse operator $[I - A(k)]^{-1}$ is bounded, then $\sigma_H(A) = \hat{B} \backslash \rho(A)$. It is known (Gohberg and Krein, 1965) that all $k \in \sigma_H(A)$ are the characteristic numbers of the operator-valued function $I - A(k)$ (because $A(k)$ is a kernel operator), namely, for every $k \in \sigma_H(A)$ there exists a non-trivial solution of eqn. 1.14 (and, hence, a non-trivial solution of eqn. 1.7 in l_2).

1.3 The equivalence theorem and fundamental frequency discreteness for open circular resonators

Let us prove the equivalence of the problem on characteristic numbers for the matrix operator-valued function $A(k)$ and the spectral problem given by eqns. 1.1–1.5.

We will formulate the properties of the matrix operator-valued function $A(k)$ in the form of several lemmas.

Lemma 1.2: Assume that there exist such a $k_0 \in B$ that eqn. 1.17 has a non-trivial solution $\mathbf{x} \in l_2^M$. Then the following asymptotic estimations hold for $\mathbf{x}^n = \{x_p^n\}_{p=-\infty}^{\infty}$

(i) $\quad x_p^n = \dfrac{1}{p} \exp[jp(\psi_n + \pi)][C_1^n P_{n-1}(u_n) + C_2^n P_n(u_n)] + O(|p|^{-\frac{5}{2}})$

$\quad p \to \infty \ (n \geq N)$

(ii) $\quad |x_p^n| < C_3^n \dfrac{|\epsilon_{n-N} - \epsilon_0|}{|\epsilon_{n-N} + \epsilon_0|} |p| \gamma_n^p$

$\quad p \to \infty, \quad 0 < \gamma_n < 1 \ (n > N)$

The proof is based on the estimations for $a_{pm}^{ns}(k)$.

The next assertions are the corollaries from lemma 1.2.

Lemma 1.3: Assume that the estimations from lemma 1.2 hold for $\{x_p^n\}_{p=-\infty}^{\infty}$. Then the function $u(\mathbf{x})$ defined by eqn. 1.6 satisfy the Meixner condition and conditions (A) and (B) formulated in Section 1.2.

Lemma 1.4: Every solution of eqn. 1.17 (or, equally, of eqn. 1.12) satisfies the system of coupled functional equations given by eqns. 1.9, 1.11 and eqn. 1.7. And vice versa, every solution of eqns. 1.7, 1.9 and 1.11 satisfying the estimations of lemma 1.2 is a solution of eqn. 1.17.

This lemma gives a rigorous justification for the method of the Riemann-Hilbert problem applied for the considered family of open resonators. The proof may be easily obtained from the previous lemmas.

Now one can formulate and prove the equivalence theorem.

Theorem 1.1: The spectral problem given by eqns. 1.1–1.5 and the problem on characteristic numbers for the matrix operator-valued function $A(k)$ are equivalent. The eigenoscillations spectrum of every open resonator with the cross-section formed by a domain from the family CB_c^ε coincides with the set of characteristic numbers of the operator-valued function $I - A(k)$.

The equivalence is understood in the following way. Assume that $k_0 \in \sigma_H(A)$ and

$$\mathbf{x} = \left\| \begin{array}{c} \mathbf{x}^1 \\ \vdots \\ \mathbf{x}^M \end{array} \right\| \qquad \mathbf{x}^n = \{x_m^n\}_{m=-\infty}^\infty \quad (n = 1, 2, \ldots, M)$$

is a non-trivial solution to eqn. 1.17 for $k = k_0$. Then the function

$$u(\mathbf{r}) = \begin{cases} \displaystyle\sum_{n=1}^M \sum_{m=-\infty}^\infty x_m^n G_m(\hat{k}r_n, \hat{a}a_n) \exp(jm\phi_n) \\ \qquad \mathbf{r} \in R^2 \setminus (S \cup \Gamma_\epsilon) \\ \displaystyle\sum_{m=-\infty}^\infty x_m^n J_m(\hat{k}a_{Ns+n}) J_m(kr_{Ns+n}\sqrt{\epsilon_n\mu_n})(\Delta_m^n)^{-1} \exp(jm\phi_{Ns+n}) \\ \qquad \mathbf{r} \in \Omega_\epsilon \end{cases}$$

$$(1.18)$$

where

$$\Delta_m^n = \frac{1}{2} j\hat{k}a_{Ns+n} \pi [J_n'(\hat{k}a_{Ns+n}) J_m(ka_{Ns+n}\sqrt{\epsilon_n\mu_n}) - w_n J_m(\hat{k}a_{Ns+n}) J_m'(ka_{Ns+n}\sqrt{\epsilon_n\mu_n})]$$

is not identically zero and $u(\mathbf{r})$ is a solution to the problem given by eqns. 1.1–1.5 for $k = k_0$, i.e. k_0 is a fundamental frequency. And vice versa, if the function in the form given by eqn. 1.18 is a solution to the problem given by eqns. 1.1–1.5 for a certain $k = k_0$, then $k_0 \in \sigma_H(A)$, i.e. k_0 is a characteristic number of the operator-valued function $I - A(k)$.

Let us note that the formulae in eqn. 1.18 are valid when perfectly conducting screens are outside dielectric inclusions. Analogous representations hold in the E-case when the screens are situated both inside dielectric pivots and on their boundaries.

Let us proceed to the proof of the equivalence theorem. Assume that $k_0 \in \sigma_H(A)$ and

$$
\mathbf{x} = \left\| \begin{array}{c} \mathbf{x}^1 \\ \vdots \\ \mathbf{x}^M \end{array} \right\| \qquad \mathbf{x}^n = \{x_m^n\}_{m=-\infty}^{\infty} \quad (n = 1, 2, \ldots, M)
$$

is a (non-trivial) solution to eqn. 1.17 for $k = k_0$. Let us construct the function $u(\mathbf{r})$ with the help of the formulae given by eqn. 1.18. In accordance with the proof from Section 1.2, $u(\mathbf{r})$ satisfies radiation conditions (eqn. 1.5) and eqn. 1.1 everywhere in $R^2 \setminus (S \cup \Gamma_\epsilon)$. The results of lemmas 1.2 and 1.3 guarantee that $u(\mathbf{r})$ satisfies also eqn. 1.4 under the fulfillment of conditions (A) and (B) from section 2. Finally, the construction procedure carried out in Section 1.2 for $u(\mathbf{r})$ and application of lemma 1.4 allow us to assert that $u(\mathbf{r})$ satisfies boundary value conditions given by eqn. 1.2 and conjugation conditions given by eqn. 1.3. The inverse assertion is valid: if for a certain k_0 $u(\mathbf{x})$ given by eqn. 1.18 is a solution to the problem given by eqns. 1.1–1.5, then $k_0 \in \sigma_H(A)$.

Hence the spectral problem given by eqns. 1.1–1.5 and the problem on characteristic numbers for the operator-valued function $I - A(k)$ defined by formulae 1.8 and 1.13 are equivalent. The eigenoscillations spectrum $\sigma(\Delta)$ coincides with the set $\sigma_H(A)$ of the characteristic numbers of $I - A(k)$.

Let us prove the discreteness of the set $\sigma(\Delta)$ by investigating analytical properties of the matrix operator-valued function $A(k) = \|A^{ns}(k)\|_{n,s=1}^{M}$.

Lemma 1.5: $A^{ns}(k) : l_2 \to l_2$ are analytical operator-valued functions when $k \in C_k$ and $n \leq N_S$ or when $k \in B$ and $n > N_S$ and finite-meromorphic operator-valued functions when $k \in C_k$ and $n > N_S$.

Let $L : l_2 \to C$ be an arbitrary bounded functional. Analytical property of $A^{ns}(k)$ is equivalent to the fact that the function $L(A^{ns}(k)\mathbf{x})$ is analytical with respect to k for every $\mathbf{x} \in l_2$ (Sobolev, 1966). Due to the Riesz theorem

$$L(A^{ns}(k)\mathbf{x}) = \sum_{p,m} a_{pm}^{ns}(k) x_m y_p \tag{1.19}$$

where $\mathbf{y} = (y_p)_{p=-\infty}^{\infty} \in l_2$ is uniquely determined by the functional L. It follows from eqn. 1.8 that the properties of cylindrical functions and integral representations given by eqn. 1.13 for the matrix elements, $a_{pm}^{ns}(k)$ are analytical functions when $k \in C_k$, $n \leq N_S$ or when $k \in \hat{B}$, $n > N_S$ and meromorphic functions when $k \in C_k$, $n > N_S$. The poles of $a_{pm}^{ns}(k)$ $(n > N)$ coincide with the points from B $(\hat{B} = C_k \setminus B)$. Therefore, in order to prove that $A^{ns}(k)$ are analytical operator-valued functions, it is sufficient to establish the uniform convergence of the series in eqn. 1.19 in every compactum $k \in F \subset \hat{B}$ (for $n \leq N_S$, $F \subset C_k$). Let us constrict ourselves to the case $n \leq N_S$, $s \leq N_S$ because the case $n > N_S$ is treated in a similar way. The estimations obtained for $a_{pm}^{ns}(k)$ yield inequalities

$$\sum_{p,m} |a_{pm}^{ns}(k) x_m y_p| \leq \|\mathbf{x}\|\|\mathbf{y}\| \sum_{p,m} |a_{pm}^{ns}(k)|$$

$$< D\|\mathbf{x}\|\|\mathbf{y}\| \left(|a_{00}^{ns}(k)| + \sum_{p \neq 0, m \neq 0} |m||p|^{-\frac{3}{2}} \gamma_{(ns)}^{m} \right)$$

where $D = \text{const}$, $0 < \gamma_{(ns)} < 1$. Hence the series $\sum_{p \neq 0, m \neq 0} |m||p|^{-\frac{3}{2}} \gamma_{(ns)}^{m})$ converges what is sufficient for the proof of the required uniform convergence of the series in eqn. 1.19.

Assume now that $n > N_S$ and $k_0 \in B$ and let us consider only the case $s \leq N_S$. Then, in accordance with eqn. 1.8, we have

$$a_{pm}^{ns}(k) = \frac{R_m(\hat{k}a_s) T_{m-p}(\hat{k}l_{ns})}{H_p^{(1)}(\hat{k}a_n)} \frac{\epsilon_{n-N_S} - \epsilon_0}{\epsilon_{n-N_S} + \epsilon_0} \exp[j(m-p)\psi_{sn}] \ (n \neq s)$$

$$k_0 = \frac{\nu_{pq}}{a_n \epsilon_0 \mu_0}$$

where ν_{pq} is a root of the equation $H_p^{(1)}(z) = 0$. It is known (Watson, 1949) that all such roots have unit multiplicity and lie in the lower half-plane $\text{Im } z < 0$ $(p = 0, 1, 2, \ldots,)$. Hence for a given k_0 there exist such whole numbers p_0 and n_0 that the following expansions are valid in a sufficiently small vicinity of k_0

$$a_{p_0 m}^{n_0 s}(k) = (k - k_0)^{-1} b_{p_0 m}^{n_0 s} + c_{p_0 m}^{n_0 s}(k) \quad (m = 0, \pm 1, \pm 2, \ldots, \ s = 1, 2, \ldots, N_S)$$

where $c_{pom}^{nos}(k)$ are analytical in the vicinity of k_0. It follows from lemma 1.1 and the estimations obtained for $|a_{pm}^{ns}|$ that the matrices $\|b_{pom}^{nos} \delta_{po}^{p}\|_{p,m=-\infty}^{\infty}$, $\|d_{pm}^{nos}(k)\|_{p,m=-\infty}^{\infty}$, $d_{pom}^{nos}(k) = c_{pom}^{nos}(k)$, $d_{pm}^{nos}(k) = a_{pm}^{nos}(k)$, $p \neq p_0$, determine finite-dimensional operator-valued functions B^{nos} and kernel operator-valued functions $D^{nos}(k)$, respectively, acting in l_2.

Hence the following representation holds for the operator-valued functions $A^{nos}(k)$

$$A^{nos}(k) = (k - k_0)^{-1} B^{nos} + D^{nos}(k)$$

in a vicinity of every point $k_0 \in B$ and the required property of $A^{nos}(k)$ stated in lemma 1.5 follows then from the general results proved by Gohberg and Krein (1965) for finite-meromorphic operator-valued functions.

Now we can assert that $I - A(k)$ is a canonical Fredholm operator-valued function, $k \in \hat{B}$, and due to the meromorphic Fredholm theorem (Reed and Simon, 1978) $I - A(k)$ may have only isolated characteristic numbers. In other words, if $\alpha(k) = \dim \text{Ker}(I - A(k))$, then

$$\alpha(k) = \alpha_0 = \text{const} \tag{1.20}$$

everywhere in the domain where $A(k)$ is an analytical operator-valued function of k (i.e. when $k \in \hat{B}$) except of a set of isolated points where $\alpha_0 \neq \alpha(k) < \infty$. The existence theorem for the excitation (inhomogeneous) problem proved in chapter two for the considered family of open resonators yields $\{k : \text{Im}\, k > 0\} \subset \rho(A) \neq \emptyset$, i.e. $\alpha_0 \neq 0$ and the following assertion is valid.

Theorem 1.2: The operator-valued function (resolvent) $R(k) = [I - A(k)]^{-1}$ exists for $k \in \sigma_H(A) \setminus B$ and may be continued as an operator-valued function analytical in $\hat{B} \setminus \sigma_H(A)$ and finite-meromorphic in $\hat{B}$.

Theorems 1.1 and 1.2 enable us to conclude that the spectrum of two-dimensional eigenoscillations for the considered family of open circular resonators has a finite multiplicity and forms a discrete set of points in $\{k : \text{Im}\, k < 0\}$: for every positive R_1, R_2, $0 < R_1 < R_2$ there may be not more than a finite number of fundamental frequencies (eigenvalues of the problem given by eqns. 1.1–1.5) in every domain $\{k : \pi < \arg k < 2\pi, R_1 < |k| < R_2\}$.

This general result may be specified. In fact, due to the properties of $a_{pm}^{ns}(k)$, the norm of $A(k)$ tends to zero when $k \to 0$, $\mu_n \to \mu_0$ and $\epsilon_n \to \epsilon_0$ $(n = N_S + 1, N_S + 2, \ldots, M - N_S)$. Hence there always exists a sufficiently small vicinity of the origin $\{k : 0 < |k| < \delta_0\}$ (when dielectric inclusions are absent, $\epsilon_n = \epsilon_0$, $\mu_n = \mu_0$, $n = 1, 2, \ldots$) where $\|A(k)\| < 1$. It is clear that such a vicinity is preserved and this inequality holds also for sufficiently "weak" dielectric inclusions (when $\epsilon_n \sim \epsilon_0$, $\mu_n \sim \mu_0$, $n = 1, 2, \ldots$) proving thus the existence of $R(k)$ and the absence of eigenvalues (fundamental frequencies) for sufficiently small $|k|$.

In addition to this, it is easy to show that the eigenvalues spectrum $\sigma(\Delta)$ of the problem given by eqns. 1.1–1.5 is symmetric with respect to the imaginary axis: if $k \in \sigma(\Delta)$, then $k^* \in \sigma(\Delta)$ where * denotes the complex conjugation.

Let us note also that the family considered CB_c^ϵ essentially differs from open resonators studied by Lax and Phillips (1972) and Sanchez-Palencia (1980). In particular, it will be shown that open circular resonators may have high Q-factor eigenoscillations corresponding to the fundamental frequencies situated close to the real axis.

We will prove in what follows that the spectrum discreteness and the absence of eigenvalues in the upper half-plane of the complex frequency spectral parameter are the general properties characteristic for open cylindrical structures and remain valid for open resonators whose transversal cross-sections are formed by the domains belonging to the families NB_c and BD_c. But in order to prove these results, one has to use a different approach based, in particular, on the solution to the boundary-value problems for the Helmholtz equation in the generalised statement (see Sections 1.8 and 2.5).

1.4 Implicit operator-valued functions: generalised dispersion equations

In this Section we will give a brief introduction to the methods of operator-valued functions for solving some nonlinear eigenvalue problems of electrodynamics. Determination of fundamental frequencies and normal waves in slot lines and resonators are considered as basic examples. Some principal features of this technique are given together with mathematical background for computational and analytical results.

From the mathematical point of view the spectral characteristics (or the spectral data) of wave fields in resonance and guiding structures, namely, the propagation constants of eigenwaves (or normal waves) and fundamental frequencies of eigenoscillations (or resonance oscillations), are directly connected with the eigenvalues of certain boundary value problems and the questions inevitably arise concerning the existence of these eigenvalues, their properties as functions of various parameters and effective ways of their calculation.

The basic idea of the approach developed recently in the number of books and articles (see Sanchez-Palencia, 1980, Shestopalov, 1980a, Ilyinsky and Shestopalov, 1989, Shestopalov, 1991a, Shestopalov, 1994) is to collect spectral data of guiding systems and resonators by solving the boundary eigenvalue problems of determining the spectra of normal waves and fundamental frequencies. They are reduced to the operator eigenvalue problems nonlinear with respect to the spectral parameter $z = \gamma$ or $z = \omega$, where ω is a frequency of electromagnetic field and γ is a longitudinal wavenumber, represented in the most general form as the problem on characteristic numbers or the generalised dispersion equation

$$K(z, \mathbf{a})\phi = 0$$

where $K(z, \mathbf{a}) : X \rightarrow Y$ is a multi-parameter operator-valued function describing mathematically the properties of the considered structure, acting, in general, on the pair of Banach spaces X, Y, $\mathbf{a}$ is a vector of non-spectral parameters and this homogeneous operator equation determines $z = z(\mathbf{a})$ implicitly. Knowledge of spectral data in such a form enables one to reconstruct then all the most important characteristics of the considered object.

In fact, for the majority of the following problems it will be proved that K is a Fredholm operator and, consequently, its characteristic numbers form a countable set of (complex) isolated points. Hence one can obtain a sufficient volume of the spectral data in the form of the numbered families of implicit functions $\{z_\nu(\mathbf{a})\}_{\nu=1}^{M}$ (the generalised dispersion curves).

In many cases it is convenient to vary only one non-spectral parameter. In particular, one of the $\mathbf{a}$ components (a_0) may serve as a small parameter (e.g. a_0 may be chosen as the diameter of the boundary irregular perturbation) and the spectral data are obtained as asymptotic series with respect to a_0 (see Section 1.9).

The typical example here is the case when K is a scalar or matrix integral operator-valued function

$$K(z, a_0)\phi \equiv \int_L K(z; t, s)\phi(s)\,\mathrm{d}\,s = 0 \quad t \in L$$

where the kernel $K(z; t, s)$ depends on the spectral parameter and the vector of non-spectral parameters consists of one component $a_0 = \mathrm{diam}\,L$.

Another well-studied situation is when K acts in the Hilbert space (l_2) of infinite complex sequences and defined by an infinite matrix $\|k_{mn}(z, \mathbf{a})\|_{m,n=1}^\infty$.

The main advantages of the analytical technique presented for determining $z = z(\mathbf{a})$ by functional methods are as follows:

(i) it gives results in the domains of variation of the parameters where numerical methods cannot be used and in this way the justified asymptotics complement the numerical data;

(ii) it enables one to obtain the reliable results in the form of explicit formulae suitable for the direct engineering calculations, which guarantee the accuracy within the given domain of the parameters variation.

We will establish in Chapters 1 and 3 that in the problems of fundamental frequencies and normal waves in the cylindrical slot lines and resonators K turns out to be a finite-meromorphic operator-valued function:

$$K = K(\lambda) = P_n(\lambda) + U_n(\lambda) = T(\lambda) + \frac{A_n(\lambda)}{\lambda - \lambda_n} + U_n(\lambda)$$

where $\lambda = z^2$, $A_n(\lambda)$ is a finite-dimensional operator, $T(\lambda)$ is invertible and $U_n(\lambda)$ is completely continuous. Such an explicit form of the operator K enables us to prove the existence of characteristic numbers in the vicinities of poles λ_n and to construct approximate formulae for calculating the characteristic numbers with given accuracy (in particular, in the form of the small parameter asymptotic series). Here the families of implicit functions $\{z_\nu(\mathbf{a})\}_{\nu=1}^M$ may be naturally numbered with respect to (known) poles λ_n.

This approach also opens the way to the studies of three-dimensional and inverse problems (Shestopalov, 1991a, Lozhechko and Shestopalov, 1994b) because in many cases one can establish a one-to-one correspondence between the domains of variation of the spectral and non-spectral parameters by means of the functions $z(\mathbf{a})$ and to reconstruct the latter with appropriate spectral data.

We may conclude that mathematical modelling of the wave and resonance phenomena may be conditionally divided within the frame of the approach developed in this book into the following stages.

1. Explicit construction of operators K and generalised dispersion equations for the particular (required) types of the structures (problems); mathematical treatment, including regularisation, explicit semi-inversion of the operators, determination of spectra localisation; obtaining the spectral data by the explicit asymptotic formulae which are used both for direct calculations and as a "source" of zero approximations for numerical computations.

2. Working out numerical procedures and computer software based on the specially constructed algorithms for solving generalised dispersion equations; the results are obtained in the form of the families of generalised dispersion curves.

Both stages will result in obtaining the required spectral data and other characteristics with guaranteed accuracy for the types of structures considered.

Exact and reliable analytical results applied in combination with justified numerical algorithms are of principal importance for practical calculations of normal waves, fundamental frequencies and other characteristics of guiding systems and resonators. They allow one to guarantee the accuracy and to decrease computation time up to more than two orders (from minutes to seconds). These results naturally form the basis of several applied programs for numerical analysis of transmission lines and other types of guiding, scattering and resonance systems. Various examples are presented in Chapters 2, 4 and 5.

1.5 Generalised potentials and the conditions at infinity

In the theory of boundary-value problems the integrals

$$u(\mathbf{r}) = \int\limits_C E(\mathbf{r}, \mathbf{q}) \xi(\mathbf{q}) \, dl_q, \quad v(\mathbf{r}) = \int\limits_C \frac{\partial}{\partial \mathbf{n}_q} E(\mathbf{r}, \mathbf{q}) \eta(\mathbf{q}) \, dl_q$$

are called the potentials. Here $\mathbf{r} = (x_1, x_2)$, $\mathbf{q} = (y_1, y_2) \in R^2$; $E(\mathbf{r}, \mathbf{q})$ is the fundamental solution of the second order elliptic differential operator, $\frac{\partial}{\partial \mathbf{n}_q}$ is the normal derivative in the point $\mathbf{q}$ of the closed piece-wise smooth boundary C of the domain in R^2, $\xi(\mathbf{q})$, $\eta(\mathbf{q})$ are sufficiently smooth functions defined on C. In the case of the Helmholtz operator $\mathcal{L}(k) = \Delta + k^2$ one can choose $E(\mathbf{r}, \mathbf{q})$ in the form

$$E(\mathbf{r}, \mathbf{q}) = \mathcal{E}(\mathbf{r} - \mathbf{q}) = \frac{j}{4} H_0^{(1)}(k|\mathbf{r} - \mathbf{q}|)$$

where $\mathcal{E}(\mathbf{r} - \mathbf{q})$ satisfies Sommerfeld's radiaton conditions in the domain $\Omega^+ = \{\mathbf{r} : x_2 > 0\}$ for real values of k; $H_0^{(1)}(z) = -2jg(z) + h(z)$ is the Hankel function, $\frac{1}{2}g(\mathbf{r} - \mathbf{q}) = \frac{1}{2\pi} \ln \frac{1}{|\mathbf{r} - \mathbf{q}|}$ is the kernel of two-dimensional single layer potential and the second derivative of $h(z)$ has logarithmic singularity.

The study of potentials and the method of boundary integral equations with strongly elliptic Fredholm operators for 'transmission' boundary value problems for Helmholtz equation in Sobolev spaces is carried out by von Petersdorff (1989) and in other papers [a detailed list of references is also presented by von Petersdorff (1989)].

We will consider here some properties of generalised potentials and Green's potentials in two-dimensional domains of a certain class together with application of the method of generalised potentials for transmission-type boundary eigenvalue problems arising in the electrodynamics of guided waves and resonators.

Let us denote by $S_\Pi = S_\Pi(L)$ the simply-connected domains bounded by closed or unclosed piece-wise smooth contours $\Gamma = \partial S_\Pi$, such that

$$L \subset L_0 \subset \Gamma \qquad L_0 \subset \{x_2 = 0\}$$

i.e. $L_0 = [c, d]$, $L = (a, b)$, $c < a < b < d$. L_0 and $L \subset L_0$ may consist of a finite number of nonintersecting finite intervals $L_0 = \cup_{l=1}^{N_L}[a_{l-1}, a_l]$, $L = \cup_{l=1}^{N_L}(d_{l-1}, d_l)$ where d_l $(l = 1, 2, \ldots N_L)$ are called the endpoints (edges) of L.

Let us say that the functions $U_l(\mathbf{r})$ are the generalised single layer potential (SL) $(l = 1)$ or the double layer potential (DL) $(l = 2)$ if

$$U_l(\mathbf{r}) = \int_L K_l(\mathbf{r}, t)\phi_l(t)\,dt, \quad \mathbf{r} \in S_\Pi(L) \tag{1.21}$$

where

$$K_l(\mathbf{r}, t) = g_l(\mathbf{r}, t) + F_l(\mathbf{r}, t) \quad (l = 1, 2)$$

$$g_1(\mathbf{r}, t) = g(\mathbf{r}, \mathbf{q}^0) = \frac{1}{\pi} \ln \frac{1}{|\mathbf{r} - \mathbf{q}^0|}$$

$$g_2(\mathbf{r}, t) = \frac{\partial}{\partial y_2} g(\mathbf{r}, \mathbf{q}^0) \quad [\mathbf{q}^0 = (t, 0)]$$

$F_{1,2}$ are smooth functions and we shall assume that in every closed domain $S_{0\Pi}(L) \subset S_{\Pi}(L)$

(i) $F_1(\mathbf{r}, t)$ is once continuously differentiable with respect to the variables of $\mathbf{r}$ and continuous in t;

(ii) $F_2(\mathbf{r}, t)$ and

$$F_2^1(\mathbf{r}, t) = \frac{\partial}{\partial x_2} \int_q^t F_2(\mathbf{r}, s) \, \mathrm{d}s, \quad q \in R^1$$

are continuous.

We shall also assume that the densities of the generalised potentials $\phi_1 \in X_a(L)$; $\phi_2 \in X_0(L)$, where the functional spaces X_a, X_0 are defined as:

$$X_0(L) = \{\psi : \psi \in C(\overline{L}), \psi' \in \widetilde{H}_\mu^1 \, \psi(d_l) = 0 \quad (l = 1, 2, \ldots, 2N_L)\}$$

$$X_a(L) = \widetilde{H}_\mu(L) = \{\psi : \psi R \in H_\mu(L)\}$$

$$\widetilde{H}_\mu^1(L) = Y_a(L) = \{\phi : \phi \in C(\overline{L}) \cap C^1(\overline{L}_\delta) \, R\phi' \in H_\mu(L)\} \tag{1.22}$$

$$\overline{L}_\delta = \cup_{l=1}^{N_L}[d_{2l-1} + \delta, d_{2l} - \delta] \quad R(t) = \prod_{l=1}^{N_L} \sqrt{|d_l - t||t - d_{l-1}|}$$

where $H_\mu(L)$ is a class (Banach space) of Hölder functions defined on the intervals L with constant $\mu \in (0, 1)$, so that X_a, Y_a are the Hölder classes with the weight $R(t)$.

The following lemmas contain the assertions which are known for the logarithmic potentials. They can be proved straightforwardly for the generalised potentials considered in the S_Π-domains.

Lemma 1.6:

$$\frac{\partial}{\partial x_2} \int_L g_2(\mathbf{r}, t) \phi_2(t) \, \mathrm{d}t = \int_L \frac{\partial}{\partial x_1} g_1(\mathbf{r}, t) \frac{\mathrm{d}\phi_2}{\mathrm{d}t}(t) \, \mathrm{d}t$$

$$\mathbf{r} \in S_\Pi, \quad \phi_2(t) \in X_0(L)$$

Lemma 1.7:

$$\lim_{\mathbf{r}\to\mathbf{r}^0\in L}\frac{\partial}{\partial x_2}\int_L g_2(\mathbf{r},t)\phi_2(t)\,\mathrm{d}\,t = \int_L \lim_{\mathbf{r}\to\mathbf{r}^0\in L}\frac{\partial}{\partial x_1}g_1(\mathbf{r},t)\frac{\mathrm{d}\,\phi_2}{\mathrm{d}\,t}\,\mathrm{d}\,t = \mathcal{S}_L\frac{\mathrm{d}\,\phi_2}{\mathrm{d}\,t}$$

$$\mathcal{S}_L f(t) \equiv \frac{1}{\pi}\int_L \frac{f(t')}{t'-t}\,\mathrm{d}\,t', \quad \phi_2(t) \in X_0(L)$$

Lemma 1.8: The properties of the normal and tangential derivatives of SL and of the trace of DL on L coincide with the ones of logarithmic potentials.

Lemma 1.9: There exists the limiting value (trace) of the normal derivative of DL in S_Π-domain and it admits the representation as a singular integral:

$$\lim_{\mathbf{r}\to\mathbf{r}^0\in L}\frac{\partial U_2}{\partial x_2} = \lim_{\mathbf{r}\to\mathbf{r}^0\in L}\int_L \left[\frac{\partial g_1}{\partial x_1} + F_2^1\right](\mathbf{r},t)\frac{\mathrm{d}\,\phi_2}{\mathrm{d}\,t}\,\mathrm{d}\,t$$

$$= (\mathcal{S}_L + K_2^1)\frac{\mathrm{d}\,\phi_2}{\mathrm{d}\,t}, \quad \phi_2(t) \in X_0(L)$$

where

$$K_2^1 f = \int_L F_2^1(\mathbf{r},t)f(t)\,\mathrm{d}\,t$$

The proof of the last lemma follows from the previous lemmas and definition of DL, under the condition that the density of DL is from $X_0(L)$.

Details of the proofs of these lemmas are given by Shestopalov (1990).

All the assertions of lemmas 1.6–1.9 are valid for a more general family of the domains $S_\Pi^c(L)$ bounded by a closed piece-wise smooth contours $\Gamma = \partial S_\Pi^c$ such that

$$L \subset L_0 \subset \Gamma \qquad L_0 \subset \cup\{x_2 = \text{const}\}$$

$S_\Pi \subset S_\Pi^c$ and L_0 and $L \subset L_0$ consist of a finite number of nonintersecting finite rectilinear intervals

$$L_0 = \cup_{x_2=\text{const}} \cup_{l=1}^{N_L} \{\mathbf{r} : x_1 \in [a_{l-1}, a_l], x_2 = \text{const}\}$$

$$L = \cup_{x_2=\text{const}} \cup_{l=1}^{N_L} \{\mathbf{r} : x_1 \in (d_{l-1}, d_l), x_2 = \text{const}\}$$

and the endpoints ("edges") of L have the x_1-co-ordinates d_l $(l = 1, 2, \ldots N_L)$.

The definition of the generalised single layer and double layer potentials is thus connected with the domains from a certain class ($S_{\Pi}^c(L)$-domains) whose boundaries contain several rectilinear intervals (L).

Now it is easy to see that the domains Ω_1 and Ω_b forming the cross-sections of the families of the domains NB_p, NB_b^r, NB_u^r BD_b^r, and BD_u^r introduced in Section 1.1 belong to the class $S_{\Pi}^c(L)$ where L denotes the "slots" coinciding with rectilinear intervals $\Gamma_S^I = \bigcup_{i=1}^{M_S} \Gamma_{Si}^I$, $\Gamma_{Si}^I \subset \{r : x_2 = \text{const}\}$, $i = 1, 2, \ldots, M_S$ (in particular the domains of the family NB_p belong to the class $S_{\Pi}(L)$).

Let us introduce Green's functions $G^i = G^i(\mathbf{r}, \mathbf{q}; k)$ of the ith boundary-value problem ($i = 1, 2$) for the Helmholtz equation $\mathcal{L}(k)U = 0$ of a (bounded or unbounded) domain $\Omega \in S_{\Pi}^c$ (under the assumption of its existence). In particular,

$$h^l(\mathbf{r}, \mathbf{q}; k) = \mathcal{E}(\mathbf{r} - \mathbf{q}) + (-1)^l \mathcal{E}(\mathbf{r} - \mathbf{q}^*)$$
$$\mathbf{q} = (y_1, y_2), \quad \mathbf{q}^* = (y_1, -y_2) \tag{1.23}$$

which is Green's function of the first ($l = 1$) or the second ($l = 2$) boundary-value problem for the Helmholtz equation of the half-plane $\Omega^+ = \{r : x_2 > 0\}$ is the kernels of SL for $l = 2$ and

$$\frac{\partial}{\partial y_2} h^1(\mathbf{r}, \mathbf{q}^0; k)$$

is the kernel of DL for $l = 1$ in every $S_{\Pi}^c(L)$-domain such that $S_{\Pi}^c(L) \subset \{r : x_2 > 0\}$.

We will call the integrals

$$u(\mathbf{r}) = \int_L G^2(\mathbf{r}, \mathbf{q}^0; k)\phi(t)\,dt \quad \mathbf{r} \in \Omega \tag{1.24}$$
$$\phi(t) =\in \widetilde{H}_\mu(L), \quad \mathbf{q}^0 = (t, 0) \in L$$

$$v(\mathbf{r}) = \int_L \mathcal{K}_1(\mathbf{r}, \mathbf{q}^0; k)\psi(t)\,dt, \quad \mathbf{r} \in \Omega, \quad \mathbf{q}^0 = (t, 0) \in L \tag{1.25}$$
$$\mathcal{K}_l(\mathbf{r}, \mathbf{q}^0; k) = \frac{\partial}{\partial y_2} G^1(\mathbf{r}, \mathbf{q}^0; k), \quad \psi(t) =\in X_0(L)$$

the single layer and double layer Green's potentials.

Theorem 1.3: The single layer and double layer Green's potentials given by eqns. 1.24 and 1.25 are the generalised single layer and double layer potentials, respectively, in the $S_\Pi^c(L)$-domains. The functions

$$G^2(\mathbf{r}, \mathbf{q}^0; k), \quad \frac{\partial}{\partial y_2} G^1(\mathbf{r}, \mathbf{q}^0; k)$$

admit representations as the kernels of SL and DL.

The proof for a (bounded) domain $S_\Pi(L)$ is based on the study of the integral representations for the solutions of two auxiliary problems

$$(\Delta + k^2)u = 0, \quad (\Delta + k^2)v = 0, \qquad \mathbf{r} \in S_\Pi(L)$$

$$\left. \frac{\partial u}{\partial \mathbf{n}} \right|_L = \phi \in X_a(L), \quad v|_L = \psi \in X_0(L) \tag{1.26}$$

where it is assumed that $\operatorname{supp} \phi \subset L$ and $\operatorname{supp} \psi \subset L$.

To get the required representations it is sufficient to apply Green's identity to the functions $u(\mathbf{r})$, $V_2(\mathbf{r}, \mathbf{q}; k) = h_2(\mathbf{r}, \mathbf{q}; k) - G^2(\mathbf{r}, \mathbf{q}; k)$ and $v(\mathbf{r})$, $V_1(\mathbf{r}, \mathbf{q}; k) = h_1(\mathbf{r}, \mathbf{q}; k) - G^1(\mathbf{r}, \mathbf{q}; k)$ and to directly verify the fulfillment of conditions (i) and (ii) formulated in the definition of generalised potentials. Here it is taken into account that h_2 and $\frac{\partial}{\partial y_2} h_1$ are the kernels of SL and DL, respectively.

The proof for a domain $S_\Pi^c(L)$ is based on the separate consideration of integral representations in a vicinity of every rectilinear component of L (coinciding with a part of the line $x_2 = c \neq 0$) with the help of Green's functions $h_{1,2}^c$ of the half-plane Ω^+.

The direct application of the above results yields the following:

Theorem 1.4: There exist limiting values (traces) on L of the first (tangential) derivative of the single layer Green's potential u given by eqn. 1.24 and of the normal derivative of the double layer Green's potential v given by eqn. 1.25 in the S_Π-domains and they admit representations as singular integrals:

$$\lim_{\mathbf{r} \to \mathbf{r}^0 \in L} \frac{\partial u}{\partial x_1} = \left(\mathcal{S}_L + \hat{G} \right) \phi = \frac{1}{\pi} \int_L \frac{\phi(t)}{t - x_1} \, dt + \int_L \hat{g}(\mathbf{r}^0, t; k) \phi(t) \, dt$$

$$\lim_{\mathbf{r} \to \mathbf{r}^0 \in L} \frac{\partial v}{\partial x_2} = \lim_{\mathbf{r} \to \mathbf{r}^0 \in L} \int_L \left(\frac{\partial g_1}{\partial x_1} + g_2^1 \right) (\mathbf{r}, t; k) \frac{d\phi_2}{dt} \, dt = (\mathcal{S}_L + G_2^1) \frac{d\phi_2}{dt}$$

$$\tag{1.27}$$

where $\phi_2 \in X_0(L)$, $\phi \in X_a(L)$,

$$G_2^1 f = \int_L g_2^1(\mathbf{r}^0, t; k) f(t) \, dt$$

$$g_2^1(\mathbf{r}^0, t; k) = \frac{\partial}{\partial x_2}\bigg|_{\mathbf{r}=\mathbf{r}^0 \in L} \int_q^t \left[\frac{\partial}{\partial y_2} G^1(\mathbf{r}, \mathbf{q}^0; k) - \frac{\partial}{\partial y_2} h^1(\mathbf{r}, \mathbf{q}^0; k) \right] ds, \quad q \in R^1$$

$\hat{g}(\mathbf{r}^0, t; k)$ is once continuously differentiable and $g_2^1(\mathbf{r}^0, t; k)$ is continuous with respect to $x_1, t \in L$, $\mathbf{r}^0 = (x_1, 0)$, $\mathbf{q}^0 = (s, 0)$.

One can also prove this assertion for an arbitrary S_Π^c-domain.

In order to obtain representations for the solutions to boundary-value problems with the generalised Reichardt-Sveshnikov conditions at infinity in (unbounded) S_Π^c-domains in the form of Green's potentials let us prove the following lemma.

Lemma 1.10: Let the functions $U_l(\mathbf{r})$ satisfy the Helmholtz equation $\mathcal{L}(k)U_l = 0$ $(l = 1, 2)$ in the half-plane Ω^+ and the Reichardt-Sveshnikov conditions at infinity given by eqn. 1.5 with $p_n(\phi) = \cos n\phi$ for $l = 2$ and $p_n(\phi) = \sin n\phi$ for $l = 1$. Then

$$\int_{C_R^+} \left[U_l(\mathbf{q}) \frac{\partial h_l}{\partial \mathbf{n}_q}(\mathbf{r}, \mathbf{q}) - \frac{\partial U_l}{\partial \mathbf{n}_q} h_l(\mathbf{r}, \mathbf{q}) \right] dl_q = 0, \quad \mathbf{r} \in \Omega^+ \setminus B_O^R \quad (l = 1, 2)$$

$$\tag{1.28}$$

$$C_R^+ = \partial B_O^R \cap \Omega^+, \quad B_O^R = \{\mathbf{r} : |\mathbf{r}| = \sqrt{x_1^2 + x_2^2} < R\}$$

for sufficiently big $R > 0$.

We will prove this lemma limiting ourselves to the case $l = 2$.

First of all note that due to the addition theorem for cylindrical functions Green's function $h_2(\mathbf{r}, \mathbf{q}; k)$ admits representation

$$h_2(\mathbf{r}, \mathbf{q}; k) = j \sum_{n=0}^{\infty} H_n^{(1)}(k, r) \theta_n(R, \phi_0, \phi) = \sum_{n=0}^{\infty} H_n^{(1)}(k, r) \eta_n(R, \phi_0) \cos n\phi$$

$$\theta_n = J_n(k, R) \cos n\phi \cos n\phi_0$$

$$\eta_n = j J_n(k, R) \cos n\phi_0 \quad (n = 0, 1, 2, \ldots)$$

$$\tag{1.29}$$

for $R > 0$, $\mathbf{r}, \mathbf{q} \in \Omega^+$, $\mathbf{r} = (r, \phi) \neq \mathbf{q} = (R, \phi_0)$.

Let the functions $U(r, \phi)$ and $V(r, \phi)$ defined for $0 \leq \phi \leq \pi$, $r \geq R$ and continuously differentiable with respect to r, ϕ are expanded in the absolutely and uniformly convergent Fourier series admitting term-wise differentiation

$$U(r, \phi) = \sum_{n=0}^{\infty} U_n(r) \psi_n(\phi)$$

$$V(r, \phi) = \sum_{n=0}^{\infty} V_n(r) \psi_n(\phi)$$

$$\psi_n(\phi) = \cos n\phi \quad (0 \leq \phi \leq \pi, \ r \geq R)$$

Let us form the Wronskian

$$W_r[U, V] = \sum_{n=0}^{\infty} U_n \psi_n \sum_{n=0}^{\infty} V'_n \psi_n - \sum_{n=0}^{\infty} U'_n \psi_n \sum_{n=0}^{\infty} V_n \psi_n$$

$$= \sum_{n=0}^{\infty} \psi_n^2 W_r[U_n, V_n] + \sum_{n=0}^{\infty} \sum_{s=0}^{n} \psi_s \psi_{n-s} W[U_s, V_{n-s}]$$

Integration of this equality with respect to $\phi \in [0, \pi]$ yields

$$W[U, V](r) = \int_0^{\pi} W_r[U, V] \, d\phi = \delta \sum_{n=0}^{\infty} W[U_n, V_n]$$

where $\delta = $ const and the series converges uniformly for $r \geq R$. In particular, if

$$V_n(r) = c_n U(r), \quad c_n = \text{const} \quad (n = 0, 1, 2, \ldots)$$

then

$$W[U, G] \equiv 0 \quad \text{for} \quad r \geq R \tag{1.30}$$

Assume now that the functions $U_2(\mathbf{r})$ satisfy the Reichardt-Sveshnikov conditions at infinity given by eqn. 1.5 and let us calculate the Wronskian

$$W[U_2, h_2] = U_2(\mathbf{y}) \frac{\partial h_2}{\partial \mathbf{n}_q}(\mathbf{r}, \mathbf{q}) - \frac{\partial U_2}{\partial \mathbf{n}_q} h_2(\mathbf{r}, \mathbf{q})$$

$$= \sum_{n=0}^{\infty} [C_n \eta_n(R, \phi)]^2 W_r \left[H_n^{(1)}, H_n^{(1)} \right]$$

$$+ \sum_{n=0}^{\infty} \sum_{n=0}^{\infty} \psi_s(\phi) \psi_{n-s}(\phi) W \left[H_s^{(1)}(kr), H_{n-s}^{(1)}(kr) \right]$$

$$\times C_s \eta_s(R, \phi_0) C_{n-s} \eta_{n-s}(R, \phi_0)$$

Integration over $\phi \in [0, \pi]$ with an account of the orthogonality of ψ_n and fulfillment of the condition given by eqn. 1.30 yields

$$W[U_2, h_2](r) = \int_0^\pi W_r[U_2, h_2]d\phi = 0 \quad (r > R)$$

for sufficiently big $R > 0$, which completes the proof of lemma 1.10.

The case $l = 1$ is considered in a similar way.

The result of lemma 1.10 may be easily generalised for open domains which differ from the half-plane inside a finite ball.

Lemma 1.11: Let $G^i = G^i(\mathbf{r}, \mathbf{q}; k)$ be the Green's functions of the ith boundary-value problem ($i = 1, 2$) for the Helmholtz equation $\mathcal{L}(k)U = 0$ of an arbitrary unbounded domain $\Omega \in S_\Pi^c$ coinciding with an "upper" domain Ω_1 forming the cross-section corresponding to one of the families NB_p, NB_b^r or NB_u^r introduced in Section 1.1 and the functions $U_l(\mathbf{x})$ satisfy the Helmholtz equation $\mathcal{L}(k)U_l = 0$ ($l = 1, 2$) in Ω and the Reichardt-Sveshnikov conditions at infinity given by eqn. 1.5 with $p_n(\phi) = \cos n\phi$ for $l = 2$ and $p_n(\phi) = \sin n\phi$ for $l = 1$. Then

$$\int_{C_R^+} \left[U_l(\mathbf{q}) \frac{\partial G^l}{\partial \mathbf{n}_q}(\mathbf{r}, \mathbf{q}) - \frac{\partial U_l}{\partial \mathbf{n}_q} G^l(\mathbf{r}, \mathbf{q}) \right] dl_q = 0, \quad \mathbf{r} \in \Omega \setminus B_O^R \quad (l = 1, 2)$$

$$(1.31)$$

$$C_R^+ = \partial B_O^R \cap \Omega, \quad B_O^R = \{\mathbf{r} : |\mathbf{r}| < R\}$$

for sufficiently large $R \geq R_0 > 0$ such that the ball $B_O^{R_0}$ contains the (irregular) part of the boundary of Ω which differs from the straight line.

Let us again consider only the case $l = 2$. Green's functions $G^2 = G^2(\mathbf{r}, \mathbf{q}; k)$ satisfy the Helmholtz equation $\mathcal{L}(k)_\mathbf{r} G^2 = 0$ ($\mathbf{r} \neq \mathbf{q}$) in every domain $B_{R_1}^{R_2} = \Omega^+ \cap \{\mathbf{r} : R_1 < |\mathbf{r}| < R_2\}$, $R_1 \geq R_0$ Hence the variables separation (in polar co-ordinates) applied in $B_{R_1}^{R_2}$ for G^2 as a function of $\mathbf{r}$ (for fixed $\mathbf{q}$) yields the representation similar to that given by eqn. 1.29 (and by eqn. 1.5) and the further proof of lemma 1.11 literally repeats the one of lemma 1.10.

Theorem 1.5: The solutions $u(\mathbf{r})$ and $v(\mathbf{r})$ to each of the problems given by eqn. 1.26 considered in an arbitrary unbounded domain $\Omega \in S_\Pi^c$ (coinciding with an "upper" domain Ω_1 forming the cross-section corresponding to one of the families NB_p, NB_b^r or NB_u^r) which satisfy the Reichardt-Sveshnikov conditions at infinity given by eqn. 1.5 with $p_n(\phi) = \cos n\phi$ for the function u and $p_n(\phi) = \sin n\phi$ for the function v admit the representations as single layer (for u) and double layer (for v) Green's potentials given by eqns. 1.24 and 1.25 (which are the generalised single layer and double layer potentials satisfying the conditions stated by theorem 1.4 given in eqn. 1.27).

In order to prove this theorem, it is sufficient to apply the second Green identity in the domain $\Omega \cup B_O^R$ for Green's functions $G^l = G^l(\mathbf{r}, \mathbf{q}; k)$ and solutions $u(\mathbf{q})$ for $l = 2$, $v(\mathbf{q})$ for $l = 1$ ["eliminating" a ball-type vicinity $\mathbf{q} : |\mathbf{q} - \mathbf{r}| < \delta$ of the observation point $\mathbf{r}$ and taking R such that the ball $B_O^{R_0}$ contains the (irregular) part of the boundary of Ω] and to use the result of lemma 1.11 given by eqn. 1.31.

1.6 Elements of the spectral theory of integral operator-valued functions

In this chapter we present a short survey on the methods of integral operator-valued functions considering only specific types of integral operators which are typical for electrodynamic problems treated in this book.

For the general definitions and notions of the spectral theory of operator-valued functions which are used below (such as the resolvent set, spectrum, characteristic number, finite-meromorphic and Fredholm operator-value functions etc.) one may refer, for example, to Gohberg and Sigal (1974), Sanchez-Palencia (1980), Ilyinsky and Shestopalov (1989) or to any book on spectral theory.

We begin with the analysis of basic properties of the integral operators with logarithmic singularity of the kernel considering first the integral equation

$$l\phi \equiv \frac{1}{\pi} \int_{-a}^{a} \ln \frac{1}{|t - s|} \phi(s) ds = f(t) \tag{1.32}$$

where $f \in Y_a$ and the functional (Banach) spaces $Y_a = Y_a(-a, a)$, $X_a = X_a(-a, a)$ are defined in Section 1.5.

The well-known results of the classical theory of singular integral operators allow us to state that the operator $l : X_a \to Y_a$ is bounded for $0 < \mu < 0.5$, there exists the (continuous) inverse operator $l^{-1} : Y_a \to X_a$, bounded for $0 < \mu < 0.5$ for $a \neq 2$ which is defined with the help of the explicit formula obtained by Muskhelishvili (1962) giving a unique solution of eqn. 1.32 for every function $f(t) \in Y_a$

$$\phi(t) = -\frac{1}{\pi R(t)} \int_{-a}^{a} \frac{R(s)f'(s)}{s-t} \, d s + \frac{C}{R(t)} \tag{1.33}$$

where

$$\text{const} = C = \frac{f(t) + \frac{1}{\pi} \int_{-a}^{a} h(t,s) \, d s \int_{-a}^{a} \frac{R(\tau)f'(\tau)}{\tau - s} \, d \tau}{\int_{-a}^{a} h(t,s) \, d s}$$

$$R(t) = \sqrt{a^2 - t^2}, \quad h(t,s) = -\frac{\ln|t-s|}{\pi R(s)}$$

when $a \neq 2$, that is when

$$C_0(a) = \int_{-a}^{a} h(t,s) \, d s = \ln\frac{2}{a} \neq 0$$

Lemma 1.12: The operator $l : X_a \to Y_a$ is of Fredholm type for every $a \neq 0$.

The proof is obvious for $a \neq 2$. For $a = 2$ it is sufficient to put $s_1 = ps$ in eqn. 1.32 and represent l as the sum of invertible and completely continuous (one-dimensional) operators, taking into account that the spaces X_a, Y_a are isomorphic for all $p \neq 0$.

Now let us consider the general form of the integral operator with logarithmic singularity of the kernel

$$K\phi = (l + N)\phi \equiv \int_{-a}^{a} \left[\frac{1}{\pi} \ln \frac{1}{|t-s|} + N(t,s)\right] \phi(s) \, d s$$

where $N(t,s)$ is a smooth function and we shall demand that

$$\frac{\partial N}{\partial t}(\gamma;t,s) = \frac{H(\gamma;t,s) - H(\gamma;t,t)}{t-s} \tag{1.34}$$

$H(\gamma;t,s)$ satisfies a Hölder condition with respect to the variables $t,s \in (-a, a)$.

The condition given by eqn. 1.34 is sufficient for the integral operator with the kernel N to be completely continuous on the pair of Banach spaces X_a, Y_a

(Chernokozhin and Shestopalov, 1982). The next statement follows from the index perturbation theorem (Gohberg and Sigal, 1974).

Theorem 1.6: The operator $K : X_a \to Y_a$ is of Fredholm type for $0 < \mu < 0.5$.

Theorem 1.7: The integral equation

$$K\phi = f(t), \quad f \in Y_a$$

has a unique solution for sufficiently small values of $a > 0$ in a weighted Hölder class X_a.

In order to prove this theorem we will introduce the norm in the Banach space $X_a = X_a(L);\ L = (-a, a)$:

$$\|\phi\|_{X_a} = \|\phi\|_a = \|\phi R\|_{H_\mu}$$

$$\|\phi\|_{H_\mu} = \|\phi\|_{C(L)} + h_\mu(\phi)$$

where

$$h_\mu(\phi) = \sup_{t,s \in L} \frac{|\phi(t) - \phi(s)|}{|t - s|}$$

Using the explicit form of the inverse operator l^{-1} one can get the following estimate, assuming that $a < 1$:

$$\|l^{-1} N\phi\|_a \leq C_K(a)\|\phi\|_a \qquad (1.35)$$

$$C_K(a) = C_S \left[\pi a^{1-\mu}\|N_t\|_a + \pi \ln^{-1}\frac{2}{a}\left(\|N\|_{C(L)} - a \ln a\|N_t\|_a\right) \right]$$

where C_S is the norm of the singular Cauchy operator in Hölder spaces. It is clear that

$$\|l^{-1}N\|_a < 1$$

for sufficiently small values of a and $C_K(a) \to 0$ as $a \to 0$.

Let us proceed to the integral operator-valued function with logarithmic singularity of the kernel

$$K(\gamma)\phi = (c\mathcal{L} + N(\gamma))\phi \equiv \int_{-a}^{a} [c \ln|t - s| + N(\gamma; t, s)]\phi(s)\,d s \qquad (1.36)$$

where $c = $ const, $N(\gamma; t, s)$ is a smooth function satisfying the condition given by eqn. 1.34 and

$$N(\gamma; t, s) = \frac{a_0(\gamma) n_0(t) m_0(s)}{\gamma - \gamma_0} + N_0(\gamma; t, s) \equiv \frac{A_{-1}(\gamma; t, s)}{\gamma - \gamma_0} + N_0(\gamma; t, s) \quad (1.37)$$

N, N_0 and A_{-1} are holomorphic functions in the neighborhood of γ_0 for every point $(t, s) \in \Pi_a = (-a, a) \times (a, a)$.

In accordance with the definition (Ilyinsky and Shestopalov, 1989) $K(\gamma)$ is finite-meromorphic in γ_0 since the one-dimensional integral operator A_{-1} has the degenerated kernel $A_{-1}(\gamma_0; t, s)$.

We shall say that $K(\gamma)$ admits separation of the simple pole pencil $P_0(\gamma)$ at the point γ_0 if

$$K(\gamma) = P_0(\gamma) + N_0(\gamma) \equiv c\mathcal{L} + \frac{A_{-1}}{\gamma - \gamma_0} + N_0(\gamma) \quad (1.38)$$

If $P_0(\gamma), N_0(\gamma)$ are real-valued functions for real values of γ, we say that $K(\gamma)$ admits separation of the simple-real-pole pencil.

According to the above results $\mathcal{L}$ is invertible ($a \neq 2$) and N, N_0 are completely continuous (integral) operators acting on the pair of Banach spaces $X_a = \widetilde{H}_\lambda(-a, a) \to Y_a = \widetilde{H}_\mu^1(-a, a)$, $\lambda < \mu < 1/2$ so that $K(\gamma)$ is a Fredholm $(\Phi-)$ operator.

Theorem 1.8: Let $K(\gamma)$ admit separation of the simple pole pencils in different points $Q_K = \{\gamma_i, i = 1, 2, \ldots, M\}$ in a domain G of the complex plane γ. Then the resolvent set $\rho_K(G)$ is $G \backslash T_K$ where T_K is a set of isolated points in G.

The estimate given by eqn. 1.35 can be obtained for the integral operator-valued function $K(\gamma)$ given by eqn. 1.36. Actually, assuming that $K(\gamma)$ is holomorphic in the domain $\mathcal{D}$ in the complex $\gamma-$plane one can see that in this case $C_K = C_K(a, \gamma)$ which admits uniform estimate on every compactum $\mathcal{D}_0 \subset \mathcal{D}$ Hence, we may assert that there exists a sufficiently small value of a such that every compactum $\mathcal{D}_0 \subset \mathcal{D}$ belongs to the resolvent set of $K(\gamma)$.

In other words, for every fixed $\gamma \neq \gamma_i$ we may choose such a small value of a that

$$\|\mathcal{L}^{-1} N(\gamma)\| < 1 \quad (\gamma \in G \backslash Q_K)$$

Then there exists only a trivial solution of the equation $K(\gamma)\phi = 0$ and therefore $\gamma \in \rho_K(G)$, since $\rho_K(G)$ is an open set, so we conclude that $\alpha(K(\gamma)) = \dim \ker K(\gamma) = 0, \gamma \in G\backslash T_K$.

Let us obtain explicit formulae for the characteristic number and inverse operator for the abstract pole pencil

$$P(\gamma) = \Upsilon + A_{-1}(\gamma - \gamma_0)^{-1} = \Upsilon + \langle ., x \rangle \psi(\gamma - \gamma_0)^{-1}$$

where χ is a continuous functional [in the case of the integral operator given by eqns. 1.36 and 1.38 $\psi = n_0, \langle ., \chi \rangle = (., m_0)$], $A_{-1} : X \to R^1$ is an one-dimensional operator and Υ is an invertible operator on a pair of Banach spaces $X, Y; u = \Upsilon^{-1}\psi \in X, u \neq 0$.

Lemma 1.13: The pole pencil $P(\gamma)$ has the unique characteristic number γ^*. If $\langle u, \chi \rangle \neq 0$ then $\gamma^* \neq \gamma_0$.

Actually, the equation

$$P(\gamma)\phi = \Upsilon\phi + \langle \phi, \chi \rangle \psi(\gamma - \gamma_0)^{-1} = 0 \qquad (1.39)$$

which states the problem of characteristic numbers for the pole pencil, is equivalent to the following equation

$$\phi + \langle \phi, \chi \rangle u(\gamma - \gamma_0)^{-1} = 0. \qquad (1.40)$$

One can substitute ϕ from eqn. 1.40 to eqn. 1.39:

$$\langle \phi, \chi \rangle \psi + \langle (\gamma - \gamma_0)^{-1} \langle \phi, \chi \rangle u, \chi \rangle \psi = 0 \qquad (1.41)$$

Assume that $\gamma \neq \gamma_0$. Then eqn. 1.41 is equivalent to the equation

$$\langle \phi, \chi \rangle [1 + \langle u, \chi \rangle (\gamma - \gamma_0)^{-1}] = 0 \qquad (1.42)$$

If $\langle \phi, \chi \rangle = 0$, then eqn. 1.39 has only the trivial solution and none of the $\gamma \neq \gamma_0$ can be a characteristic number.

Let us assume now that $\langle u, \chi \rangle \neq 0$, then

$$\gamma^* = \gamma_0 - \langle u, \chi \rangle$$

is the characteristic number of $P(\gamma)$ with corresponding eigenvector u.

Lemma 1.14: The multiplicities of the pole and the characteristic number of the pole pencil $P(\gamma)$ each are equal to one.

We will prove this lemma only for the case of pole multiplicity. Let us give an explicit form of the inverse operator $P^{-1}(\gamma)$. It is obvious that γ_0 is the pole of $P(\gamma)$ and the characteristic number of $P^{-1}(\gamma)$. To represent the latter let us consider the equation

$$\Upsilon\phi + \frac{\langle\phi,\chi\rangle\psi}{\gamma - \gamma_0} = f$$

which is equivalent to

$$\phi + \frac{\langle\phi,\chi\rangle u}{\gamma - \gamma_0} = \Upsilon^{-1}f$$

After substitution of ϕ from this equation into the initial one and assuming that $\gamma \neq \gamma_0$ we obtain

$$\frac{\langle\phi,\chi\rangle u}{\gamma - \gamma_0} = \frac{\langle\Upsilon^{-1}f,\chi\rangle u}{\gamma - \gamma_0} - \frac{\langle\phi,\chi\rangle\langle u,\chi\rangle u}{(\gamma - \gamma_0)^2}$$

$$\langle\phi,\chi\rangle = \frac{\langle\Upsilon^{-1}f,\chi\rangle}{1 + (\gamma - \gamma_0)^{-1}\langle u,\chi\rangle}$$
$$= \frac{\langle\Upsilon^{-1}f,\chi\rangle}{1 - (\gamma - \gamma_0)^{-1}(\gamma^* - \gamma_0)} = \frac{(\gamma - \gamma_0)\langle\Upsilon^{-1}f,\chi\rangle}{\gamma - \gamma^*}$$

One can see that $\gamma^* = \gamma_0$ only if $\langle u,\chi\rangle = 0$. Then

$$\phi = \Upsilon^{-1}f - (\gamma - \gamma^*)^{-1}\langle\Upsilon^{-1}f,\chi\rangle u$$

and we get the required form:

$$P^{-1}(\gamma)f = \Upsilon^{-1}f - (\gamma - \gamma^*)^{-1}\langle\Upsilon^{-1}f,\chi\rangle u$$

where

$$u = \Upsilon^{-1}\psi \qquad \gamma^* = \gamma_0 - \langle u,\chi\rangle$$

corresponding to the above notations and γ^* is the characteristic number of $P(\gamma)$.

The assertion of this lemma is easily deduced from the explicit representations of the operators P, P^{-1}.

Now we can prove the main result for the finite-meromorphic integral operator-valued functions with logarithmic singularity of the kernel which enables one to state sufficient conditions of the existence of characteristic numbers is contained in the following theorem.

Theorem 1.9: Let $K(\gamma)$ satisfy the conditions from Theorem 1.8. Then there exists a positive value of a_0, such that if $a \in (0, a_0)$ there lies in the neighbourhood of every point $\gamma \in Q_K$ at least one characteristic number of $K(\gamma)$.

Let $\mathcal{K}(\gamma_0; p)$ be the circle $\{|\gamma - \gamma_0| = p\}$, where $\gamma_0 \in Q_K, 0 < p < \min_{\gamma_0 \neq \gamma \in Q_K} |\gamma - \gamma_0|$. Using the above estimates we may see that

$$\|\mathcal{L}^{-1} N_0(\gamma)\|_{X_a} \to 0, \quad \|\mathcal{L}^{-1}\psi\|_{X_a} \to 0, \ a \to 0 \tag{1.43}$$

uniformly on every compactum $G_0 \subset G\backslash Q_K$, hence the characteristic number γ_P^* of the simple pole pencil $P_0(\gamma)$ in γ_0 taken from lemma 1.13 tends to γ_0. It means that for a sufficiently small positive value of a there are one pole and one characteristic number of $P(\gamma)$ in the circle $\mathcal{K}(\gamma_0; p)$ with corresponding radius p, therefore the total multiplicity of $P(\gamma)$ in $\mathcal{K}(\gamma_0; p)$ is equal to zero. It also follows from lemma 1.13 that for the fixed value of a,

$$|\gamma - \gamma_P^*(a)| \geq \delta > 0, \qquad \gamma \in \mathcal{K}(\gamma_0; p)$$

and we have estimates

$$\sup_{\gamma \in \mathcal{K}(\gamma_0;p)} \|P^{-1} N_0(\gamma)\|_{X_a} \leq \sup \|\mathcal{L}^{-1} N_0(\gamma)\| + \sup \frac{|\langle \mathcal{L}^{-1} N_0(\gamma), x\rangle| \|\mathcal{L}^{-1}\phi\|}{|\gamma - \gamma_0|}$$
$$\leq C_1(a) + C_2(a)p^{-1} < 1$$

where the last inequality is true when a is sufficiently small. It is easy to see that the total multiplicities of $K(\gamma)$ and $P_0(\gamma)$ coincide, corresponding to $\mathcal{K}(\gamma_0; p)$, and they both are equal to zero. Admitting one simple pole of $K(\gamma)$ in $\mathcal{K}(\gamma_0; p)$ we finally obtain that in $\mathcal{K}(\gamma_0; p)$ the operator-valued function $K(\gamma)$ has at least one characteristic number.

As another important example of an integral operator-valued function we will consider

$$K(\gamma)\phi = \int_{-a}^{a} \mathcal{K}(\gamma; t, s)\phi(s)\,\mathrm{d}s$$

$$\mathcal{K}(\gamma; t, s) = \sum_{n=0}^{\infty} \beta_n(\gamma)\psi_n(t)\psi_n(s) \tag{1.44}$$

in the Hilbert space X, where the system of functions $\{\psi_n\}$ is assumed to be complete; the formal operator series

$$\sum_{n=0}^{\infty} \beta_n(\gamma)(., \psi_n)\psi_n \tag{1.45}$$

is assumed to converge in operator norm; the coefficients $\beta_n(\gamma)$ are meromorphic functions in a domain G, so that $K(\gamma)$ is a finite-meromorphic operator-valued function.

Theorem 1.10: Under the following conditions:

$$\beta_n(\gamma_0) < 0 \quad (\beta_n(\gamma_0) > 0) \quad (n = 0, 1 \ldots)$$

the point γ_0 belongs to the resolvent set $\rho_K(G)$.

To prove this theorem it is sufficient to verify that the quadratic form

$$F_K(\gamma) = (K(\gamma)\phi, \phi) = \sum_{n=0}^{\infty} \beta_n(\gamma)|(\phi, \psi_n)|^2$$

is of fixed sign, using the conditions given above.

Theorem 1.11: Assume that $\beta_n(\gamma)$ satisfies the following conditions:

(i) $\beta_0(\gamma)$ takes all real values if $\operatorname{Im}\gamma = 0$, $\operatorname{Re}\gamma \in (c, d)$;
(ii) $\beta_n(\gamma), n \geq 1$ satisfies the condition from theorem 1.10 on the segment $[c, d]$.

Then there exists at least one characteristic number of $K(\gamma)$ on (c, d).

Assume that there exists a nontrivial element $\phi \in X$ with $(\phi, \psi_n) = 0, n \geq 1$, then the roots of the equation $\beta_0(\gamma) = 0$ are the characteristic numbers looked for. If we assume that such an element does not exist then the operator

$$K_1(\gamma) = \sum_{n=1}^{\infty} \beta_n(\gamma)(., \psi_n)\psi_n$$

is invertible on (c,d). If we put $0 \neq f_1 = K_1^{-1}\psi_0$, then the equation $K(\gamma)\phi = 0$ is equivalent to

$$\beta_0(\gamma)(K_1^{-1}(\gamma)\psi_0, \psi_0) + 1 = 0$$

which has roots on (c,d), according to the condition (i).

As a corollary from theorem 1.11 we find that the operator-valued function $K(\gamma)$ has at least one real (isolated) characteristic number on the interval (c,d).

Using the result of lemma 1.13, estimations given by eqn. 1.35 and the assumptions concerning the properties of functions m_0 and n_0 in eqn. 1.37 determining the degenerated kernel, it is easy to show that the pole pencil P_0 and the integral operator-valued function $K(\gamma)$ given by eqn. 1.38 has characteristic numbers γ^P and γ^K such that

$$\gamma^P = \gamma_0 - \frac{a_0}{\pi c}(l^{-1}n_0, m_0), \quad |\gamma^P - \gamma^K| < C_n a^2 |\ln a| \quad (C_n = \text{const}) \quad (1.46)$$

situated for $a = \text{diam}\, L \ll 1$ in the vicinity of the pole γ_0. γ^P may be considered as an approximate value of γ^K with the accuracy given by eqn. 1.46 which may be used for direct computations of characteristic numbers.

Applying the technique of matrix integral operator-valued functions developed by Chernokozhin and Shestopalov (1982) one can show that lemma 1.12 and theorems 1.6–1.9 remain valid for "pure" matrix operators

$$\mathcal{V} \equiv \mathcal{T} + \mathcal{N} = \|t_{ik}\|_{i,k=1}^M + \|N_{ik}\|_{i,k=1}^M$$

considered on the direct product of appropriate combination of spaces X_0, X_a and Y_a when the set of integration (L) consists of an arbitrary (finite) number of nonintersecting intervals. Each of t_{ik} is either the "logarithmic" operator l given by eqn. 1.32 or the singular integral operator S_L, and N_{ik} are integral operators with smooth kernels $(i, k = 1, 2, \ldots, M)$. The limited size of this book and Chapter does not allow us to present the complete proof of these results for matrix operators so when considering matrix operator-valued functions we will refer to the detailed proofs presented by Chernokozhin and Shestopalov (1982), Ilyinsky and Shestopalov (1989) and Shestopalov (1990, 1991a).

1.7 Reduction of the problem of fundamental frequencies to the Fredholm boundary integral equation. The fundamental frequency discreteness for two-dimensional open resonators

Let us consider the problems E and H on fundamental frequencies given by eqn. 1.1–1.5 with $\lambda = k^2$ taken as a spectral parameter (corresponding to the cases of the E and H oscillations, respectively) for open resonators whose cross-sections $\Omega = \Omega^1 \cup \Omega^2$ are formed by the domains from the families NB_p, NB_b^r, NB_u^r BD_b^r or BD_u^r. Using the notation of Section 1.1, assume that $N_2 = 1$, i.e. $\Omega^2 = \Omega_1^2$ is always a (one) simply-connected bounded or unbounded $S_\Pi(L)$-domain where L are the "slots" coinciding with rectilinear intervals $\Gamma_S^I = \bigcup_{i=1}^{M_S} \Gamma_{Si}^I$, $\Gamma_{Si}^I \subset \{(x_1, x_2) : x_2 = \text{const}\}$, $i = 1, 2, \ldots, M_S$. We will begin with the case $M_S = 1$, $L = (-a, a) \subset \mathbf{r} : \{x_2 = 0\}$ corresponding to the family NB_p and then generalise the results for arbitrary $S_\Pi^c(L)$-domain and natural M_S (for the families NB_b^r and NB_u^r).

We will assume in what follows that $\epsilon = \epsilon(\mathbf{r}) = \epsilon_m = \text{const}$, $\mu(\mathbf{r}) = \mu = 1$, $\mathbf{r} \in \Omega^m$ $(m = 1, 2)$.

Let us introduce Green's functions $G_m^i = G_m^i(\mathbf{r}, \mathbf{q}; \lambda)$ of the ith boundary-value problem $(i = 1, 2)$ for the Helmholtz operator $\Delta + \lambda\epsilon$ of the (bounded or unbounded) domains Ω^m (under the assumption of its existence). In particular, $\Omega^1 = \Omega^+ = \{\mathbf{r} : x_2 > 0\}$ for the family NB_p and, consequently, in this case $G_1^i(\mathbf{r}, \mathbf{q}; k) = h^i(\mathbf{r}, \mathbf{q}; k)$ where h^i are given by eqn. 1.23.

It is known (Reed and Simon, 1980) that for two-dimensional domains bounded by piece-wise smooth simply-connected contours Green's functions $G_m^i(\mathbf{r}, \mathbf{q}; \lambda)$ admit analytical continuation to a complex domain D_m^i as meromorphic functions of λ on the whole complex plane C_λ with real finite-order poles (which form a set P_m^i without finite accumulation points), that is, here $D_m^i = C_\lambda \setminus P_m^i$ where P_m^i is a countable set of isolated points $(i, m = 1, 2)$.

There is no general information on the structure of D_m^i for arbitrary unbounded domains with noncompact boundaries except for certain particular cases, like the half-plane Ω^+ when Green's functions $G_1^{1,2} = h^{1,2}$ given by eqn. 1.23 admit analytical continuation to the infinite-sheet Riemann surface $\mathcal{D}^{(0)} = \Lambda$ (of the function $\ln \lambda$) with one logarithmic branch point $\lambda = 0$. [as well as the fundamental solution $\mathcal{E}(\mathbf{r} - \mathbf{q}) = \frac{i}{4}H_0^{(1)}(\sqrt{\lambda}|\mathbf{r} - \mathbf{q}|)$]. In this way we will assume that for the considered unbounded domains (from the families NB_b^r and NB_u^r) Green's functions $G_m^i(\mathbf{r}, \mathbf{q}; \lambda)$ together with the limiting values ("traces") of

$$G_m^2(\mathbf{r}, \mathbf{q}; \lambda)\big|_{\mathbf{q} \in L}, \quad G_m^2(\mathbf{r}, \mathbf{q}; \lambda)\big|_{\mathbf{r}, \mathbf{q} \in L}$$

and

$$\frac{\partial G_m^1}{\partial x_2}\bigg|_{\mathbf{q} \in L} (\mathbf{r}, \mathbf{q}; \lambda), \quad \frac{\partial G_m^1}{\partial x_2}\bigg|_{\mathbf{r}, \mathbf{q} \in L} (\mathbf{r}, \mathbf{q}; \lambda)$$

admit analytical continuations to one and the same complex manifolds D_m^i such that

(i) $D_m^i \subseteq \mathcal{D}^{(0)}$ (where $m = 1, 2$ for the family NB_u^r and $m = 1$ for the family NB_u^r);

(ii) $\emptyset \neq D^i = \bigcap_{m=1}^{2} D_m^i \subseteq \mathcal{D}^{(0)}$, $i = 1, 2$ (for the family NB_u^r);

(iii) D_m^i and D^i contain a strip $\lambda : |\operatorname{Im} \lambda| < d$ for certain $d > 0$ with the exception of not more than a countable quantity of cuts and singular points.

Let us note that $D^1 = D^2 = \mathcal{D}^{(0)}$ for the domains of the family NB_p and in one specific (canonical) case related to the family NB_u^r when $\Omega^1 = \{\mathbf{r} : x_2 > 0\}$, $\Omega^2 = \{\mathbf{r} : x_2 < 0\}$.

In addition, since the "traces" of Green's functions on L for bounded domains are assumed to be meromorphic functions of λ on the whole complex plane C_λ, then here $D_m^i = C_\lambda \setminus P_m^i$ (and $D^i = C_\lambda \setminus P^i = \bigcup_{m=1}^{2} P_m^i$) where P_m^i is a countable set of isolated points $(i, m = 1, 2)$.

These assumptions on D_m^i may be treated as indirect limitations defining a class of unbounded domains from the families NB_b^r and NB_u^r and they will be overcome in Section 2.6 with the help of an alternative approach based on solving the spectral problems in the generalised statement.

Let us denote by $u = u_m(\mathbf{r})$ and $v = v_m(\mathbf{r})$, $\mathbf{r} \in \Omega^m$, (nontrivial) solutions of the problems E and H (under the assumption of its existence). In accordance with the results of theorems 1.4 and 1.5, u_m and v_m admit representations in each of the domains Ω^m as Green's potentials given by eqns. 1.24 and 1.25 with the kernels determined in terms of Green's functions G_m^i and the densities

$$\begin{aligned} \phi &= \phi_m(t) = \frac{\partial u_m}{\partial x_2}(\mathbf{q}^0) \in \widetilde{H_\mu}(L) \\ \psi &= \psi_m(t) = v_m(\mathbf{q}^0) \in X_0(L), \quad \mathbf{q}^0 = (t, 0) \in L \quad (m = 1, 2) \end{aligned} \tag{1.47}$$

We take into account that, as follows from Green's identity, theorem 1.4 and the classical theorem of the potential theory concerning the behaviour of the normal derivative of a simple layer logarithmic potential on smooth boundaries, eqn. 1.47 may be treated as identities on L (and simultaneously as definitions of the functions ϕ_m and ψ_m). So we may put $\psi_1(t) = \psi_2(t)$, $\phi_2(t) = \epsilon_2/\epsilon_1\phi_1(t)$ and it fulfils the second conjugation condition given by eqn. 1.3 for u and the first conjugation condition for v. To satisfy the first conjugation condition for u and the second for v we will apply the results of theorem 1.4 for eqns. 1.24 and 1.25 assuming that the densities are given by eqn. 1.47 and the kernels are expressed by G_m^i and let $\mathbf{r}$ tend to L from the domains Ω^1, Ω^2, respectively. Finally we will obtain the required boundary integral equations

$$K_E(\lambda)\psi \equiv \int_L \mathcal{K}_E(\lambda; t, s)\psi_1'(s)\,d\,s = 0, \quad t \in L$$

$$\mathcal{K}_E(\lambda; t, s) = \sum_{m=1}^{2} \frac{\partial}{\partial x_2}\bigg|_{\mathbf{r}=\mathbf{r}^0 \in L} \int_q^s \frac{\partial}{\partial y_2} G_m^1(\mathbf{r}, \mathbf{q}^0; \lambda)\,d\,s' \qquad (1.48)$$

$$q \in R^1, \quad \mathbf{r}^0 = (t, 0), \quad \mathbf{q}^0 = (s', 0)$$

where $\psi = \psi_1' \in \widetilde{H_\mu}(L)$ and

$$K_H(\lambda)\phi_1 \equiv \int_L \mathcal{K}_H(\lambda; t, s)\phi_1(s)\,d\,s = 0, \quad t \in L$$

$$\mathcal{K}_H(\lambda; t, s) = \sum_{m=1}^{2} \epsilon_m G_m^2(\mathbf{r}^0, \mathbf{q}^0; \lambda) \qquad (1.49)$$

where $\phi_1 \in \widetilde{H_\mu}(L)$.

The inverse assertion is true: if we take any nontrivial solutions $\psi, \phi_1 \in \widetilde{H_\mu}(L)$, $(\psi, 1) = \int_L \psi(s)\,d\,s = 0$ of eqns. 1.48, 1.49 and put $\phi_2 = \epsilon_2/\epsilon_1\phi_1$, $\psi_1(s) = \int_{-a}^{s} \psi(s')\,d\,s'$ then Green's potentials $v_m(\mathbf{r})$ and $u_m(\mathbf{r})$ given by eqns. 1.24 and 1.25 define the corresponding (nontrivial) solutions of the initial spectral problem on fundamental frequencies given by eqn. 1.1–1.5. Really, they satisfy the Helmholtz equation in Ω^m, boundary-value conditions given by eqns. 1.2 and 1.3 (because, e.g. $\phi_m(t) \equiv \frac{\partial u_m}{\partial x_2}|_L$). One can also show using the properties of singular integrals (Muskhelishvili, 1962) that $v_m(\mathbf{r})$ and $u_m(\mathbf{r})$ satisfy the "edge" conditions given by eqn. 1.4 and the radiation conditions in unbounded domains (of the families NB_p) given by eqn. 1.5 (due to theorem 1.5). Hence we have proved

Theorem 1.12: The problems E and H are equivalent for $\lambda \in D^i$ $(i = 1, 2)$ to the integral equations 1.48 and 1.49 where $K_E(\lambda) : X_a \to X_a$ and $K_H(\lambda) :$ $X_a \to Y_a$ are Fredholm operator-valued functions

The equivalence is based on the following: treat eqns. 1.48 and 1.49. as (nonlinear) eigenvalue problems or the problems on characteristic numbers for the integral operator-valued functions $K_E(\lambda)$ and $K_H(\lambda)$ of the complex variable λ, where λ is also the spectral parameter of the problems E and H. In the case when both domains Ω^m are bounded (the families BD_u^r and BD_b^r), $D^i = C_\lambda \backslash P^i = \bigcup_{m=1}^2 P_m^i$ where P^i are countable sets of isolated points [poles of Green's functions coinciding with the real eigenvalues of the Dirichlet $(i = 1)$ and Neumann $(i = 2)$ boundary eigenvalue problems for the Laplace equation in Ω^m]. Hence here the integral operator-valued functions $K_E(\lambda)$ and $K_H(\lambda)$ may have, in accordance with the above assumption concerning the analytical continuation of the "traces" of Green's functions, only isolated singularities (poles) and they are meromorphic functions on C_λ. In such a case the corresponding problems on characteristic numbers may be considered on the whole λ-plane and one can put $D^i = C_\lambda$ $(i = 1, 2)$ in theorem 1.12.

The Fredholm property of $K_E(\lambda)$ and $K_H(\lambda)$ directly follows from the index perturbation theorem (Gohberg and Krein, 1965) and theorem 1.4 which enables one to "completely" separate the kernel singularities in eqns. 1.48 and 1.49 (given by eqn. 1.27), eliminating the invertible (l, S_L) and completely continuous parts of the integral operators.

In accordance with the notation and results of Section 1.3, we may conclude, using the principle of the discreteness of the set of characteristic numbers for Fredholm meromorphic operator-valued functions (Gohberg and Krein, 1965, Sanchez-Palencia, 1980), that the following theorem holds which establishes the spectrum discreteness of two-dimensional E and H eigenoscillations for the considered families of open circular resonators.

Theorem 1.13: The spectrum $\sigma_{E,H}(\Delta)$ of the problems E and H given by eqns. 1.1–1.5 consists of the points having finite multiplicities and forms a discrete set in D^i, i.e. there may be not more than a finite number of fundamental frequencies (eigenvalues of the problem given by eqns. 1.1–1.5) in every compactum $Q \subset D^i$ $(i = 1, 2)$.

One can specify the structure of D^i: $D^i = C_\lambda$ for the families BD_u^r, BD_b^r; $D^i = \mathcal{D}^{(0)}$ for the family NB_p.

Application of the technique of generalised potentials and matrix integral operator-valued functions described in Section 1.6 enables one to generalise the main result stated in theorem 1.13 for the families NB_b^r and NB_u^r when Ω^2 consists of several subdomains $(M_S > 1)$ by introducing the "narrowing" of solutions and Green's functions onto each subdomain of a multi-connected domain Ω^2.

1.8 Existence of fundamental frequencies

Consider first the canonical structure for the families BD_u^r and BD_b^r introduced in Section 1.1, the rectangular shielded slot resonator, whose cross-section in the plane $x_3 = 0$ is formed by two rectangular domains $\Omega^1 = \Pi_{ab_1} = \{\mathbf{r} = (x_1, x_2) : 0 < x_1 < a, \ 0 < x_2 < b_1\}$ and $\Omega^2 = \Pi_{ab_2}\{\mathbf{r} = (x_1, x_2) : 0 < x_1 < a, \ -b_2 < x_2 < 0\}$ with the common part of the boundaries $\Gamma = \partial\Omega^1 \cap \partial\Omega^2 = \{\mathbf{r} : x_2 = 0, \ 0 \le x_1 \le a\}$ containing an interval $L = \{\mathbf{r} : x_2 = 0, \ \frac{a}{2} - w = d_1 < x < d_2 = \frac{a}{2} + w\}$ (a slot) with endpoints ("edges") $\partial L = \{A_1 = d_1, \ A_2 = d_2\}$.

The mathematical statements of the problems E and H to determine the E- and H-type fundamental frequencies of such a cylindrical slot resonator are given in Section 1.2 (eqns. 1.1–1.5).

In order to prove the existence of fundamental frequencies we will apply in what follows the method of integral operator-valued functions developed in Sections 1.5 and 1.6 [limiting ourselves to the case of the H oscillations and assuming that $\epsilon = \epsilon(\mathbf{r}) = \epsilon_i = \text{const}$, $\mu(\mathbf{r}) = \mu = 1$, $\mathbf{r} \in \Omega^i$ $(i = 1, 2)$].

Let us introduce Green's functions of the second boundary-value problem for the Helmholtz equation 1.1 of the rectangular domains Π_{ab_i} $(i = 1, 2)$

$$G_i^2(\mathbf{r}, \mathbf{q}; \lambda) = \frac{4}{\epsilon_i a b_i} \sum_{n=0}^{\infty} \sum_{m=0}^{\infty} \varepsilon_n \varepsilon_m \frac{\psi_n(x_1, a)\psi_n(y_1, a)\psi_m(x_2, b_i)\psi_m(y_2, b_i)}{\lambda - \lambda_{nm}^{(i)}}$$

$$(1.50)$$

where

$$\varepsilon_0 = 0.5 \ \varepsilon_n = 1 \ (n \geq 1) \quad \psi_n(t,d) = \cos \frac{\pi n t}{d}$$

$$\lambda_{nm}^{(i)} = \frac{\pi^2}{\epsilon_i} \left(\frac{n^2}{a^2} + \frac{m^2}{b_i^2} \right) \quad (i = 1, 2)$$

As shown by Ilyin (1956), if $|\mathbf{r} - \mathbf{q}| \geq \delta, \delta > 0$, the series in eqn. 1.50 converges uniformly with respect to $\mathbf{r}, \mathbf{q} \in \Pi_{ab_i}$ if one sums with respect to increasing values of $\lambda_{nm}^{(i)}$ in every inner subdomain $Q \subset \Pi_{ab_i}$ and has a logarithmic singularity:

$$G_i^2(\mathbf{r}, \mathbf{q}; \lambda) = c_i \ln |\mathbf{r} - \mathbf{q}| + N_i(\mathbf{r}, \mathbf{q}; \lambda), \quad \mathbf{r}, \mathbf{q} \in \Pi_{ab_i}$$

where $c_i = $ const, N_i are once continuously differentiable with respect to the co-ordinates of $\mathbf{r}, \mathbf{q}$ for $\lambda \notin \Lambda_i = \{\lambda_{nm}^{(i)}\}_{m,n=0}^{\infty}$ and together with G_i^2 are meromorphic functions with respect to λ with the same simple (if all of them are different) real poles $\lambda_{nm}^{(i)}$ $(m, n = 0, 1, \ldots, i = 1, 2)$. The following representations hold for Green's functions [obtained by Ilyinsky and Shestopalov (1989)]

$$G_i^2(\mathbf{r}, \mathbf{q}; \lambda)\big|_{\mathbf{q}=\mathbf{q}^0 \in L} = G_i^2(\mathbf{r}, \mathbf{q}^0; \lambda) = f(\mathbf{r}, \mathbf{q}; \lambda) + g_i(\mathbf{r}, \mathbf{q}; \lambda)$$

$$f(\mathbf{r}, \mathbf{q}; \lambda) = \frac{1}{2\pi} \ln \Phi(\mathbf{r}, \mathbf{q}^0) = \frac{1}{\pi} \ln |\mathbf{r} - \mathbf{q}^0| - \frac{\ln 2}{2\pi} + f_1(\mathbf{r}, \mathbf{q}^0) \qquad (1.51)$$

where $\mathbf{q}^0 \in L$, $g_i(\mathbf{r}, \mathbf{q}; \lambda)$ and $f_1(\mathbf{r}, \mathbf{q}; \lambda)$ are once continuously differentiable with respect to the co-ordinates of $\mathbf{r}$ in every closed domain $\overline{Q_0} \subset \Pi_{ab_i}$, $L \subset \partial Q_0$ $(i = 1, 2)$. This means that when one of the points lies on L, Green's function has the same singularity as the kernel of a two-dimensional single-layer potential, i.e. in accordance with the terminology of Section 1.5, $G_i^2(\mathbf{r}, \mathbf{q}^0; \lambda)$ $(i = 1, 2)$ are the kernels of the generalised single-layer potential. And what is more, Green's functions of the rectangular domains satisfy all the demands concerning the properties of analytical continuations and "traces" formulated in Section 1.7 for bounded domains, namely, $G_i^2(\mathbf{r}, \mathbf{q}; \lambda)$ and $G_i^2(\mathbf{r}, \mathbf{q}^0; \lambda)$ are meromorphic functions of λ on the whole complex plane C_λ with the (isolated) real simple $\lambda_{nm}^{(i)}$ $(m, n = 0, 1, \ldots, i = 1, 2)$.

The above conclusions enable us to assert, using the results of Section 1.7, that the spectral problem considered H is equivalent in C_λ to the problem of characteristic numbers for the Fredholm integral operator-valued function $K_H(\lambda)$ given by eqn. 1.47. It implies, in particular, the discreteness of the set σ_H of the characteristic numbers λ^H in C_λ, that is, there may be not more than a finite number of the points of σ_H in every finite ball $\{\lambda : |\lambda| < R\}$.

Let us prove the "global" existence of characteristic numbers of $K_H(\lambda)$.

It is easy to show that the kernel of $K_H(\lambda)$ admits the following representation:

$$K_H(\lambda; t, s) = \beta_0(\lambda) + \sum_{n=1}^{\infty} \beta_n(\lambda)\psi_n(t)\psi_n(s) \tag{1.52}$$

where

$$\beta_0(\lambda) = -\frac{1}{a}[f(\lambda q_1) + f(\lambda q_2)]$$

$$\beta_n(\lambda) = \frac{2}{a}\{\epsilon_1 b_1 f[q_1(n^2 r_1 - \lambda)] + \epsilon_2 b_2 f[q_2(n^2 r_2 - \lambda)]\}$$

$$q_i = \epsilon_i b_i^2, \quad r_i = \frac{\pi^2}{\epsilon_i a^2} \quad (i = 1, 2)$$

$$f(z) = \begin{cases} -\dfrac{\coth\sqrt{-z}}{\sqrt{-z}}, & z < 0 \\[2mm] -\dfrac{\cot\sqrt{z}}{\sqrt{z}}, & z > 0 \end{cases}$$

$K_H(\lambda; t, s)$ is a meromorphic function with the same set of poles $\Lambda = \Lambda_1 \cup \Lambda_2$ as Green's functions $G_i^2, i = 1, 2$ have. If $\lambda \notin \Lambda$ then the series in eqn. 1.52 converges uniformly for $s, t \in L, |s - t| \geq \delta, \delta > 0$. The series given by eqn. 1.52 converges also in $L_2(L)$ and $L_{2,R}(L)$ where $L_{2,R}(L)$ is the weighted L_2-space with the weight $R(t) = \sqrt{(t - d_1)(d_2 - t)}$ and has the logarithmic singularity:

$$K_H(\lambda; t, s) = g_0(\lambda; t, s) + N(\lambda; t, s) \tag{1.53}$$

where

$$g_0(\lambda; t, s) = \frac{\epsilon_1 + \epsilon_2}{\pi}\left[\ln\left(2\sin\frac{\pi}{2a}|s - t|\right) + \ln\left(2\sin\frac{\pi}{2a}|s + t|\right)\right]$$

$$= \frac{\epsilon_1 + \epsilon_2}{\pi}\ln|s - t| + N_1(\lambda; t, s)$$

$$N(\lambda; t, s) = \sum_{n=0}^{\infty}\left[\beta_n(\lambda) - (\epsilon_1 + \epsilon_2)\frac{a}{\pi n}\right]\psi_n(t)\psi_n(s)$$

where N is a meromorphic function with the same poles as K_H has, the series for $N(\lambda; t, s)$ converges absolutely and uniformly for $s, t \in L$ and admits term-wise differentiation with respect to these variables. The function β_0 has simple poles at the points

$$B_0 = \{\beta_{0m}^{(i)}\}_{m=0}^{\infty} \quad (i = 1, 2)$$

where

$$\beta_{om}^{(1)} = \frac{\pi^2}{q_1}m^2, \quad \beta_{0m}^{(2)} = \frac{\pi^2}{q_2}m^2 \quad (m = 0, 1, \dots)$$

The functions $\beta_n(\lambda)$ have poles at the points

$$B_n = \{\beta_{nm}^{(i)}\}_{m=0}^{\infty} \quad (i = 1, 2)$$

where

$$\beta_{nm}^{(1)} = \beta_{0m}^{(1)} + n^2 r_1, \quad \beta_{nm}^{(2)} = \beta_{0m}^{(2)} + n^2 r_2 \quad (m = 0, 1, \dots \ n = 1, 2, \dots)$$

Let $B_n = \{\beta_{nk}\}_{k=0}^{\infty}$ be the sequences of real non-negative poles of $\beta_n(\lambda)$ ordered in accordance with increasing values. Then in every interval $\Lambda_{nk} = (\beta_{nk}, \beta_{nk+1})$ the function $\beta_n(\lambda)$ takes all real values and

$$\beta_n(\lambda) < 0 \quad \text{for} \quad \lambda < \min_{k \geq 0} \beta_{nk} = n^2 \min_{i=1,2} r_i = n^2 r$$

Therefore, under the condition

$$r > \min(\beta_{01}^{(1)}, \beta_{01}^{(2)})$$

which is equivalent to

$$\min_{i=1,2} \epsilon_i^{-1} > a^2 \min_{i=1,2} \epsilon_i b_i^2$$

all $\beta_n(\lambda) < 0$ for $n = 1, 2, \dots, \lambda \in \Lambda_{00}$, and $\beta_0(\lambda)$ takes all real values in the interval Λ_{00}.

It is easy to show using the explicit form of the coefficients that for every $\beta_n(\lambda), n \geq 1$, and for every first interval Λ_{n1} where $\beta_n(\lambda)$ takes all real values the following assertions hold:

(i) all successive coefficients $\beta_k(\lambda), k > n$, are of fixed sign;

(ii) all previous coefficients $\beta_k(\lambda), 0 \leq k < n$, are of the same fixed sign on a certain subinterval $\Lambda_{n1}^0 \subset \Lambda_{n1}$.

In order to apply theorem 1.10, take $L_{2,R}(L)$ as the Hilbert space X and consider the operator-valued function $K_H(\lambda)$ acting in X, where the system of functions $\{\psi_n(t)\}$ is complete. The series in the corresponding quadratic form given by eqn. 1.46 converges absolutely on every compactum $Q_0 \subset \{\lambda | Im\lambda = 0\}$, which does not contain poles from Λ, because $\beta_n(\lambda) \sim \frac{1}{n}$ and (ϕ, ψ_n) are the Fourier coefficients of the function $\phi \in L_{2,R}$.

Now we may state the required "global" existence of the fundamental frequencies.

Theorem 1.14: The problem H of the fundamental frequencies for the rectangular slot resonator has real isolated eigenvalues on every interval $\Lambda_{nk}; n, k = 0, 1, \ldots$, between every two different neighbouring poles of the operator-valued function $K(\lambda)$ of the problem H.

Using the technique developed in Section 1.6, it is not difficult to show the "local" existence of the H-type fundamental frequencies for a shielded resonator whose cross-section is formed by an arbitrary domain from the families BD_u^r or BD_b^r introduced in Section 1.1.

Consider in this way such a shielded slot resonator whose cross-section in the plane $x_3 = 0$ is formed by two arbitrary S_Π-domains Ω^1 and Ω^2 with the common part of the boundaries $\Gamma = \partial\Omega^1 \cap \partial\Omega^2$ containing an interval $L = \{\mathbf{r} : x_2 = 0, \frac{a}{2} - w = d_1 < x < d_2 = \frac{a}{2} + w\}$ (a "slot").

In accordance with the result of Section 1.5, Green's functions of the second boundary-value problem for the Helmholtz equation 1.1 of the domains Ω^i $(i = 1, 2)$ admit the representations as kernels of the generalised single-layer potentials:

$$G_i^2(\mathbf{r}, \mathbf{q}; \lambda) = c_i \ln|\mathbf{r} - \mathbf{q}| + N_i(\mathbf{r}, \mathbf{q}; \lambda), \quad \mathbf{r}, \mathbf{q} \in \Omega_0^i \subset \Omega^i \qquad (1.54)$$

where Ω_0^i is any closed simple-connected subdomain of Ω^i such that $L \subset \partial\Omega_0^i$, $c_i = $ const, N_i are once continuously differentiable with respect to the co-ordinates of $\mathbf{r}, \mathbf{q}$. Green's functions of the bounded S_Π^c domains admit representations in the form of bilinear expansions

$$G_i^2(\mathbf{r}, \mathbf{q}; \lambda) = \sum_n \frac{u_n^{(2i)}(\mathbf{r}) u_n^{(2i)}(\mathbf{q})}{\lambda - \lambda_n^{(i)}} \qquad (1.55)$$

where $u_n^{(2i)}(\mathbf{r})$ are the eigenfunctions of the second boundary-value problem of the Laplace equation of the domain Ω^i corresponding to the eigenvalues $\lambda_n^{(i)}$ and the series converge in $L_2(\Omega_i)$, $\mathbf{r} \neq \mathbf{q}$. Such expansions together with eqn. 1.54 hold when $\mathbf{r}, \mathbf{q} \in \Omega^i$ and define G_i^2 as meromorphic functions of λ.

In order to prove the "local" existence of eigenvalues of the problem given by eqns. 1.1–1.5 we will make the assumptions which narrows the class of domains (corresponding to the families BD_u^r and BD_b^r) in such a way that eqns. 1.55 would hold also for the "traces" of Green's functions on L. These assumptions indirectly separate (define) subfamilies S_Π^{LE} (of bounded domains) of the families S_Π and S_Π^c introduced in Section 1.1.

Assume that Green's functions G_i^2 satisfy all the demands concerning the properties of analytical continuations formulated in Section 1.7 for bounded domains, namely,

(i) $G_i^2(\mathbf{r}, \mathbf{q}; \lambda)$, $G_i^2(\mathbf{r}, \mathbf{q}^0; \lambda)$ and $G_i^2(\mathbf{r}^0, \mathbf{q}^0; \lambda)$ $(\mathbf{r}^0, \mathbf{q}^0 \in L)$ are meromorphic functions of λ on the whole complex plane C_λ with the (isolated) real simple poles $\lambda_n^{(i)}$ $(n = 0, 1, \ldots, i = 1, 2)$;

(ii) the following representation holds for the "traces"

$$
\begin{aligned}
G_i^2(\mathbf{r}^0, \mathbf{q}^0; \lambda) &= \frac{u_n^{(i)}(\lambda; t) u_n^{(i)}(\lambda; s)}{\lambda - \lambda_n^{(i)}} + N_i^{2n}(\lambda; t, s) \\
N_i^{2n}(\lambda; t, s) &= c_i \ln |s - t| + N_{i1}^{2n}(\lambda; t, s) \\
\mathbf{r}^0 &= (t, 0) \in L, \quad \mathbf{q}^0 = (s, 0) \in L
\end{aligned}
\tag{1.56}
$$

in a vicinity of every pole $\lambda_n^{(i)}$ where the relations given by eqn. 1.43 hold for $u_n^{(i)}$, N_{i1}^{2n} for $t, s \in \bar{L}$ uniformly with respect to λ, N_i^{2n} and N_{i1}^{2n} are the meromorphic functions of λ with the same poles as Green's functions G_i^2 have (except for $\lambda_n^{(i)}$ in the case of N_i^{2n}), N_i^{2n} is holomorphic in a vicinity of $\lambda_n^{(i)}$ and N_{i1}^{2n} is once continuously differentiable with respect to $t, s \in L$ $(n = 0, 1, \ldots, i = 1, 2)$.

Let us note that we do not generally assume $u_n^{(i)}$ to coincide with the "traces" (limiting values) of the eigenfunctions $u_n^{(2i)}(\mathbf{r})$, $\mathbf{r} \to L$ and the assumption (ii) is thus "weaker" than the demand for the series in eqn. 1.55 to converge (for $\mathbf{r} \neq \mathbf{q}$) in the closed domains $\bar{\Omega}^i$ $(i = 1, 2)$.

We will call the domains whose Green's functions G_i^2 satisfy the conditions (i) and (ii) the S_Π^{LE}-domains.

The properties of Green's functions and the above assumptions enable us to literally repeat the procedure of reduction to the integral equation 1.49 (described in Section 1.7) and to state that $K_H(\lambda)$ given by eqn. 1.49 is a finite-meromorphic integral operator-valued function which admits separation of the simple-real-pole pencil in a vicinity of every pole $\lambda_n^{(i)}$ $(n = 0, 1, \ldots, i = 1, 2)$. Now the direct application of theorem 1.9 [under the assumptions (i) and (ii)] yields the required "local" existence of the H-type fundamental frequencies.

Theorem 1.15: Assume that resonator's cross-section is formed by arbitrary bounded S_Π^{LE}-domains Ω^1 and Ω^2. Then there exists a positive value a_0 such that if $a\mathrm{diam}L \in (0, a_0)$ there lies in a vicinity of every point $\lambda_n^{(i)}$ $(n = 0, 1, \ldots, i = 1, 2)$ at least one eigenvalue of the problem given by eqns. 1.1–1.5 [or a characteristic number of $K_H(\lambda)$].

When $\lambda \in \sigma_K$ where σ_K is the set of characteristic numbers of the operator-valued function $K_H(\lambda)$, $\omega = \sqrt{\lambda}$ is the H-type fundamental frequency of the resonator considered.

Of course, rectangular domains $\Omega^i = \Pi_{ab_i}$ $(i = 1, 2)$ are the S_Π^{LE}-domains and the "local" existence holds for a rectangular shielded slot resonator.

Let us consider the boundary integral equation 1.47 in the case when resonator's cross-section is formed by arbitrary bounded S_Π^{LE}-domains. Using the result of lemma 1.13, it is easy to show that the pole pencil $P = P_n$ of the operator-valued function K_H given by eqn. 1.47 corresponding to every pole $\lambda_n^{(i)}$ and K_H itself have characteristic numbers λ_{ni}^P and λ_{ni}^K such that

$$\lambda_{ni}^P = \lambda_n^{(i)} - \frac{1}{\pi(c_1 + c_2)}(l^{-1}u_n^{(i)}, u_n^{(i)}) \quad (n = 0, 1, \ldots i = 1, 2) \qquad (1.57)$$

and

$$|\lambda_{ni}^P - \lambda_{ni}^K| < C_{ni}a^2|\ln a| \quad (C_{ni} = \mathrm{const}) \qquad (1.58)$$

for $a = \mathrm{diam}\, L \ll 1$. Eqn. 1.57 gives reliable approximate values of fundamental frequencies with known accuracy given by eqn. 1.58 and may be used for direct computations in the case of arbitrary cross-sections because one has to know only "weak" values $(\cdot, u_n^{(i)})$ of eigenfunctions on L.

Explicit approximate formulae for the squared values λ_{mk}^{iH} of the (higher order) fundamental frequencies of the rectangular shielded slot resonator considered in this Section may be also represented with the help of singular integrals:

$$\lambda_{mk}^{iH} = \lambda_{mk}^{(i)} - \int_L \psi_m(t, b_i) R^{-1}(t) \left[\frac{m}{a} F_m^i(t) + C_m \right] dt$$

$$R(t) = \sqrt{\left(t - \frac{a}{2} + w \right) \left(\frac{a}{2} + w - t \right)}$$

$$F_m^i(t) = \int_L \frac{R(s) \psi_m(s, b_i)}{s - t} ds \qquad (1.59)$$

$$C_m = \frac{1}{\ln \frac{1}{w}} \left[\psi_m(t, b_i) - \frac{1}{\pi^2} \int_L R^{-1}(s) \ln |t - s| F_m^i(s) ds \right]$$

$$L = \left(\frac{a}{2} - w, \frac{a}{2} + w \right) \quad (m, k = 0, 1, \ldots \; i = 1, 2)$$

The accuracy in eqn. 1.59 is determined by the estimate given by eqn. 1.58.

An alternative proof of the "local" existence together with the analysis and classification of eigenoscillations may be carried out by the perturbation theory methods developed in Section 1.9 which enable one, in addition, to obtain the data which is more precise than that given by eqn. 1.59.

The assertion of theorem 1.15 may be generalised for arbitrary "coupled" slot resonators with cross-sections formed by arbitrary bounded S_{Π}^{LE}-domains whose boundaries contain N rectilinear slots ($N > 1$), if to apply the technique of matrix integral operator-valued functions developed by Chernokozhin and Shestopalov (1982) (see the remark at the end of Section 1.6). Of course, the multiplicities of eigenvalues (and fundamental frequencies) will essentially depend on N, i.e. there may be at least N different eigenvalues in a vicinity of any (simple) pole of Green's function. The limited size of this book does not allow us to present the complete proofs of these results and we again refer to the detailed analysis carried out by Chernokozhin and Shestopalov (1982) and Ilyinsky and Shestopalov (1989).

We may conclude that the spectra of the II-type fundamental frequencies of the shielded slot resonators from the families BD_u^r and BD_b^r consists (at least for sufficiently narrow slots L) of the regular (countable) component $\Lambda_{\text{reg}}^H = \cup_{i=1}^2 \cup_{m,k} \lambda_{mk}^{iH}$, $m^2 + k^2 \neq 0$, "perturbed" with respect to the higher order eigenoscillations spectra of cylindrical resonators with homogeneous filling (which form the slot resonators) and irregular component Λ_{irr}^H containing not more than a finite number of the fundamental frequencies (corresponding to the so called slot resonances) "perturbed" with respect to the zero eigenvalue

$\lambda_{00}^{(1)} = \lambda_{00}^{(2)} = 0$ of the Neumann boundary eigenvalue problem for the Laplace equation in each of the domains forming the cross-section of a slot resonator.

1.9 The small parameter method. Explicit determination of fundamental frequencies in the form of asymptotic series

We will again begin with the consideration of the canonical structure of the family NB_p^r introduced in Section 1.1, an open slot resonator with non-compact boundaries, the rectangular image slot resonator, whose cross-section in the plane $x_3 = 0$ is formed by the half-plane $\Omega^1 = \Omega^+ \{\mathbf{r} = (x_1, x_2) : x_2 > 0\}$ and the rectangular domain $\Pi_{ab} = \Omega^2 = \{\mathbf{r} : -\frac{a}{2} < x_1 < \frac{a}{2} \quad -b < x_2 < 0\}$ symmetric with respect to the axis $x_1 = 0$ with an interval (a slot) $L = \{(\mathbf{r} : x_2 = 0, \ -l < x < l\}$ on the common part of the boundaries of Ω_1, Ω_2 whose endpoints (edges) $\partial L = \{A_1 = -l, \ A_2 = l\}$.

The mathematical statements of the problems E and H to determine the E- and H-type fundamental frequencies of the image slot resonator are given in Section 1.2 (eqns. 1.1–1.5) with $p_m(\phi) = \sin m\phi$ for the E-case and $p_m(\phi) = \cos m\phi$ for the H-case ($m = 0, 1, 2, \ldots$) in the radiation conditions given by eqn. 1.5.

We will denote by FF_E and FF_H the spectra of the E- and H-type fundamental frequencies (coinciding with the sets σ_E and σ_H of the squared values of the eigenvalues of the problems E and H considered with respect to the spectral parameter $\lambda = \omega^2$).

The study of problems E and H in the generalised statement carried out in Section 2.5 allows one to prove that they are of the Fredholm type what implies the discreteness of the sets σ_E and σ_H which consist of isolated points $\lambda : \operatorname{Im} \sqrt{\lambda} < 0$, i.e. in accordance with the generally accepted terminology, image slot resonators may have only scattering frequencies. This result yields the absence of fundamental frequencies in the closed upper half-plane of the main zero ("physical") sheet of the Riemann surface $\mathcal{H}$ [of the analytical continuation of the Hankel function $H_0^{(1)}(z)$] and, in particular, the absence of real fundamental frequencies.

Of course, the spectra discreteness may be proved by the method of the boundary integral operator-valued functions developed in Sections 1.5–1.8 for shielded resonators. Consider first the problem H.

Since both Ω^1 and Ω^2 are the S_Π-domains (see Section 1.5) one can show that the solution $u = u_i(\mathbf{r})$ $(\mathbf{r} \in \Omega^i,\ i = 1, 2$ to the problem H admits the representation in the form of the generalised single-layer potentials

$$
\begin{aligned}
u_1(\mathbf{r}) &= \int_L \frac{j}{2} H_0^{(1)}(\sqrt{\lambda\epsilon_1}|\mathbf{r} - \mathbf{q}^0|)\varphi(s)\,d\,s \quad (\mathbf{r} \in \Omega^1) \\
u_2(\mathbf{r}) &= \int_L G^H(\mathbf{r}, \mathbf{q}^0; \lambda)\varphi(s)\,d\,s \quad (\mathbf{r} \in \Omega^2)
\end{aligned}
\tag{1.60}
$$

where $\mathbf{q}^0 = (s, 0) \in L$, $G^H = G_2^2(\mathbf{r}, \mathbf{q}; \lambda)$ is Green's functions of the second boundary-value problem for the Helmholtz equation 1.1 of the rectangular domain Π_{ab} given by eqn. 1.50 (with the change of notation corresponding to the considered symmetric rectangular domain) and

$$
\varphi(s) = \frac{1}{\epsilon_1}\frac{\partial u_1}{\partial x_2} = \frac{1}{\epsilon_2}\frac{\partial u_2}{\partial x_2}
$$

is assumed, in accordance with the edge and transmission conditions, to belong to the weighted Hölder class $H_\mu^*(L)$ defined in Section 1.5. The first integral in eqn. 1.60 satisfies the radiation conditions given by eqn. 1.5 with $p_m(\phi) = \cos m\phi$ $(m = 0, 1, 2, \ldots)$.

The above conclusions enable one to equivalently reduce the problem H to the problem on characteristic numbers for the Fredholm integral operator-valued function with a logarithmic singularity of the kernel

$$
K(\lambda)\varphi \equiv \int_L K(\lambda; t, s)\varphi(s)\,d\,s = 0 \quad t \in L
$$

$$
\begin{aligned}
K(\lambda; t, s) &= \epsilon_1 \frac{j}{2} H_0^{(1)}(\sqrt{\lambda\epsilon_1}|t - s|) + \epsilon_2 G^H(\mathbf{r}^0, \mathbf{q}^0; \lambda) \\
&= \frac{\epsilon_1 + \epsilon_2}{\pi} \ln \frac{1}{|t - s|} + g^H(\mathbf{r}^0, \mathbf{q}^0; \lambda)
\end{aligned}
$$

where $\mathbf{r}^0 = (t, 0) \in L$ and, due to the results of Section 1.5, $g^H(\mathbf{r}, \mathbf{q}^0; \lambda)$ is once continuously differentiable in each closed domain $\bar{\Omega}_2^0 \subset \Omega^2$, $L \subset \bar{\Omega}_2^0$. Such equivalent reduction gives an alternative proof of the spectrum discreteness.

In the case of the rectangular image slot resonator one can directly show that K is the finite-meromorphic operator-valued function and

$$K(\lambda; t, s) = \epsilon_1 \frac{j}{2} H_0^{(1)}(\sqrt{\lambda \epsilon_1}|t - s|) + \epsilon_2 \sum_{k=0}^{\infty} q_k(\lambda) \psi_k(t) \psi_k(s)$$

$$= P_n(\lambda; t, s) + U_n(\lambda; t, s)$$

$$P_n(\lambda; t, s) = \frac{\epsilon_1 + \epsilon_2}{\pi} \ln \frac{1}{|t - s|} + q_n(\lambda) \psi_n(t) \psi_n(s)$$

$$\psi_k(t) = \cos \frac{\pi k t}{c}$$

the coefficients $q_n(\lambda) = (\lambda - \lambda_{nm})^{-1} q_{nm}^1(\lambda)$ are meromorphic functions of λ and each of them has the infinite sequence of poles $\lambda_{nm} = \pi^2 \epsilon_2^{-1}(n^2 c^{-2} + m^2 b^{-2})$, (see eqn. 1.52), P_n and U_n define the pole-pencil operator-valued function

$$P_n = \frac{\epsilon_1 + \epsilon_2}{\pi} \Upsilon + q_n(\cdot, \psi_n)$$

and integral operator-valued function with differentiable kernel holomorphic in the vicinity of λ_{nm}, respectively $(m, n = 0, 1, 2, \ldots)$, Υ denotes the integral operator-valued function with logarithmic kernel (like in eqn. 1.39).

Using the notation for the weighted Hölder spaces introduced by eqn. 1.2 of Section 1.5, we will consider $K(\lambda) = K(\lambda, l) : X_a(-l, l) \to Y_a(-l, l)$ as a family of the Fredholm integral operator-valued function taking $l =$ diam$L$ as a (small) parameter and represent $K(\lambda) = \tilde{K}(\lambda, l)$ in the vicinity of a fixed pole λ_{NM} after the change of variables $t = lt, s = ls$ in the form

$$\tilde{K}(\lambda, l)\varphi = K_{NM}(\lambda, l)\varphi + \frac{4}{bc} \frac{\delta_N \delta_M}{\lambda_{NM} - \lambda}(\varphi, \psi_N)\psi_N,$$

where $\delta_0 = 0.5, \delta_m = 1, m = 1, 2, \ldots,$

$$\varphi_n(t) = \begin{cases} \cos \frac{\pi n t l}{c}, & n = 2k \\ \sin \frac{\pi n t l}{c}, & n = 2k + 1 \end{cases}$$

the scalar product is defined as

$$(\varphi, \varphi_n) = \int_{-1}^{1} \varphi(t)\varphi_n(t) \, dt$$

and K_{NM} is expressed by the formula

$$K_{NM}(\lambda, l)\varphi = L_1 \varphi + l^2 \ln l \hat{K}_{NM}(\lambda, l)\varphi$$

$$L_1\varphi = (\epsilon_1 + \epsilon_2)\Upsilon\varphi + \left[\frac{\epsilon_1 + \epsilon_2}{\beta} + C_{NM}(\lambda)\right](\varphi, 1)$$

$K_{NM}(\lambda, l)$ which is an analytical operator-valued function in the vicinity of λ_{NM} and the kernel of $\hat{K}_{NM}$ is uniformly bounded as a function of l and is represented as follows:

$$K_{NM}(\lambda, l)\varphi = (\epsilon_1 + \epsilon_2)L + \frac{\epsilon_1 + \epsilon_2}{\pi}\ln\frac{1}{l}(\varphi, 1)\cdot 1 + C_{NM}(\lambda)(\varphi, 1)\cdot 1$$

$$+ l^2\ln l(\epsilon_1^2 + \epsilon_2^2)\frac{\lambda}{4\pi}P\varphi - jl^2\frac{\lambda\epsilon_1^2}{8}P\varphi$$

$$+ l^2 Q_{NM}(\lambda, l)\varphi + l^4\ln l R(\lambda, l)\varphi,$$

where

$$C_{NM}(\lambda) = \frac{\epsilon_1}{2\pi}\ln\frac{1}{\lambda\epsilon_1} + \frac{\epsilon_1}{\pi}\ln\frac{2}{\gamma} + \frac{j\epsilon_1}{2} - \frac{\epsilon_2}{\pi}\ln\frac{\pi}{c} - \frac{\epsilon_2}{\pi}\ln 2$$

$$+ \frac{\lambda\epsilon_2^2}{\pi}\frac{c^2 S_3}{2\pi^2} + \frac{\lambda\epsilon_2^2}{\pi}\frac{c^2}{2\pi^2}T_3 + \frac{\epsilon_2}{\pi}\sum_{n=1}^{\infty}\frac{\exp(-\frac{2\pi nb}{c})}{n\sinh\frac{2\pi nb}{c}}$$

$$+ \frac{2\lambda\epsilon_2^2 c^2}{\pi^3}\sum_{n=1}^{\infty}\frac{\exp(-\frac{2\pi nb}{c})}{(2n)^3\sinh\frac{2\pi nb}{c}} + \frac{\lambda\epsilon_2^2 bc}{\pi^2}\sum_{n=1}^{\infty}\frac{1}{(2n)^2\sinh\frac{2\pi nb}{c}}$$

$$+ \frac{4}{bc}\sum_{n=1}^{\infty}\sum_{m=0}^{\infty}\delta_m\left(\frac{1 - \delta_{2nmNM}}{\lambda_{2nm} - \lambda} - \frac{1}{\lambda_{2nm}} - \frac{\lambda}{\lambda_{2nm}^2}\right)$$

$$+ \frac{2}{bc}\sum_{m=0}^{\infty}\frac{\delta_m}{\lambda_0 m - \lambda}$$

$$S_3 = 1.20205690..., \quad T_3 = -0.90154268..., \quad P\varphi = \int_{-1}^{1}(t - s)^2\varphi(s)\,ds$$

and $Q_{NM}(\lambda, l)$ and $R(\lambda, l)$ are the operator-valued functions analytical with respect to λ in the vicinity of λ_{NM} and uniformly bounded with respect to l.

Below, we will also introduce the "exponentially" small parameter

$$\beta = \left(\frac{1}{\pi}\ln\frac{1}{l}\right)^{-1}$$

It is possible to generalise lemma 1.14 and to prove that there exist the inverse operators L_1^{-1} and K_{NM}^{-1} which are given for $l, \beta \ll 1$ by the following asymptotic formulae:

$$L_1^{-1}f = \frac{\Upsilon^{-1}f}{\epsilon_1 + \epsilon_2} - \frac{\ln 2}{\pi}\left[1 - \beta\frac{\ln 2}{\pi}\right.$$

$$\left. + \beta^2\frac{\ln 2}{\pi}\left(\frac{\tilde{C}_{NM}}{\epsilon_1 + \epsilon_2} + \frac{\ln 2}{\pi}\right) + O(\beta^3)\right]\frac{(\Upsilon^{-1}f, 1)}{\epsilon_1 + \epsilon_2}\Upsilon^{-1}1$$

$$\tag{1.61}$$

and

$$K_{NM}^{-1}(\lambda, l)f = L_1^{-1}f + O(l^2 \ln l)$$

$$= \frac{\Upsilon^{-1}f}{\epsilon_1 + \epsilon_2} - \frac{\ln 2}{\pi}\left[1 - \beta\frac{\ln 2}{\pi} + \beta^2\frac{\ln 2}{\pi}\left(\frac{\tilde{C}_{NM}}{\epsilon_1 + \epsilon_2}\right.\right.$$

$$+ \left.\left.\frac{\ln 2}{\pi}\right)\right]\frac{(\Upsilon^{-1}f, 1)}{\epsilon_1 + \epsilon_2}L^{-1}1 + O(\beta^3) \tag{1.62}$$

if $(\Upsilon^{-1}f, l) \neq 0$; if $(\Upsilon^{-1}f, 1) = 0$, then

$$L_1^{-1}f = \frac{L^{-1}f}{\epsilon_1 + \epsilon_2}$$

and, consequently,

$$K_{NM}^{-1}(\lambda, l)f = \frac{L^{-1}f}{\epsilon_1 + \epsilon_2} - l^2 \ln l \frac{\epsilon_1^2 + \epsilon_2^2}{(\epsilon_1 + \epsilon_2)^2}\frac{\lambda}{4\pi}L^{-1}PL^{-1}f$$

$$+ jl^2\frac{\lambda}{8}\frac{\epsilon_1^2}{(\epsilon_1 + \epsilon_2)^2}L^{-1}PL^{-1}f \tag{1.63}$$

$$- \frac{l^2}{(\epsilon_1 + \epsilon_2)^2}L^{-1}Q_{NM}(\lambda)L^{-1}f + O(l^4 \ln^2 l)$$

here $\tilde{C}_{NM} = \operatorname{Re} C_{NM}(\lambda_{NM})$.

In accordance with the assertions of Sections 1.6 and 1.8 concerning the "local" existence, $\tilde{K}(\lambda, l)$ has exactly one characteristic number (which coincides with the one of $K(\lambda)$) in a vicinity of λ_{MN} for sufficiently small l. In order to obtain an approximate representation for this characteristic number in the form of a segment of an asymptotic series with respect to β, we will "asymptotically" regularise the problem of characteristic numbers for $\tilde{K}(\lambda, l)$ with the help of K_{NM}^{-1} yielding the equivalent equation with respect to λ

$$1 + \frac{4}{ab}\frac{\delta_N\delta_M}{\lambda_{NM} - \lambda}\left(K_{NM}^{-1}\psi_N, \psi_N\right) = 0 \tag{1.64}$$

which has two families of solutions essentially differing for the even and odd indices N. In fact, $(\Upsilon^{-1}\psi_{2m+1}, 1) = 0$ and, therefore, the asymptotic series for the "even" (λ_{NM}^e) and "odd" (λ_{NM}^o) characteristic numbers have the form

$$\lambda_{NM}^e = \lambda_{NM} + \beta\frac{4}{ab}\frac{\delta_M}{\epsilon_1 + \epsilon_2} - \beta^2\frac{4}{ab}\frac{\delta_M}{\epsilon_1 + \epsilon_2}\left[\frac{\tilde{C}_{NM}}{\epsilon_1 + \epsilon_2} + \frac{\ln 2}{\pi}\right]$$

$$- j\beta^2\frac{2}{ab}\frac{\delta_M\epsilon_1}{(\epsilon_1 + \epsilon_2)^2} + O(\beta^3) \tag{1.65}$$

$$\lambda_{NM}^o = \lambda_{NM} + l^2 \frac{2\delta_M \pi^3 N^2}{ab^3(\epsilon_1 + \epsilon_2)} + l^4 \ln l \frac{\lambda_{NM} \delta_M \pi^3 N^2}{2ab^3} \frac{\epsilon_1^2 + \epsilon_2^2}{(\epsilon_1 + \epsilon_2)^2}$$
$$- jl^4 \frac{\lambda_{NM} \delta_M}{4ab^3} \frac{\epsilon_1^2}{(\epsilon_1 + \epsilon_2)^2} \pi^4 N^2 + r(l)$$

(1.66)

where

$$\operatorname{Re} r(l) = O(l^4)$$
$$\operatorname{Im} r(l) = O^*(l^6 \ln^2 l)$$

$K_{NM}^{-1} \psi_N$ are the eigenfunctions corresponding to λ_{NM}^e and λ_{NM}^o which together with the fundamental frequencies $\omega_{NM}^{e,o} = \sqrt{\lambda_{NM}^{e,o}}$ are situated in the lower half-planes λ and ω.

Hence, we have proved the following theorem.

Theorem 1.16: There exists a positive value a_0 such that if $\operatorname{diam} L \in (0, a_0)$ there lies in the vicinity of every point $\lambda_{nm} = \pi^2 \epsilon_2^{-1}(n^2 c^{-2} + m^2 b^{-2})$ $(n, m = 0, 1, \dots)$ at least one complex eigenvalue [a characteristic number of $K(\lambda)$] with a strictly negative imaginary part of the problem given by eqns. 1.1–1.5 for the rectangular image slot resonator.

It is easy to see that the odd H-type oscillations have much higher Q-factors than the even ones and the ratio of their Q-factors has the order $O(l^4 \ln^2 l)$.

The analytical representations obtained may be used as canonical data in the modelling of open image-type resonators. Since Green's functions are meromorphic with respect to λ, this method may be generalised under natural restrictions for an arbitrary domain Ω^2.

In the case of E-polarisation it is convenient to reduce the boundary-value problem given by eqns. 1.1–1.5 with the help of the known procedure described by Hönl *et al.* (1962) to the integro-differential equation with respect to the function $\psi(s) \in X_0(-l, l)$ coinciding with the longitudinal component of the electric field (the functional space X_0 is defined in Section 1.5):

$$\left(\frac{d^2}{dt^2} + k_1^2\right) \int_{-l}^{l} \frac{j}{2} H_0^{(1)}\left(k_1|t - s|\right) \psi(s) \, ds$$
$$+ \left(\frac{d^2}{dt^2} + k_2^2\right) \int_{-l}^{l} g^E(t, s; \lambda)\psi(s) \, ds = 0 \quad (1.67)$$

where $k_i^2 = \lambda \epsilon_i$ $(i = 1, 2)$,

$$g^E(t, s; \lambda) = \frac{1}{\pi} \ln \frac{1}{|t - s|} - \frac{1}{\pi} \ln \frac{\sin \frac{\pi|t-s|}{2c}}{|t - s|} + \frac{1}{\pi} \ln \cos \frac{\pi(t - s)}{2c}$$

$$+ \frac{2}{\pi} \sum_{n=1}^{\infty} \frac{\exp\left(-\frac{\pi b n}{c}\right)}{n \sinh \frac{\pi b n}{c}} \psi_n(t) \psi_n(s) + \frac{4}{b c \epsilon_2} \sum_{n,m=1}^{\infty} \frac{\lambda}{\lambda_{nm}^2} \psi_n(t) \psi_n(s)$$

$$+ \frac{4}{b c \epsilon_2} \sum_{n,m=1}^{\infty} \left(\frac{1}{\lambda_{nm} - \lambda} - \frac{1}{\lambda_{nm}} - \frac{\lambda}{\lambda_{nm}^2} \right) \psi_n(t) \psi_n(s) \qquad (1.68)$$

$$\psi_n(t) = \begin{cases} \cos \frac{\pi n t}{c}, & n = 2k \\ \sin \frac{\pi n t}{c}, & n = 2k + 1 \end{cases}$$

Let us rewrite the integro-differential equation 1.67 in the form of the problem of characteristic numbers

$$K_1(\lambda)\psi = 0$$

$K_1(\lambda)$ may be represented in the form

$$K_1(\lambda) = 2 \frac{d}{dx} S + M(\lambda)$$

where S is the Cauchy singular integral operator and

$$M(\lambda)\psi = \frac{1}{\pi} \int_{-l}^{l} M(t, s; \lambda)\psi(s) \, ds$$

where $M(t, s; \lambda)$ is an integrable function of the variables t and s. The operator

$$\frac{d}{dx} S : Y_a(-l, l) \to X_a(-l, l), \quad \mu < \frac{1}{2},$$

is continuous and continuously invertible (Muskhelishvili, 1962).

The inverse operator $\left(\frac{d}{dx}S\right)^{-1} : X_a(-l, l) \to Y_a(-l, l)$ is defined by the formula

$$\left(\frac{d}{dx}S\right)^{-1} f = -\frac{\sqrt{l^2 - t^2}}{\pi} \int_{-l}^{l} \frac{F(s)}{\sqrt{l^2 - s^2}} \frac{ds}{s - t}$$

$$F(s) = \int_{0}^{s} f(t') \, dt'$$

The operators $M(\lambda) : X_0(-l, l) \to X_a(-l, l)$ is completely continuous; therefore, $K_1(\lambda)$ is a Fredholm operator-valued function and we will consider

$K_1(\lambda)$ for $\lambda \in \Lambda$, where Λ denotes the complex plane cut along the line $\operatorname{Re}\lambda < 0$, $\operatorname{Im}\lambda = 0$.

It is easy to show that $K_1(\lambda)$ is finite-meromorphic in the domain Λ with the poles

$$\lambda_{nm} = \frac{\pi^2}{\epsilon_2}\left(\frac{n^2}{c^2} + \frac{m^2}{b^2}\right) \quad (n, m = 1, 2, \ldots)$$

and, similar to the H-case analysed above, $K_1(\lambda)$ may be considered as a family of the Fredholm integral operator-valued function with l as a (small) parameter and after the change of variables $t = lt, s = ls$ eqn. 1.67 takes the form

$$\tilde{K}_1(\lambda, l)\psi = 0 \tag{1.69}$$

where

$$\tilde{K}_1(\lambda, l)\psi = \frac{2}{\pi}\frac{\mathrm{d}}{\mathrm{d}t}\int_{-1}^{1}\frac{\psi(s)\,\mathrm{d}s}{s - t} + l^2\int_{-1}^{1}M(lt, ls; \lambda)\psi(s)\,\mathrm{d}s$$
$$t \in (-1, 1)$$

For all $\lambda \in \Lambda \setminus \{\lambda_{nm}\}$ and $l > 0$ $\tilde{K}_1(\lambda, l) : Y_a(-1, 1) \to X_a(-1, 1)$ is a Fredholm holomorphic operator-valued function. The spectra of $\tilde{K}_1(\lambda, l)$ and $K_1(\lambda)$ coincide.

Choosing a certain pole λ_{NM} we will represent $\tilde{K}_1(\lambda, l)$ in its vicinity in the form

$$K_{NM}(\lambda, l) = 2\frac{\mathrm{d}}{\mathrm{d}x}S + l^2\ln l N_2(\lambda)$$
$$+ l^2 N_3(\lambda) + l^4 N_4(\lambda, l)$$

where

$$N_i(\lambda)\psi = \int_{-1}^{1}\tilde{N}_i(t, s; \lambda)\psi(s)\,\mathrm{d}s \quad (i = 2, 3, 4)$$

are the integral operators with the kernels $\tilde{N}_i(t, s; \lambda)$,

$$\tilde{N}_2(t, s; \lambda) = -\frac{\lambda}{2\pi}(\epsilon_1 + \epsilon_2)$$

$$\tilde{N}_3(t,s;\lambda) = \frac{\lambda}{4\pi}\epsilon_1 - \frac{\lambda}{2\pi}\epsilon_1 \ln \frac{\gamma\sqrt{\lambda\epsilon_1}|t-s|}{2} + \frac{j}{4}\lambda\epsilon_1 + \frac{\lambda}{2\pi}\epsilon_2 \ln \frac{1}{|t-s|}$$

$$+ \frac{\pi}{3(2c)^2} - \frac{\lambda}{\pi}\epsilon_2 \ln \frac{\pi}{2c} - \frac{\pi}{(2c)^2}$$

$$+ \frac{1}{\pi}\sum_{n=1}^{\infty}\xi_n \frac{\exp\left(-\frac{\pi n b}{c}\right)}{n \sinh \frac{\pi n b}{c}} + \frac{\lambda^2\epsilon_2^2 c^2}{\pi^3}\cdot 1.3147495\ldots$$

$$+ \frac{\lambda\epsilon_2 c^2}{2\pi^3}\sum_{n=1}^{\infty}\xi_n \frac{\exp\left(-\frac{\pi n b}{c}\right)}{n^3 \sinh \frac{\pi n b}{c}} + \frac{\lambda\epsilon_2 bc}{2\pi^2}\sum_{n=1}^{\infty}\xi_n \frac{1}{n^2 \sinh^2 \frac{\pi n b}{c}}$$

$$+ \frac{\lambda^2\epsilon_2^2 c^4}{\pi^5}\cdot 0.7533927\ldots$$

$$+ \frac{3}{8}\frac{\lambda^2\epsilon_2 bc^3}{\pi^4}\sum_{n=1}^{\infty}\xi_n \frac{1}{n^4 \sinh^2 \frac{\pi n b}{c}}$$

$$+ \frac{3}{8}\frac{\lambda^2\epsilon_2^2 c^4}{\pi^5}\sum_{n=1}^{\infty}\xi_n \frac{1}{n^5 \sinh \frac{\pi n b}{c}} + \frac{2\lambda^2\epsilon_2^2 b^2 c^2}{\pi^3}\sum_{n=1}^{\infty}\xi_n \frac{\coth \frac{\pi n b}{c}}{n^3 \sinh^2 \frac{\pi n b}{c}}$$

$$+ \frac{2}{bc\epsilon_2}\sum_{n,m=1}^{\infty}\xi_n \frac{\lambda^3(1-\delta_{n-N,m-M})}{\lambda_{nm}^3(\lambda_{nm}-\lambda)}$$

$$\tilde{N}_4(t,s,\lambda) = \frac{\lambda\epsilon_2}{4b}|t-s| - \frac{\lambda\epsilon_2(t+s)}{4b} - \frac{\lambda^2\epsilon_2^2 c^2(t+s)}{8b}$$

Here

$$\xi_n = \left(\lambda\epsilon_2 - \frac{\pi^2 n^2}{c^2}\right)\alpha_n, \quad \alpha_n = 1-(-1)^n$$

Obviously,

$$\frac{1}{\epsilon_2}\left(\frac{d^2}{dx^2} + l^2 k_2^2\right)\psi_N(lx) = l^2(\lambda - \lambda_{N0})\psi_N(lx),$$

where $\lambda_{N0} = \frac{\pi^2 N^2}{\epsilon_2 c^2}$.

In the vicinity of the point λ_{NM} eqn. 1.69 may be written as follows

$$K_{NM}(\lambda,l)\psi + l^2 \frac{\lambda-\lambda_{N0}}{\lambda_{NM}-\lambda}\frac{4}{ab}(\psi,\psi_N)\psi_N = 0 \qquad (1.70)$$

where $K_{NM}(\lambda;l)$ is invertible for sufficiently small l (since it may be expressed as the sum of an invertible operator and a "small" operator) and

$$K_{NM}^{-1}\psi = \left(2\frac{\mathrm{d}}{\mathrm{d}x}S + l^2 \ln l N_2 + l^2 N_3 + l^4 \ln l N_4\right)^{-1}\psi$$
$$= \left(I + \frac{l^2 \ln l}{2}S_1 N_2 + \frac{l^2}{2}S_1 N_3 + \frac{l^4 \ln l}{2}S_1 N_4\right)^{-1}\frac{1}{2}S_1\psi$$
$$= \frac{1}{2}\left(I - \frac{l^2 \ln l}{2}S_1 N_2 - \frac{l^2}{2}S_1 N_3 + l^4 \ln l N_4\right)S_1\psi \qquad (1.71)$$

where

$$S_1 = \left(\frac{\mathrm{d}}{\mathrm{d}x}S\right)^{-1}$$

Applying the operator $K_{NM}^{-1}(\lambda, l)$ to both sides of eqn. 1.70 we obtain

$$\psi + \frac{4}{bc}l^2\frac{\lambda - \lambda_{N0}}{\lambda_{NM} - \lambda}(\psi, \psi_N)K_{NM}^{-1}(\lambda, l)\psi_N = 0 \qquad (1.72)$$

It follows from eqn. 1.72 that if a characteristic number λ_{NM}^* of $K(\lambda, l)$ exists in a vicinity of λ_{NM}, it must be a solution of the equation

$$1 + \frac{4}{bc}l^2\frac{\lambda - \lambda_{N0}}{\lambda_{NM} - \lambda}(K_{NM}^{-1}\psi_N, \psi_N) = 0$$

The equivalent form of this equation is

$$\lambda = \lambda_{NM} + \frac{4}{bc}l^2(\lambda - \lambda_{N0})(K_{NM}^{-1}\psi_N, \psi_N) = 0$$

Taking into account eqn. 1.71 we get

$$\lambda_{NM}^* = \lambda_{NM} + l^2\frac{2}{bc}(\lambda_{NM} - \lambda_{N0})(R_{NM}S_1\psi_N, \psi_N) + O(l^6 \ln l)$$

where

$$R_{NM} = I - \frac{l^2 \ln l}{2}S_1 N_2(\lambda_{NM}) - \frac{l^2}{2}S_1 N_3(\lambda_{NM})$$

and, consequently,

$$\operatorname{Re}\lambda_{NM}^* = \lambda_{NM} + l^2\frac{2}{bc}(\lambda_{NM} - \lambda_{N0})(S_1\psi_N, \psi_N) + O(l^4 \ln l)$$
$$\operatorname{Im}\lambda_{NM}^* = -l^4\frac{1}{bc}(\lambda_{NM} - \lambda_{N0})\frac{\lambda\epsilon_1}{2} \cdot (S_1(S_1\psi_N, 1), \psi_N) + O(l^6 \ln l)$$

$$(1.73)$$

Let N be odd. Then $\psi_N = 1 + O(l^2)$ and

$$S_1\psi_N = S_1 1 + O(l^2)$$

$$= -\frac{1}{\pi}\sqrt{1-t^2}\int_{-1}^{1}\frac{s}{\sqrt{1-s^2}}\frac{ds}{s-t} + O(l^2)$$

$$= -\sqrt{1-t^2} + O(l^2)$$

$$(S_1\psi_N, \psi_N) = (S_1\psi_N, 1) + O(l^2) = -\frac{\pi}{2} + O(l^2)$$

$$(S_1(S_1\psi_N, 1), \psi_N) = \frac{\pi^2}{4} + O(l^2)$$

The resulting formulae are

$$\operatorname{Re}\lambda_{NM}^* = \lambda_{NM} - l^2\frac{\pi}{bc}(\lambda_{NM} - \lambda_{N0}) + O(l^4\ln l)$$

$$= \frac{\pi^2}{\epsilon_2}\left(\frac{N^2}{c^2} + \frac{M^2}{b^2}\right) - \frac{l^2\pi^3 M^2}{b^3 c\epsilon_2} + O(l^4\ln l) \qquad (1.74)$$

$$\operatorname{Im}\lambda_{NM}^* = -l^4\frac{\epsilon_1\pi^2}{8bc}(\lambda_{NM} - \lambda_{N0})\lambda_{NM} + O(l^6\ln l)$$

$$= -l^4\frac{\epsilon_1\pi^6}{8\epsilon_2^2 b^3 c}M^2\left(\frac{N^2}{c^2} + \frac{M^2}{b^2}\right) + O(l^6\ln l)$$

and the characteristic number λ_{NM}^* is situated in the lower half-plane of the complex plane λ.

In the case of even N

$$(S_1\psi_N, 1) = 0$$

one cannot obtain the expression for $\operatorname{Im}\lambda$ by eqn. 1.73 and it is necessary to use greater powers of l. Here

$$\psi_N = \sin\frac{\pi N l x}{c} = \frac{\pi N l x}{c} + O(l^3)$$

$$S_1\psi_N = -\sqrt{1-t^2}\frac{\pi N l x}{2c} \quad \text{and} \quad (S_1\psi_N, \psi_N) = -\frac{\pi^3 N^2 l^2}{16c^2}$$

Therefore,

$$\operatorname{Re}\lambda_{NM}^* = \lambda_{NM} - \frac{\pi^3 N^2 l^4}{4a^3 b}(\lambda_{NM} - \lambda_{N0}) + O(l^6\ln l)$$

$$= \frac{\pi^2}{\epsilon_2}\left(\frac{N^2}{a^2} + \frac{M^2}{b^2}\right) - \frac{l^4\pi^5 M^2 N^2}{8a^3 b^3\epsilon_2} + O(l^6\ln l) \qquad (1.75)$$

$$\operatorname{Im}\lambda^*_{NM} = -l^6 \frac{\lambda^2 \epsilon_2^2}{64}(S_1 N_5 S_1 \psi_N, \psi_N) + O(l^{10}\ln l)$$

where N_5 is the integral operator with the kernel $N_5(s) = s$.

Calculating the scalar product in the right-hand side we obtain the asymptotic representation

$$\begin{aligned}
\operatorname{Im}\lambda^*_{NM} &= -\frac{l^8 \lambda^2_{NM} \epsilon_1^2 \pi^6 M^2 N^2}{2^{12} a^3 b^3 \epsilon_2} + O(l^{10}\ln l)\\
&= -\frac{l^8 \epsilon_1^2 \pi^{10} M^2 N^2}{2^{12} a^3 b^3 \epsilon_2^3}\left(\frac{N^2}{a^2} + \frac{M^2}{b^2}\right)^2 + O(l^{10}\ln l)
\end{aligned}$$

$$(1.76)$$

This characteristic number is also situated in the lower half-plane as follows from the general results presented in Section 2.5.

Eqns. 1.74–1.76 give asymptotic expansions for the squared values of the E-type fundamental frequencies of the rectangular image slot resonator. Similar formulae may be obtained for an image resonator with arbitrary domain $\Omega^2 \in S_\Pi(L)$ if to use the representations given by eqns. 1.57 and 1.58. Of course, the order of accuracy in corresponding asymptotic series will be less than those obtained for the (rectangular) domain with given geometry.

1.10 The Morse critical points of empty open resonators. Evolution equations

It is known after the works of Shteinshleiger (1955) that interaction of oscillations takes place in open resonators for certain dependences between spectral and non-spectral parameters, when their small variations lead to sharp changes of resonator's properties: a considerable rise or fall of diffraction loss, emergence of intertype (hybrid) fields etc. Such phenomena are usually called intertype oscillations and the model proposed by Shteinshleiger (1955) for closed resonators describes them as intertype oscillations of an ideal resonator perturbed by a small inhomogeneity when eigenoscillations with a double frequency degeneration interact due to the presence of this inhomogeneity.

In this section we will construct a mathematical model of intertype oscillations in empty open circular resonators whose transversal cross-sections are formed by the domains from the family CB_c^ε on the basis of the theory of functions of several complex variables and the theory of singularities.

As shown in Section 1.3, the methods of spectral theory allow us to determine for an open cylindrical resonator the characteristic values of the spectral parameter $\kappa = ka$ for which there exist nontrivial solutions to homogeneous Maxwell's equations, satisfying the boundary-value conditions on the resonator surfaces, the Meixner conditions and the Reichardt conditions at infinity. The mathematically rigorous solution of this problem has been obtained on the basis of the spectral theory of non-self-adjoint holomorphic Fredholm matrix operator-valued functions considered in the Cartesian product $H = \bigotimes_{n=1}^{M} l_2$ of the M copies of the Hilbert space l_2.

Analytical properties of the operator-valued function $A(\kappa)$ corresponding to the considered boundary eigenvalue problems allow us to prove that the set (spectrum) of its characteristic numbers is discrete and each characteristic number has a finite multiplicity. The problem on characteristic numbers may be reduced to the dispersion equation

$$F(\kappa) \equiv \det\{I - A(\kappa)\} = 0 \qquad (1.77)$$

Eqn. 1.77 contains the infinite characteristic determinant of the kernel operator $A(\kappa)$ which depends also on resonator's non-spectral parameters [the distance between the mirrors (l), the radius (a) and the aperture (b) of the mirrors for an open symmetrical resonator which is transformed for $l = l_0 = 2[a - (a^2 - b^2)^{1/2}]$ to a closed resonator]. Let us take only one of them, namely, $\chi = \frac{b}{a}$. We will rewrite the operator-valued function in the form $A = A(\kappa, \chi)$ assuming that χ varies in the domain $D_\chi \overline{\subset} C$ where $A(\kappa, \chi)$ is holomorphic as an operator-valued function of χ (here C is the complex plane). We note also that $A(\kappa, \chi)$ is a kernel operator-valued function holomorphic with respect to two complex variables κ, χ for $(\kappa, \chi) \in \Lambda \times D_\chi$. The parameter χ corresponds to resonator's adjustment. We will assume also that eigenvalues $\kappa \subset \Lambda_0$, where $\Lambda_0 = \{\kappa \in C : -\pi < \arg \kappa < \pi; \ \kappa \neq 0\}$ is the zero sheet of Λ.

Since $A(\kappa, \chi)$ is a kernel and holomorphic operator-valued function, the zeros of the dispersion equation $F(\kappa, \chi) = 0$ coincide with the characteristic numbers of $I - A(\kappa, \chi)$ forming the set $\sigma(\kappa)$ ($\sigma(\kappa)$ is defined on the subset $D = \Lambda_0 \times D_\kappa \subset C^2$). It is important to establish what happens to $\sigma(\kappa)$ when eigenvalues κ vary in D_κ.

With this purpose we will introduce the analytical set D having the form $\sigma_0 = \{(\kappa, \chi) \subset D : F(\kappa, \chi) = 0\}$ and consider $F(\kappa, \chi)$ as a mapping $F : C^2 \to C$ defined in the domain D (here we assume, of course, that χ is a complex value). If an isolated singularity (κ_0, χ_0) of the mapping F is situated close to D then, as it follows from the general results of the theory of functions of several complex variables (Shabat, 1976) and of the theory of singularities of smooth mappings (Arnold *et al.*, 1982), the local structure of σ_0 in the vicinity of (κ_0, χ_0) is defined by the type and position of (κ_0, χ_0). Far from (κ_0, χ_0) this set has a local structure of a hyper-plane $(\bar{\alpha}\kappa + \bar{\beta}\chi + \bar{\gamma} = 0)$. It is clear that small variations of χ in the vicinity of (κ_0, χ_0) lead to small variations of $\sigma(\kappa)$.

The phenomenon of intertype oscillations has well-known and commonly accepted features and, first of all, the dispersion law which regulates the special behaviour of two fundamental frequencies as functions of χ when $\operatorname{Im}\chi = 0$ (the Wien graphs). The natural problem appears to describe in terms of regular and singular points of the mapping F the conditions, which correspond to the dispersion law characteristic for intertype oscillations.

The local structure of σ_0 in the form of a hyper-plane does not correspond to the dispersion law. Therefore, let us consider the set of critical points

$$\sigma_K = \left\{ (\kappa, \chi) \in D : \operatorname{grad} F = \left(\frac{\partial F}{\partial \kappa}, \frac{\partial F}{\partial \chi} \right) = 0 \right\}$$

and the set of isolated Morse critical points

$$\sigma_{MK} = \left\{ (\kappa, \chi) \in \sigma_K : \frac{\partial^2 F}{\partial \kappa^2} \frac{\partial^2 F}{\partial \chi^2} - \left(\frac{\partial^2 F}{\partial \kappa \partial \chi} \right) \neq 0 \right\}$$

Let (κ_0, χ_0) be an isolated Morse critical point situated close to σ_0 [i.e. $F(\kappa_0, \chi_0) = \delta$ is sufficiently small]. Then one can write the dispersion equation in the vicinity of (κ_0, χ_0) in the form

$$\delta + \frac{1}{2} \frac{\partial^2 F}{\partial \kappa^2} \bigg|_{\substack{\kappa=\kappa_0 \\ \chi=\chi_0}} (\kappa - \kappa_0)^2 + \frac{\partial^2 F}{\partial \kappa \partial \chi} \bigg|_{\substack{\kappa=\kappa_0 \\ \chi=\chi_0}} (\kappa - \kappa_0)(\chi - \chi_0)$$
$$+ \frac{1}{2} \frac{\partial^2 F}{\partial \chi^2} \bigg|_{\substack{\kappa=\kappa_0 \\ \chi=\chi_0}} (\chi - \chi_0)^2 + O^3 = 0 \tag{1.78}$$

where O^3 denotes the small cubic terms. Since $(\kappa_0, \chi_0) \in \sigma_{MK}$, there exists the linear change of variables $\tilde{\kappa} = \psi_1(\kappa, \chi)$, $\tilde{\chi} = \psi_2(\kappa, \chi)$ (Milnor, 1971) which transforms eqn. 1.78 into

$$\tilde{\kappa}^2 + \tilde{\chi}^2 + \delta = 0 \tag{1.79}$$

what gives $\tilde{\kappa} = \pm\sqrt{-\tilde{\chi}^2 - \delta}$. For $\delta = 0$ we get $\tilde{\kappa} = \pm j\tilde{\chi}$. Qualitative properties of the solutions to eqns. 1.78 and 1.79 coincide. Eqn. 1.79 contains the dispersion law (i.e. it determines implicitly the spectral parameter as a function of non-spectral parameters), which is characteristic for intertype communication and the value of $\delta = F(\kappa_0, \chi_0)$ determines the degree of this communication. There is a double degeneration of $F(\kappa, \chi)$ for $\delta = 0$ in the point (κ_0, χ_0) with respect both to κ and χ.

It is sufficient to study $\operatorname{Re}\kappa(\chi)$ and $\operatorname{Im}\kappa(\chi)$ as solutions of eqn. 1.78 in the physical domain of variation of the non-spectral parameter determining the resonator's adjustment (when $\operatorname{Re}\chi = 0$) for $\operatorname{Re}\tilde{\kappa}(\tilde{\chi})$ and $\operatorname{Im}\tilde{\kappa}(\tilde{\chi})$ from eqn. 1.79 when $\tilde{\chi}$ varies along the straight lines $\operatorname{Re}\tilde{\chi} = \xi$, $\operatorname{Im}\tilde{\chi} = \alpha\xi + \beta$ where ξ and α are the real parameters of the line, $\beta \neq 0$ for $\operatorname{Im}\chi_0 \neq 0$. Finally we get

$$\tilde{\kappa}(\xi) = \pm\left[\left(\frac{\sqrt{a^2 + b^2} + a}{2}\right)^{1/2} + j\operatorname{sign}b\left(\frac{\sqrt{a^2 + b^2} - a}{2}\right)^{1/2}\right] \qquad (1.80)$$

where

$$a = (\alpha\xi + \beta) - \xi^2 - \delta_1, \quad b = -\delta_2 - 2\xi(\alpha\xi + \beta), \quad \delta = \delta_1 + i\delta_2, \quad \delta_1, \delta_2 \in \mathbb{R}^1$$

The two-parameter families of the curves $\operatorname{Re}\tilde{\kappa}(\xi)$ and $\operatorname{Im}\tilde{\kappa}(\xi)$ (with respect to α and β, see Figure 1.2) obtained from eqns. 1.78 and 1.79 describe all possible qualitative situations occurring in the Morse critical points for different values of α and β. We will call these curves the communication diagrams. The functions $\tilde{\kappa}(\xi)$ for $\beta = 0$ are shown in Figure 1.3 by the bold lines and the perturbed curves for $\beta \neq 0$ are shown by dot lines. Similar communication diagrams may be obtained by substituting $\operatorname{Re}\tilde{\kappa} \leftrightarrow \operatorname{Im}\tilde{\kappa}$.

Hence the presence of the Morse critical point (κ_0, χ_0) of the characteristic determinant $F(\kappa, \chi)$ yields two solutions $\kappa_+(\xi)$, $\kappa_-(\xi)$ of the equation $F(\kappa, \chi) = 0$ in the vicinity of (κ_0, χ_0) and their behaviour with respect to the resonator's adjustment χ is completely determined by eqn. 1.79 and describes intertype oscillations of an open resonator. In particular, it determines sharp changes of the Q-factor (and diffraction loss) due to small variations of the geometrical parameters.

Now one can expect that all the dispersion equations obtained in electrodynamics (and also in fusion, plasma and solid state physics etc.) may be considered in order to prove the presence (or absence) of the Morse critical

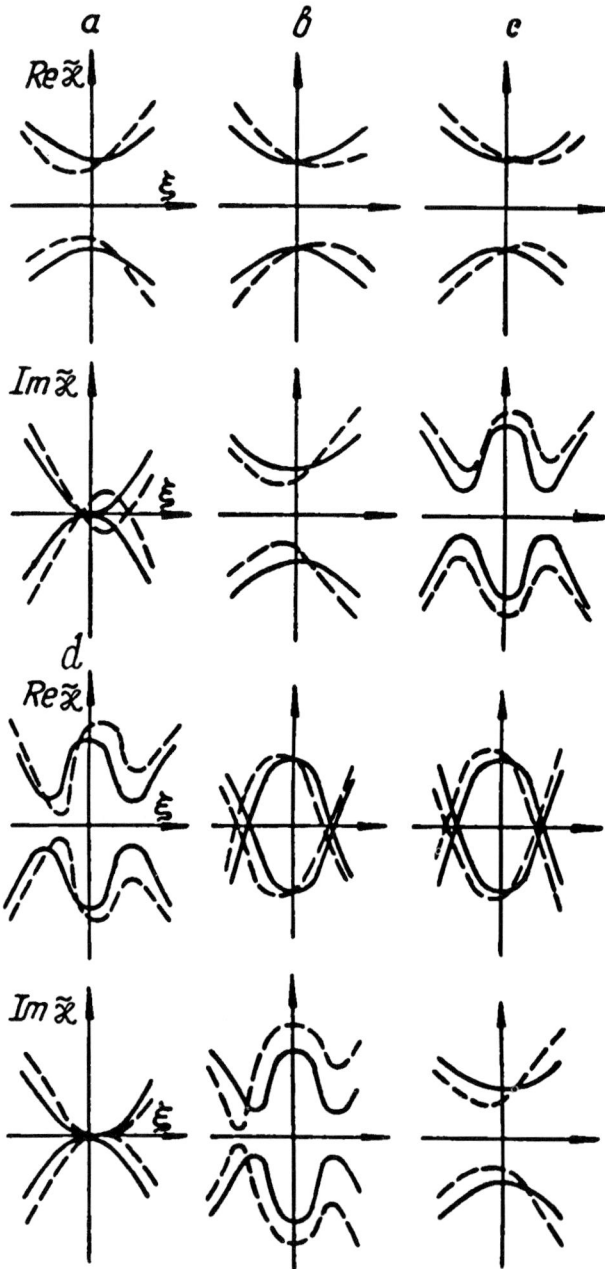

Figure 1.3   *Dispersion curves of an open resonator near the Morse critical point*

points and, hence, to check the possibilities of the existence of intertype oscillations. Here we choose the frequency as a spectral parameter and every other parameter of the problem may be taken as a non-spectral one. It is important that the dispersion equation has the form $F(\kappa, \chi) = 0$ where κ is a frequency, and χ is one of the non-spectral parameters (κ and χ are the complex values).

Consideration of the Morse critical points in the case when the function $F(\kappa, \chi)$ in the dispersion equation $F(\kappa, \chi) = 0$ depends on two complex variables κ and χ may be expanded for the case of three independent variables κ, Φ and χ. Such problems are connected with a simultaneous study of the oscillation (with respect to the spectral parameter κ) and wave (with respect to the spectral parameter Φ) processes when a non-spectral parameter (denoted by χ) vary. It is especially interesting to study the properties of dispersion equations close to the Morse critical points for non-stationary nonlinear processes.

The quadratic dispersion law given by eqns. 1.78 and 1.79 enables one to obtain linear and nonlinear differential equations and to investigate with the help of these equations the local evolution processes occurring close to the Morse critical points. The frequency and the wave vector projection are replaced by the operators $j\frac{\partial}{\partial \tau}$, $j\frac{\partial}{\partial \tau}$, respectively, where τ and ξ are linear combinations of the real time and a spatial co-ordinate (Bass *et al.*, 1993) and one can then describe the behaviour of a physical system "in the entry", "in the exit" and in the point of its "catastrophic" destruction by introducing to the evolution equations an appropriate control parameter (Shestopalov, 1994). It is clear that these very "catastrophic" changes (anomalous dispersion) are the most interesting for modelling physical objects and processes since the "normal" evolution corresponding to the regular points of dispersion equations (normal dispersion) gives a non-perturbed picture of the system's behaviour with respect to the time parameter. Auto-model solutions of nonlinear evolution equations close to the Morse critical points may have the forms of solitons, antisolitons or kinks (Bass *et al.*, 1994).

Assume that the dispersion equation $F(\kappa, \chi) = 0$ has been obtained as a result of the solution to a spectral problem where κ and χ denote the complex frequency spectral parameter and the wave vector projection on the Cartesian axis Ox (the non-spectral parameter), respectively. Let (κ_0, χ_0) be an isolated

Morse critical points of this dispersion equation determined by the conditions

$$\frac{\partial F}{\partial \kappa} = 0, \quad \frac{\partial F}{\partial \chi} = 0$$

$$\frac{\partial^2 F}{\partial \kappa^2} \frac{\partial^2 F}{\partial \chi^2} - \left(\frac{\partial^2 F}{\partial \kappa \partial \chi}\right) \neq 0$$

Then in the vicinity of (κ_0, χ_0) the dispersion equation may be represented as an equivalent quadratic form

$$A\tilde{\kappa}^2 - B\tilde{\chi}^2 + C = 0 \tag{1.81}$$

where

$$A = \frac{1}{2}\{a + c + [(a + c)^2 - 4w]^{\frac{1}{2}}\} \qquad B = -\frac{1}{2}\{a + c - [(a + c)^2 - 4w]^{\frac{1}{2}}\}$$

$$C = (\kappa_0, \chi_0) \qquad w = AC - B^2$$

$$a = \left.\frac{\partial^2 F}{\partial \kappa^2}\right|_{\substack{\kappa=\kappa_0 \\ \chi=\chi_0}}$$

$$c = \left.\frac{\partial^2 F}{\partial \chi^2}\right|_{\substack{\kappa=\kappa_0 \\ \chi=\chi_0}}$$

$$\tag{1.82}$$

Let us proceed with the help of eqn. 1.81 from the dispersion equation to the corresponding linear partial differential equation taking into account that

$$\tilde{\kappa} = j\frac{\partial}{\partial \tau} \quad \tilde{\chi} = j\frac{\partial}{\partial \tau} \tag{1.83}$$

τ and ξ are linear combinations of the time co-ordinate (t) and the spatial co-ordinate (x_1), $\xi = \alpha_1 x_1 + \beta_1 t$, $\tau = \alpha_2 x_1 + \beta_2 t$ where the coefficients $\alpha_{1,2}$ and $\beta_{1,2}$ are determined separately in every specific case.

The required partial differential equation

$$A\frac{\partial^2 u}{\partial \xi^2} - B\frac{\partial^2 u}{\partial \tau^2} = Cu \tag{1.84}$$

has the form of the Klein-Gordon equation whose type essentially depends on the signs of A and B.

Let us introduce a nonlinearity in the usual way: assume following Cornman (1973) that A, B and C are the functions of $|u|$ and sign $A =$ sign B. Then eqn. 1.84 is a hyperbolic equation describing the wave propagation.

Consider first the auto-model case when $\eta = \xi - v\tau$ and eqn. 1.84 is reduced to the form

$$\frac{d^2 u}{d\eta^2} = \frac{C|u|}{A(|u|) - v^2 B(|u|)} u \tag{1.85}$$

We will look for the solution of eqn. 1.84 in the form of a monochromatic wave $u(\xi, \tau) = u(\xi) \exp(-jv\tau)$ which yields

$$\frac{d^2 u}{dx^2} = \frac{B(|u|) + C(|u|)}{A|u|} u \tag{1.86}$$

When the solution of eqn. 1.86 is sought in the form of a plane wave $u = u(\tau) = u(\tau) \exp(jk)$, eqn. 1.84 takes the form

$$\frac{d^2 u}{d\tau^2} = \frac{k^2 A(|u|) - C(|u|)}{B|u|} u \tag{1.87}$$

Hence, in all three cases we get the equations of the type

$$\frac{d^2 u}{d\eta^2} = W(|u|)u \tag{1.88}$$

Let us consider the case of a small nonlinearity in eqn. 1.84 taking only the zero terms of the expansions for A and B in the power series with respect to $|u|$ and two terms for C, $C = C_0 + C_1|u|^2$, and looking for the solution in the form $u(\xi, \tau) = v(\xi, \tau) \exp[-j(q\xi - v\tau)]$ where $Bv^2 - Aq^2 = C_0$. Then it is easy to show (without taking into account the nonlinear interaction of intertype oscillations) that v satisfies the nonlinear Schrödinger equation

$$j\frac{\partial v}{\partial \tau} + \gamma \frac{\partial^2 v}{\partial \eta^2} + \bar{\delta} C_1 |v|^2 v = 0 \tag{1.89}$$

where

$$v = v(\bar{\eta}, \tau), \quad \bar{\eta} = \xi + v_G \tau$$

$$v_G = \frac{\pm Aq}{\sqrt{B(C_0 + Aq^2)}}$$

is the group velocity,

$$\gamma = \frac{\pm AC_0}{2\sqrt{B(C_0 + Aq^2)^3}}$$

$$\bar{\delta} = \frac{\pm C_1^2}{2\sqrt{B(C_0 + Aq^2)}} \tag{1.90}$$

the signs $+$ and $-$ in eqn. 1.90 correspond to different types of oscillations.

It is known that the Cauchy problem for eqn. 1.89 with rather general initial conditions may be solved by the inverse scattering problem method and, hence, multi-soliton and finite-zone solutions may be found close to the Morse critical points.

Let us construct evolution equations in the form

$$F(\kappa, \chi, L) = 0 \tag{1.91}$$

occurring in the vicinities of the Morse critical points and depending, in addition to the spectral parameters κ (the real time) and χ (the spatial coordinate), on the non-spectral parameter L which is used for the control of the considered evolution process.

Let us determine the (isolated and non-degenerate) Morse critical point from eqn. 1.91. We will assume that in its vicinity having the form of a hyperplane $\Omega(\kappa, \chi, L) = \{(\kappa, \chi, L) : F(\kappa, \chi, L) = 0\}$ where

$$F_\kappa = F_\chi = F_L = 0, \quad \det J(\mathbf{Z}) \neq 0, \ \mathbf{Z} = (z_1, z_2, z_3) = (\kappa, \chi, L) \tag{1.92}$$

the function F in eqn. 1.92 is considered as a mapping $F(\mathbf{Z}) : C^3 = C \times C \times C \to C$ (Milnor, 1971) and $J = ||J_{il}||_{i,l=1}^3$, $J_{il} = F_{z_i z_l}$ is its Hesse matrix (all the derivatives are calculated in the Morse critical point $\mathbf{Z}_0 = (\kappa_0, \chi_0, L_0) = (z_{10}, z_{20}, z_{30})$).

Taking only the quadratic terms of the Taylor expansion, the equation of the hyperplane $\Omega(\kappa, \chi, L)$ in the vicinity of the Morse critical point may be represented as a quadratic form

$$I_1 = A\tilde{\kappa}^2 + B\tilde{\chi}^2 + C\tilde{L}^2 + 2(D\tilde{\kappa}\tilde{\chi} + K\tilde{\kappa}\tilde{L}P\tilde{\chi}\tilde{L}) + 2\delta \tag{1.93}$$

where

$$\delta = F(\mathbf{Z}_0), \quad \tilde{\mathbf{Z}} = \mathbf{Z} - \mathbf{Z}_0$$
$$A = J_{11}, \ B = J_{22}, \ C = J_{33}, \ D = J_{12}, \ K = J_{13}, \ P = J_{23} \tag{1.94}$$

and one can describe in an analytical form the variation of the eigencharacteristics of the considered physical system using the variables (κ, χ) and the control parameter L.

Let us reduce I_1 in eqn. 1.93 with the help of eqn. 1.94 to the canonical form

$$I_2 = \lambda_1 \hat{\kappa}^2 + \lambda_2 \hat{\chi}^2 + \lambda_3 \hat{L}^2 + \delta \tag{1.95}$$

with the corresponding diagonal matrix $K_2 = \text{diag} \|\lambda_l\|_{l=1}^3$ similar to the matrix K_1 of the quadratic form given by eqn. 1.93 $[K_2 = V^{-1} K_1 V$, V (det $V \neq 0$) is the transforming matrix] where λ_l, $l = 1, 2, 3$ are the roots of the characteristic equation

$$\lambda^3 + b_1 \lambda^2 + b_2 \lambda + b_3 = 0 \tag{1.96}$$

$$
\begin{aligned}
b_1 &= -(A + B + C) \\
b_2 &= -(P^2 + D^2 + K^2 - AB - AC - BC) \\
b_3 &= -(ABC + 2DKP - AP^2 - CD^2 - BK^2)
\end{aligned}
\tag{1.97}
$$

and eqn. 1.97 coincides with the Hessian (Milnor, 1971). After the change of variable $\lambda = y - \frac{b_1}{3}$ eqn. 1.96 will have the form

$$y^3 + py^2 + q = 0 \tag{1.98}$$

where

$$p = b_2 - \frac{b_1^2}{3}, \quad q = b_3 - \frac{b_1 b_2}{3} + \frac{b_1^2}{2}$$

and

$$y_1 = \tilde{\alpha} + \tilde{\beta}$$

$$y_{2,3} = -\frac{1}{2}(\tilde{\alpha} + \tilde{\beta}) \pm \frac{\sqrt{3}}{2}(\tilde{\alpha} - \tilde{\beta})$$

$$\tilde{\alpha} = (\sqrt{q_1^2 + p_1^3} - q_1)^{\frac{1}{3}}, \quad \tilde{\beta} = (-\sqrt{q_1^2 + p_1^3} - q_1)^{\frac{1}{3}}$$

$$q_1 = \frac{q}{2}, \quad p_1 = \frac{p}{3}$$

are the roots of eqn. 1.98 $(\tilde{\alpha}\tilde{\beta} = p_1)$.

Let us go back to eqn. 1.83 and proceed from the dispersion equation

$$\lambda_1 \hat{\kappa}^2 + \lambda_2 \hat{\chi}^2 + \lambda_3 \hat{L}^2 + 2\delta = 0$$

to the Klein-Gordon equation

$$\lambda_1 \frac{\partial^2 u}{\partial \xi^2} - \lambda_2 + \gamma(L)u = 0 \qquad (1.99)$$

with the control parameter L where

$$\gamma(L) = \lambda_3 \hat{L}^2 + 2\delta \qquad (1.100)$$

Variation of L in eqn. 1.100 controls an evolution ("catastrophic") process in the vicinity of the Morse critical point.

If the signs of $\lambda_1, \lambda_2, \lambda_3$ and δ in eqn. 1.99 are equal, then 1.99 is a hyperbolic equation describing the wave propagation which depends on the media parameters (corresponding to different values of L) and varying L one can investigate the wave evolution close to the Morse critical points.

Introducing a nonlinearity in the way described above, i.e. assuming that λ_1, λ_2 and $\gamma = \gamma(\lambda_3, L, \mathbf{Z}_0)$ are the functions of $|u|$, eqn. 1.99 will have the form

$$\frac{d^2 u}{d\eta^2} = W(|u|, L)u \qquad (1.101)$$

similar to eqns. 1.85–1.87 and different from eqn. 1.88 because the right-hand side of eqn. 1.101 depends on the control parameter L.

The presence of the negative and positive dispersions is a characteristic feature of the dispersion law. They are observed close to the Morse critical points and cause various nonlinear effects.

It has been shown above that in the auto-model regime when $A = A(|u|)$, $B = B(|u|)$, $C = C|u|$ are represented for small $|u|$ as the power series segments and $u(\xi, \tau) = u(\eta)$, $\eta = \xi - v\tau$ (where $v = $ const and in the case of a wave process a nonlinear wave formation $u(\eta)$ moves close to the Morse critical point with a constant velocity v) eqn. 1.84 is reduced to eqn. 1.88 with $W(u) = C|u|[A(|u|) - v^2 B(|u|)]^{-1}$. We will look for the solution of eqn. 1.88 in the form $u(\eta) = a(\eta)\exp(js\eta)$ which yields the system of two nonlinear real-valued equations for $a(\eta)$ and $s(\eta)$

$$\begin{cases} 2\dfrac{ds}{d\eta}\dfrac{da}{d\eta} + a\dfrac{d^2 s}{d\eta^2} = 0 \\[2mm] \dfrac{d^2 a}{d\eta^2} - a\left(\dfrac{ds}{d\eta}\right)^2 = aW(a) \end{cases} \qquad (1.102)$$

The solution of eqn. 1.102 is obtained by Bass and Gurevich (1975) and has the form

$$\eta = \pm \int_{a_0}^{a} \left[M - N^2 a^{-2} + 2 \int_{a_0}^{a} \alpha W(\alpha) \, d\alpha \right] d a \qquad (1.103)$$

where

$$N = a^2(0)s'(0) \quad M = [a'(0)]^2 + [s'(0)]^2 a^2(0)$$

$$a_0 = a(0), \quad s_0 = s(0), \quad a'(0) = a'(0), \quad s'_0 = s'(0)$$

and if $a(0) = 0$, then either $a(\eta) \equiv 0$ or $s(\eta) \equiv s_0$.

Limiting oneself to the first two terms in the Taylor expansion of $W(|u|)$ for sufficiently small $|u|$, one gets

$$W(|u|) = \alpha + 2\beta |u|^2 \qquad (1.104)$$

where

$$\alpha = \frac{C_0}{A_0 - v^2 B_0}, \quad \beta = \frac{C_1 A_0 - C_0 A_1 + (C_0 B_1 - C_1 B_0)v^2}{2(A_0 - v^2 B_0)} \qquad (1.105)$$

Here it is taken into account that

$$A(|u|) = A_0 + A_1 |u|^2, \quad B(|u|) = B_0 + B_1 |u|^2$$

$$C(|u|) = C_0 + C_1 |u|^2$$

and $A_0^2 - v^2 B_0 \neq 0$.

Let us consider the (bounded) solutions of eqn. 1.88 for $W(|u|)$ having the form given by eqns. 1.104 and 1.105 when $\beta \neq 0$ (for $\beta = 0$ the properties of the solution are well-known).

If $s'_0 = 0$, then $N = 0$ and $s(\eta) = s_0$. This case corresponds to the total internal reflection and it follows from eqn. 1.103 that

$$\eta = \pm \int_{a_0}^{a} \frac{d a}{\sqrt{E + 2\alpha a^2 + \beta a^4}}, \qquad (1.106)$$

where $E = a_0'^2 - \alpha a_0^2 - \beta a_0^4$.

Let us separate the bounded solutions of eqn. 1.106:

(i) In addition to the trivial solution, there exist other stationary solutions. Namely, if $\alpha\beta < 0$, then one has the solution

$$u(\eta) \equiv u_0 = a_0 \exp[j s_0(\eta)] \qquad (1.107)$$

satisfying the initial conditions $a_0 = \sqrt{-\frac{\alpha}{2\beta}}$, $a_0' = 0$

(ii) If $\alpha < 0$, $\beta > 0$, then the solution differs from the one given by eqn. 1.107 and has the form of a "kink"

$$u(\eta) = \eta \exp(js_0) \tanh \left(\frac{1}{2}\frac{n+a_0}{n-a_0} \pm \sqrt{-\frac{\alpha}{2}}\eta\right)$$

$$n = \sqrt{-\frac{\alpha}{2\beta}}$$

for the initial conditions

$$a_0' = \pm\beta\left(\frac{\alpha}{2\beta} + a_0^2\right)$$

$$a_0 < \sqrt{-\frac{\alpha}{2\beta}}$$

(iii) If $\alpha > 0$, $\beta < 0$, then the solution has the form of a soliton

$$u(\eta) = u(0)\,\mathrm{sech}\left(\frac{1}{2}\ln\frac{\sqrt{\alpha} - \sqrt{\alpha + \beta a_0^2}}{\sqrt{\alpha} + \sqrt{\alpha + \beta a_0^2}} \pm \sqrt{\alpha}\eta\right)$$

for the initial conditions

$$a_0' = \pm a_0(\alpha + 2\beta a_0^2)$$

$$a_0 < \sqrt{-\frac{\alpha}{2\beta}}$$

In particular, if

$$a_0 = \sqrt{-\frac{\alpha}{\beta}}, \quad a_0' = 0$$

then

$$u(\eta) = u(0)\,\mathrm{sech}\left(\sqrt{\alpha}\eta\right)$$

(iv) Assume that one of the two groups of conditions holds:

(a)

$$\alpha < 0, \quad \beta > 0, \quad a_0^2 + {a_0'}^2 \neq 0$$

$$a_0 < \sqrt{-\frac{\alpha}{2\beta}}$$

$$|a_0'| < -\sqrt{\beta}\left(\frac{\alpha}{2\beta} + a_0^2\right)$$

(b) $\beta < 0$ and none of the conditions A and B holds.

Then $u(\eta)$ is a periodical solution which may be represented by means of the elliptic Jacobi functions (Akhiezer, 1970). Three such solutions may be separated depending on the initial conditions. Let us introduce the notation

$$\tau_1 = \frac{-\alpha - \sqrt{\alpha - 4\beta E}}{2\beta}, \quad \tau_2 = \frac{-\alpha + \sqrt{\alpha - 4\beta E}}{2\beta}$$

$$\eta_0 = \int_0^{a_0} \frac{d a}{\sqrt{E + \alpha a^2 + \beta a^4}}$$

$$\eta_1 = \int_{\sqrt{\tau_2}}^{a_0} \frac{d a}{\sqrt{E + \alpha a^2 + \beta a^4}}$$

If the condition **(a)** holds, then

$$u(\eta) = \sqrt{\tau_1}\exp(js_0)\,\mathrm{sn}\{\sqrt{\beta\tau_2}(\eta + \eta_0\,\mathrm{sign}\,a_0')\sqrt{\frac{\tau_1}{\tau_2}}\}$$

If $\alpha > 0$, $\beta < 0$ and

$$|a_0'| > a_0\sqrt{(\alpha + \beta a_0^2)}$$

(i.e. when the condition **(b)** holds), then

$$u(\eta) = \sqrt{\frac{\tau_1\tau_2}{\tau_1 - \tau_2}}\exp(js_0)$$

$$\times \frac{\mathrm{sn}\left\{\sqrt{\beta(\tau_2 - \tau_1)}(\eta + \eta_0\,\mathrm{sign}\,a_0');\ \sqrt{\frac{\tau_1}{\tau_1 - \tau_2}}\right\}}{\mathrm{dn}\left\{\sqrt{\beta(\tau_2 - \tau_1)}(\eta + \eta_0\,\mathrm{sign}\,a_0');\ \sqrt{\frac{\tau_1}{\tau_1 - \tau_2}}\right\}} \qquad (1.108)$$

The third possible solution is obtained when **(b)** holds and at least one of the conditions indicated before eqn. 1.108 is violated:

$$u(\eta) = \sqrt{\tau_1}\exp(js_0) \times \mathrm{dn}\left\{\sqrt{|\beta|\tau_1}(\eta + \eta_0\,\mathrm{sign}\,a_0');\ \sqrt{1 - \frac{\tau_2}{\tau_1}}\right\}$$

Assume now that there is no total internal reflection, i.e. $a_0 s_0' \neq 0$. Then

$$\eta = \pm \int_{a_0^2}^{a^2} \frac{dt}{\sqrt{Q(t)}} \tag{1.109}$$

where

$$Q(t) = \bar{E}t + \alpha t^2 + \beta t^3 - N^2$$
$$\bar{E} = a_0'{}^2 - \alpha a_0^2 - \beta a_0^4 + a_0^2 s_0'{}^2$$

If $\beta < 0$, then, as well as in the previous case, all the solutions are bounded. If $\beta > 0$, then the necessary and sufficient conditions for the existence of bounded solutions are as follows:

$\alpha < 0$,

$$D(Q) = \alpha^2 \bar{E} - 4\beta \bar{E}^3 - 27\beta^2 N^4 + 4\alpha^3 N^2 - 18\alpha\beta N^2 \bar{E} \geq 0$$

and one of the two conditions

$$|a_0'| \leq a_0 \sqrt{-(\alpha + 2\beta a_0^2 + s_0'{}^2)}$$

or

$$\begin{cases} \alpha + 3\beta a_0^2 \leq 0 \\ a_0' \leq \frac{1}{3\beta}(\alpha - 3\beta a_0^2)^2 - a_0^2 s_0'{}^2 \end{cases}$$

Let us separate now all the bounded solution of eqn. 1.109:

(a) If $\alpha + 3\beta a_0^2 < 0$, $a_0' = 0$

$$s_0' = \pm \sqrt{-(\alpha + 2\beta a_0^2)}$$

and either $\beta < 0$ or $\alpha < 0$ and $\beta > 0$, then the bounded solution has the form

$$u(\eta) = u(0) \exp(js_0'\eta)$$

i.e. $u(\eta)$ is a plane monochromatic wave.

(b) If $\alpha < 0$, $\beta > 0$, $D(Q) = 0$ and the criterion for the existence of bounded solutions holds [except for the case $\alpha + 2\beta a_0^2 < 0$, $a_0' = 0$,

$$s_0' = \pm \sqrt{-(\alpha + 2\beta a_0^2)}$$

and $\beta < 0$], then

$$u(\eta) = \left[n_1 + (n_2 - n_1)\tanh^2 B\right]^{\frac{1}{2}}$$

$$\times \exp\left\{j\left[s_0 + \frac{N\eta}{n_2} + \frac{N}{n_2\sqrt{\beta n_1}}\left(\arctan\sqrt{\frac{n_2}{n_1}} - 1\,\tanh B\right.\right.\right.$$

$$\left.\left.\left. - \arctan\sqrt{\frac{a_0^2 - n_1}{n_1}}\,\right]\right\}$$

where

$$B = \frac{1}{2}\ln\frac{\sqrt{n_2 - n_1} + \sqrt{a_0^2 - n_1}}{\sqrt{n_2 - n_1} - \sqrt{a_0^2 - n_1}} \pm \sqrt{\beta(n_2 - n_1)}\eta$$

$$n_1 = \frac{N^2}{\beta n_2^2}, \quad n_2 = \frac{\alpha\bar{E} + 9\beta N^2}{6\beta\bar{E} - 2\alpha^2}$$

and in this case the amplitude is an antisoliton.

(c) If $\beta < 0$ [except for the case $\alpha + 2\beta a_0^2 < 0$, $a_0' = 0$,

$$s_0' = \pm\sqrt{-(\alpha + 2\beta a_0^2)}$$

and $\beta < 0$], or if $\alpha < 0$, $\beta > 0$, $D(Q) > 0$ and the criterion for the existence of bounded solutions holds, then $a(\eta)$ is a periodic function and the solution $u(\eta)$ is a quasi-periodic function,

$$u(\eta) = \exp[j(s_0 + U)]\sqrt{\tau_1 + (\tau_2 - \tau_1)\operatorname{sn}^2[Z_1; k]}$$

$$Z_1 = \sqrt{\beta(\tau_3 - \tau_1)}(\eta + \eta_0\operatorname{sign} a_0')$$

$$k = \sqrt{\frac{\tau_2 - \tau_1}{\tau_3 - \tau_1}}$$

$$U = \frac{c\Pi(a_n Z_1; K; k) - \Pi(a_n Z_2; K; k)}{\tau_1\sqrt{\beta(\tau_3 - \tau_1)}}$$

$$Z_2 = \sqrt{\beta(\tau_3 - \tau_1)}\eta_0\operatorname{sign} a_0'$$

$$K = \frac{\tau_2 - \tau_1}{\tau_1}$$

where τ_1, τ_2 and τ_3 are the roots of the polynomial $Q(t)$ (increasing when $\beta > 0$ and decreasing when $\beta < 0$),

$$\eta_0 = \frac{1}{2}\int_{\tau_1}^{a_0^2}\frac{dt}{\sqrt{Q(t)}}$$

$$\Pi(m; \phi; k) = \int_0^\phi \frac{dt}{(1 + m\sin^2 t)\sqrt{1 - k^2\sin^2 t}}$$

is the standard elliptic integral of the third kind, a_n is the amplitude of the standard elliptic integral

$$\eta = \int_0^{a_n} \frac{dt}{\sqrt{1 - k^2 \sin^2 t}}$$

of the first kind and $\mathrm{sn}(\eta; k) = \sin a_n \eta$.

In any other case the solution is not bounded and tends to infinity in a finite time interval (at least in one direction).

When the wave solutions of eqn. 1.84 are sought for $A = A(|u|)$, $B = B(|u|)$, $C = C(|u|)$ as plane or monochromatic waves, one gets equations similar to eqn. 1.88 and looking for their solutions in the form

$$u(\xi, \tau) = u(\xi) \exp(j\omega\tau), \quad \omega = \text{const}$$

assumimg that $A_0 \neq 0$, then

$$W(|u|) = \frac{-\omega^2 B(|u|) + C(|u|)}{A(|u|)}$$

$$\alpha = \frac{C_0 - \omega^2 B_0}{A_0}$$

$$\beta = \frac{C_1 A_0 - C_0 A_1 + (A_0 B_0 - B_1 A_0)\omega^2}{2A_0}$$

If $u(\xi, \tau) = u(\tau) \exp(jk\xi)$ $(k = \text{const})$ and $B_0 \neq 0$, then

$$W(|u|) = \frac{-k^2 A(|u|) + C(|u|)}{B(|u|)}$$

$$\alpha = \frac{C_0 - k^2 A_0}{B_0}$$

$$\beta = \frac{C_0 B_1 - C_1 B_0 + (A_0 B_1 - B_0 A_1)k^2}{2B_0}$$

All further investigations of such bounded solutions are carried out in a similar way. Eqns. 1.84 and 1.88 which are obtained by Bass *et al.* (1993, 1994) are the hybrid of the Klein-Gordon nonlinear equation and the wave nonlinear equation.

The results obtained above have a universal character and may be applied for various nonlinear problems with "small nonlinearities" and the Morse critical points. For example, the latter are connected in plasma physics with an external magnetic field and the former with the Miller potential. Analogous situations occur in the processes of different plasma wave interactions in semi-conductors with supergratings considered by Bass and Gurevich (1975) where nonlinearities are caused by non-quadratic dependence of the electron energy on the quasi-impulse. Electron pencils cause emergence of nonlinearities in the vacuum electronics.

1.11 Calculation of fundamental frequencies

The method of the solution to the spectral problems for open circular resonators with dielectric inclusions described in Section 1.2 enables us to construct and justify a numerical algorithm for calculating the fundamental frequencies. This algorithm is based on investigating the problems on characteristic numbers for families of finite-dimensional matrix operator-valued functions and on the existence of only isolated characteristic numbers of the initial "exact" spectral problem. Therefore, this algorithm may be considered not only as a tool for numerical calculations but also as a special application of the operator perturbation theory for these particular types of problems.

We introduce the sequence of projectors

$$P_L : l_2^M \to l_2^M \quad (L = 0, 1, 2, \dots) \quad P_L = \|\delta_m^n\|_{n,m=1}^M$$

$$d_L : l_2 \to l_2, \quad d_L \mathbf{x} = \mathbf{y}$$

$$\mathbf{x} = (x_m)_{m=-\infty}^{\infty}, \quad \mathbf{y} = (y_m)_{m=-\infty}^{\infty} = \begin{cases} 0 & |m| > L \\ 1 & |m| \leq L \end{cases}$$

onto the finite-dimensional space

$$P_L(l_2^M) = \left\{ \mathbf{x} = \left\| \begin{matrix} \mathbf{x}^1 \\ \vdots \\ \mathbf{x}^M \end{matrix} \right\| \quad x_m^i = 0, \ m > L, \ i = 1, 2, \dots M \right\}$$

and the sequence of finite-dimensional matrix operator-valued functions

$$A_L(k) = P_L - P_L A(k) P_L$$

approximating the operator-valued function $A(k)$ in eqn. 1.17.

Lemma 1.15: Let the sequence $(\mathbf{x}_L)_{L=0}^{\infty}$ satisfy the conditions

$$\mathbf{x}_L \in P_L(l_2^M), \quad \|\mathbf{x}_L\| < \text{const}, \quad \lim_{L \to \infty} \|P_L\mathbf{y} - A_L(k)\mathbf{x}_L\| = 0$$

Then

$$\lim_{L \to \infty} \|\mathbf{x}_L - A(k)\mathbf{x}_L - \mathbf{y}\|_2 = 0$$

The formulation of this lemma contains the definition of "eigenconvergence" for the operator sequence $A_L(k)$ proposed by Vainikko and Karma (1974).

The following inequality holds

$$\|A(k)\mathbf{x}_L + \mathbf{y} - \mathbf{x}_L\|$$
$$\leq \|A(k)\mathbf{x}_L - P_L\mathbf{y}\| + \|P_L\mathbf{y} - \mathbf{y}\| + \|\mathbf{x}_L\|\|A(k) - P_LA(k)P_L\|$$

The estimates obtained in Section 1.3 for $|a_{pm}^{ns}(k)|$ and the definition of the projectors P_L yield

$$\|A(k) - P_LA(k)P_L\| \leq A_L^M$$

$$A_L^M = \left(\sum_{s,n=1}^{M} \sum_{|p|,|m|>L} |a_{pm}^{ns}(k)|^2 \right)^{\frac{1}{2}} \to 0, \quad |P_L\mathbf{y} - \mathbf{y}\| \to 0 \; (L \to \infty)$$

which completes the proof of the eigenconvergence for the sequence of operator-valued functions $(A_L(k))_{L=0}^{\infty}$, $k \in \hat{B}$.

Lemma 1.16: The sequence of norms $\|(A_L(k))_{L=0}^{\infty}\|$ is uniformly bounded with respect to L and k in every compactum $F \subset \hat{B}$.

We have

$$\|A_L(k)\| = \|P_L - P_LA(k)P_L\| \leq \|P_L\| + \|P_L\|^2\|A(k)\| = 1 + \|A(k)\|$$

(because $\|P_L\| = 1$) and the following estimates for the operator norm

$$\|A(k)\mathbf{x}\| = \sum_{l=1}^{M} \| \sum_{l=1}^{N+M} A^{il}(k)\mathbf{x}^l \|^2$$
$$\leq \sum_{l=1}^{M} (\sum_{l=1}^{N+M} \|A^{il}(k)\|\|\mathbf{x}^l\|)^2$$
$$\leq \sum_{l=1}^{M} \|A^{il}(k)\|^2\|\mathbf{x}\|^2$$

and, consequently,

$$\|A(k)\| \le \left(\sum_{l=1}^{M} \|A^{il}(k)\|^2\right)^{\frac{1}{2}} \le \left(\sum_{i,l=1}^{M} \sum_{p,m} |a_{pm}^{il}(k)|^2\right)^{\frac{1}{2}} \qquad (1.110)$$

Eqn. 1.110 and the properties of $|a_{pm}^{il}(k)|$ yield the required uniform estimates

$$\|A_L(k)\| < 1 + R(F)(M)$$

where $R(F)$ is a constant depending on the compactum F and resonator's geometrical parameters.

Let us denote by $\sigma(A_L)$ the set of characteristic numbers of $A_L(k)$. Eigenconvergence of the sequence $(A_L(k))_{L=0}^{\infty}$ allow us to prove the fundamental result.

Theorem 1.17: If $k_0 \in \sigma(\Delta)$, i.e. if eqn. 1.17 of Section 1.2 has a non-trivial solution for $k = k_0$, then there exists a sequence of complex numbers $(k_L)_{L=0}^{\infty}$ such that $k_L \in \sigma(A_L)$, $(L \ge L_0 > 0)$ and $\lim_{L \to \infty} |k_L - k_0| \to 0$. On the other hand, if $k_L \in \sigma(A_L)$, $(L \ge L_0 > 0)$ and there exists an infinite subsequence $\{k_L'\} \subset (k_L)_{L=0}^{\infty}$ such that $\lim_{L' \to \infty} |k_L' - k_0| \to 0$, then k_0 is an eigenvalue of the spectral problem given by eqns. 1.1–1.5 (a fundamental frequency of the open resonator whose cross-section is formed by a domain from the family CB_c^{ϵ}).

Theorem 1.17 gives a rigorous justification of the following algorithm for calculating the points of $\sigma(\Delta)$. Call $k_L \in \sigma(A_L)$, $L \ge 1$ an approximate characteristic number of the operator-valued function $I - A(k)$ if for $k = k_L$ there exists a non-trivial solution of the equation

$$A_L(k)\mathbf{x} = \mathbf{o} \qquad (1.111)$$

In accordance with the projectors definition, eqn. 1.111 is a system of linear algebraic equations with a $N_L \times N_L$ matrix $(N_L = M(2L+1)M)$ and k_L are the roots of the equations

$$\Delta_L(k) \equiv \det A_L(k) = 0$$

where $\Delta_L(k)$ are analytical functions in $\hat{B}$.

The norm of $S_L(k) = A_L(k) - P_L A(k) P_L$ which tends to zero as $L \to \infty$ uniformly with respect to k in every compactum from $\hat{B}$, will characterise the "closeness" of eqns. 1.17 and 1.111. Due to Theorem 1.17, one can separate a subsequence of approximate characteristic numbers which converges to an "exact" characteristic number of $I - A(k)$ when $L \to \infty$.

The finite-dimensional operators $A_L(k)$ may be identified with the matrices

$$A_L(k) = \|B^{ns}(k)\|_{n,s=1}^M$$
$$B^{nn}(k) = \|\delta_p^m - a_{pm}^{nn}(k)\|_{p,m=-L}^L$$
$$B^{ns}(k) = \|a_{pm}^{ns}(k)\|_{p,m=-L}^L$$

where the matrix elements are given by eqns. 1.8 and 1.13 (see Section 1.2).

Let us determine the asymptotic rate of convergence of the presented numerical algorithm.

Assume that $(k_L)_{L=0}^\infty$ is a sequence of approximate characteristic numbers such that $\lim_{L \to infty} k_L = k_0$. Due to the analytical properties of the operator-valued function $A(k)$ (see lemma 1.1) and lemma 1.5, there exists the function

$$\Delta(k) = \det(I - A(k)) = \Pi_{i \geq 1}(1 - \lambda_i(k))$$

($\lambda_i(k)$ are the eigenvalues of $A(k)$) analytical in B and meromorphic in C_k [the infinite determinant of $A(k)$ introduced, e.g. by Gohberg and Krein, (1965)] and the following estimate holds

$$|k_L - k_0| < \frac{\Delta(k_L)}{|\Delta'(k_0)| + (|k_L - k_0|)} \tag{1.112}$$

Assume that $\Delta'(k_0) \neq 0$. Then the inequality

$$|\det(I - A) - \det(I - B)| < \|A - B\|_1 \exp(\|A\|_1 + \|B\|_1 + 1) \tag{1.113}$$

holds for a pair of kernel operators A and B in l_2^M (see Reed and Simon, 1978). The kernel norm

$$\|A\|_1 = \sum_{i=1}^\infty s_i(A) \tag{1.114}$$

wher $s_i(A)$ are the singular numbers of A, i.e. the eigenvalues of the operator $(A^* A)^{\frac{1}{2}}$ (A^* is the conjugated operator).

Inequalities given by eqns. 1.112 and 1.113 yield

$$|k_L - k_0| < \|A(k_L) - P_L A(k_L) P_L\|_1 \frac{\exp(\|A(k_L)\| + 1)}{|\Delta'(k_0)|} \qquad (1.115)$$

where, due to eqn. 1.114 and the properties of singular numbers (Reed and Simon, 1978)

$$\|A(k_L) - P_L A(k_L) P_L\|_1 \leq \sum_{s,n=1}^{M} \sum_{i=1}^{\infty} s_i [A^{ns}(k_L) - d_L A^{ns}(k_L) d_L]$$

The operators d_L are induced by P_L and are the projectors on the finite-dimensional sub-space of l_2

$$d_L(l_2) = \{\mathbf{x} : \mathbf{x} = (x_m)_{m=-\infty}^{\infty}, \ x_m = 0, |m| > L\}$$

Let us first obtain the estimate for

$$s_i [A^{nn}(k_L) - d_L A^{nn}(k_L) d_L]$$

It follows from eqns. 1.8 and 1.13 that

$$A^{nn}(k_L) - d_L A^{nn}(k_L) d_L = \begin{cases} V_L^n D_L^n(k_L), & n \leq N \\ B_L^n(k_L), & n > N \end{cases} \qquad (1.116)$$

where $N \leq M$, the operators V_L^n are bounded in l_2 and do not depend on k_L. $D_L^n(k_L)$ and $B_L^n(k_L)$ are the kernel operators and they admit diagonal matrix representations

$$D_L^n(k_L) = \|\delta_m(k_L a_n \sqrt{\epsilon_0 \mu_0}) \delta_m^p\|_{|m|,|p|>L}$$
$$B_L^n(k_L) = \|a_{pm}^{nn}(k_L) \delta_m^p\|_{|m|,|p|>L}$$

where $a_{pm}^{nn}(k_L)$ and δ_m are determined by eqns. 1.8 and 1.13. Eqn. 1.116 yields

$$s_i [A^{nn}(k_L) - d_L A^{nn}(k_L) d_L] \leq \begin{cases} \|V_L^n\| s_i(D_L^n(k_L)), & n \leq N \\ s_i(B_L^n(k_L)), & n > N \end{cases}$$

It is not difficult to calculate $s_i(D_L^n(k_L))$ and $s_i(B_L^n(k_L))$, since D_L^n and B_L^n are the diagonal operators. Finally we have

$$s_i [A^{nn}(k_L) - d_L A^{nn}(k_L) d_L] \leq \begin{cases} \|V_L^n\| 2|\delta_i(k_L a_n \sqrt{\epsilon_0 \mu_0})|, & n \leq N \\ 2|a_{ii}^{nn}(k_L)|, & n > N \end{cases} \qquad (1.117)$$

$$s_i[A^{nn}(k_L) - d_L A^{nn}(k_L)d_L] = 0, \quad i \leq L$$

Due to the Allakhverdiev theorem (Sobolev, 1966),

$$s_i[A^{nn}(k_L) - d_L A^{nn}(k_L)d_L] = \min_R \|A^{ns}(k_L) - d_L A^{ns}(k_L)d_L - R\|$$

where the minimum is determined on the set of all finite-dimensional operators R of order not greater than i. In our case

$$s_i[A^{nn}(k_L) - d_L A^{nn}(k_L)d_L] = \begin{cases} \|A^{ns}(k_L) - d_L A^{ns}(k_L)d_L\| & i \leq N \\ \|A^{ns}(k_L) - d_i A^{ns}(k_L)d_i\| & i > N \end{cases}$$

and taking eqn. 1.117 into account we get

$$\|A(k_L) - P_L A(k_L) P_L\|_1 < N(k_L) \tag{1.118}$$

where

$$\begin{aligned}
N(k_L) = & \sum_{n=1}^{N} \|V_L^n\| 2 \sum_{i=L+1}^{\infty} |\delta_i(k_L a_n \sqrt{\epsilon_0 \mu_0})| \\
& + \sum_{n=N+1}^{M} 2 \sum_{i=L+1}^{\infty} a_{ii}^{nn}(k_L) \\
& + \sum_{n \neq s}^{M} (L \|A^{ns}(k_L) - d_L A^{ns}(k_L)d_L\| \\
& + \sum_{i=L+1}^{\infty} \|A^{ns}(k_L) - d_i A^{ns}(k_L)d_i\|)
\end{aligned}$$

The values $\|A^{ns}(k_L) - d_i A^{ns}(k_L)d_i\|$ may be estimated with the help of inequality

$$\|A^{ns}(k_L) - d_i A^{ns}(k_L)d_i\|^2 < \sum_{|p|,|m|>i} |a_{pm}^{ns}(k_L)|^2 \tag{1.119}$$

and asymptotic properties of $|a_{pm}^{ns}(k)|$ obtained in Section 1.3. Substitution of eqn. 1.118 to eqn. 1.115 yields the required estimate

$$|k_L - k_0| < F(k_L)$$

$$F(k_L) = N(k_L) \frac{\exp(\|A(k_L)\| + 1)}{|\Delta'(k_0)|} \tag{1.120}$$

One can show with the use of eqn. 1.119 and after the series of bulky transformations that the asymptotic rate of convergence of the presented numerical algorithm has the order $F(k_L) = O(L^{-1})$, $L \to \infty$ uniformly with respect to the parameters of the problem varying in closed intervals. The use of this estimate, as well as eqns. 1.119 and 1.120, are not convenient in practical applications (besides, they turn out to be set too high) but such estimates are of principle character for the justification of numerical methods.

Chapter 2

Excitation of oscillations: physical properties of open resonators

In this chapter we will consider various scattering problems beginning with the plane momochromatic $\mathbf{E}^0(\omega)$, $\mathbf{H}^0(\omega)$ wave diffraction by the open structure formed by a finite number N_S of infinitely thin, perfectly conducting, unclosed circular cylindrical surfaces containing inclusions in the form of N_ϵ homogeneous isotropic circular dielectric pivots ($N_S + N_\epsilon = N_\epsilon$), with the axis parallel to that of the resonator's mirrors (the permittivity and permeability of the medium, ϵ and μ, depend on spatial co-ordinates). The resonator itself is placed in a homogeneous medium with complex permittivity and permeability, ε_0, μ_0 (Im $\varepsilon_0 \geq 0$, Im $\mu_0 \geq 0$).

In accordance with the terminology introduced in Section 1.1, the transversal cross-sections of these open resonators belong to the family CB_c^ϵ.

The incident field may have the form

$$\{H_{x_3}^0, E_{x_3}^0\} \sim \exp[-j\hat{k}(x_1 \cos \beta + x_2 \sin \beta)] \quad \text{(plane wave)}$$

$$\{H_{x_3}^0, E_{x_3}^0\} \sim \frac{j}{4} H_0^{(1)}(\hat{k}|\mathbf{r} - \mathbf{r}_0| \quad \text{(cylindrical wave)}$$

where $\mathbf{k} = \hat{k}(-\cos \beta \mathbf{e}_1, -\sin \beta \mathbf{e}_2)$ is the wave vector, $\mathbf{r}_0$ is the point where the source of the cylindrical wave is situated, $\hat{k} = \frac{\omega}{c}\sqrt{\epsilon_0 \mu_0}$. It is known (Hönl et al., 1962) that the vector diffraction problem is reduced to two boundary-value problems in the determination of Green's functions corresponding to two polarisations of the incident field:

$$H^0_{x_3} \neq 0, \ E^0_{x_3} \equiv 0 \quad \text{(H-case)}$$
$$E^0_{x_3} \neq 0, \ H^0_{x_3} \equiv 0 \quad \text{(E-case)}$$

The solution of these boundary-value problems is reduced with the help of variables separation in the local co-ordinate systems by the method of the Riemann-Hilbert problem (Shestopalov, 1971) to systems of the second kind operator equations in the space l_2. The equivalence theorem and the existence of Green's functions have been proved and the connection between the diffraction problem and the spectral problem given by eqns. 1.1–1.5 have been established. The numerical and analytical methods for constructing Green's functions are presented.

Sections 2.4 and 2.5 consider spectral problems and excitation of arbitrary open resonators, in particular, with the cross-sections belonging to the general family NB_c (see Section 1.1) by means of the method of generalised solutions and variational identities. Basic distinctions of this approach are, firstly, in the treatment of spectral and excitation problems in cylindrical domains whose transversal cross-sections have noncompact boundaries, and, secondly, in applying the partial radiation Sveshnikov-Riechardt conditions at infinity, the form differing from the conditions used for cylindrical scatterers with compact boundaries. The problems with the Sveshnikov-Riechardt conditions have been widely considered before for the cases either of local bodies or periodical structures [for detailed references see Ilyinsky and Shestopalov (1989)].

The well developed approaches (methods of integral equations, partial domains etc.) cannot be applied for modelling such open resonators due to the complicated character of their noncompact boundaries.

The resonators considered model perfectly conducting plane screens with arbitrary finite perturbations, including (layered) dielectric "lugs" and strips (finite arrays) placed over the screens. They have distinct resonance properties caused by the presence of cavities which enable one to effectively control their scattering characteristics. An image slot line formed by a plane screen with longitudinal slot and a rectangular waveguide "stuck" under it may serve as a canonical example of such a scatterer.

2.1 Diffraction problems and second-kind operator equations

Let us give the mathematical formulation of the plane monochromatic wave diffraction problem, considering the structure with the transversal cross-section formed by the domain belonging to the family CB_c^ϵ. Assume that the source of the cylindrical wave $\mathbf{H}^0(\omega, \mathbf{r})$, $\mathbf{E}^0(\omega, \mathbf{r})$ lies out of the dielectric inclusions [in particular, at infinity when $\mathbf{H}^0(\omega, \mathbf{r})$, $\mathbf{E}^0(\omega, \mathbf{r})$ is a plane wave]. One has to determine the solution $\mathbf{H}^g(\omega, \mathbf{r})$, $\mathbf{E}^g(\omega, \mathbf{r})$ of homogeneous Maxwell's equations

$$
\begin{cases}
\dfrac{j}{k\mu(\mathbf{r})} \operatorname{curl} \mathbf{H}^g(\omega, \mathbf{r}) = \mathbf{E}^g(\omega, \mathbf{r}) \\[3mm]
-\dfrac{j}{k\epsilon(\mathbf{r})} \operatorname{curl} \mathbf{E}^g(\omega, \mathbf{r}) = \mathbf{H}^g(\omega, \mathbf{r})
\end{cases}
\tag{2.1}
$$

satisfying the boundary value conditions on S

$$
[\mathbf{E}^g + \mathbf{E}^0, \mathbf{n}] = 0 \quad (S \cap \Gamma_\epsilon = \emptyset)
\tag{2.2}
$$

$$
[\mathbf{E}^g, \mathbf{n}] = 0 \quad (S \subset \Gamma_\epsilon)
\tag{2.2a}
$$

the conjugation conditions on Γ_ϵ

$$
[\mathbf{E}^g_- + \mathbf{E}^0, \mathbf{n}] = [\mathbf{E}^g_+, n], \quad [\mathbf{H}^g_- + \mathbf{H}^0, \mathbf{n}] = [\mathbf{H}^g_+, n],
$$

the Meixner conditions and the Maluzhinec condition (Sveshnikov, 1950). Here $k = \frac{\omega}{c}$, c is the light velocity in vacuum, $\mathbf{n}$ is the unit normal to Γ_ϵ or to S, $\mathbf{E}_\pm$, $\mathbf{H}_\pm$ are the (external for $-$ and internal for $+$) limiting values of $\mathbf{H}^g$, $\mathbf{E}^g$ on the corresponding arcs, and the functions $\epsilon(\mathbf{r})$ and $\mu(\mathbf{r})$ are introduced in Section 1.1.

The uniqueness theorem proved by Hönl *et al.* (1962) holds for the vector boundary value problem given by eqns. 2.1–2.3 which is reduced to two independent (scalar) boundary value problems with respect to $u^1 = H_z^g$, $u^2 = E_z^g$:

$$
\Delta u(\mathbf{r}) + k^2 \chi(\mathbf{r}) u(\mathbf{r}) = 0, \quad \mathbf{r} \in R^2 \setminus (S \cup \Gamma_\epsilon)
\tag{2.4}
$$

satisfying boundary conditions on S

$$
u^2 + E_{x_3}^0 = 0, \quad \frac{\partial u^1 + H_{x_3}^0}{\partial \mathbf{n}} = 0 \quad (S \cap \Gamma_\epsilon = \emptyset)
\tag{2.5}
$$

$$u^2 = 0, \quad \frac{\partial u^1}{\partial \mathbf{n}} = 0 \quad (S \subset \Gamma_\epsilon)$$

the conjugation conditions on Γ_ϵ

$$[u_-^2 + E_{x_3}^0 - u_+^2] = 0, \quad [u_-^1 + H_{x_3}^0 - u_+^1] = 0$$

$$\left[\frac{1}{\mu(\mathbf{r})} \frac{\partial u_+^2}{\partial \mathbf{n}} - \frac{1}{\mu_0} \frac{\partial u_-^2 + E_z^0}{\partial \mathbf{n}} \right] = 0 \qquad (2.6)$$

$$\left[\frac{1}{\epsilon(\mathbf{r})} \frac{\partial u_+^1}{\partial \mathbf{n}} - \frac{1}{\epsilon_0} \frac{\partial u-^1 + H_{x_3}^0}{\partial \mathbf{n}} \right] = 0$$

the Meixner conditions

$$\iint_Q \left(|u^i|^2 + \left| \frac{\partial u^i}{\partial x_1} \right|^2 + \left| \frac{\partial u^i}{\partial x_2} \right|^2 \right) d\mathbf{r} < \infty \quad (i = 1, 2) \qquad (2.7)$$

for every compactum $Q \subset R^2$ and the Reichardt radiation conditions: there exists an $R_0 > 0$ such that for all $\mathbf{r} = (r, \phi)$: $\hat{k}r \geq R_0$, where $\hat{k} = k\sqrt{\varepsilon_0 \mu_0}$, the following representation holds for the function $u^i(\mathbf{r})$

$$u^i(\mathbf{r}) = \sum_{m=-\infty}^{\infty} a_m^i H_m^{(1)}(\hat{k}r) \exp(jm\phi) \quad (i = 1, 2) \qquad (2.8)$$

which is equivalent to the Sommerfeld radiation conditions for $\operatorname{Im} \hat{k} = \operatorname{Im} k = 0$

$$\frac{1}{\sqrt{r}} \left(\frac{\partial u^{1,2}}{\partial r} - j\hat{k}u^{1,2} \right) \to 0, \quad r \to \infty$$

and to the "liquidating" conditions formulated by Sveshnikov (1950): for $\operatorname{Im} \epsilon_0 > 0, \operatorname{Im} \mu_0 \geq 0$,

$$|u^{1,2}(\mathbf{r})| \to 0, \quad r \to \infty.$$

Here $\mathbf{n}$ is the unit normal to the boundary of the domain $S \cup \Gamma_\epsilon$, $u_\pm^{1,2}$ and $\frac{\partial u_\pm^{1,2}}{\partial \mathbf{n}}$ are the limiting values of u and $\frac{\partial u^{1,2}}{\partial \mathbf{n}}$ on the boundary, where "+" denotes the values taken from inside the domain and "−" from outside the domain.

Other components of the E^g- and H^g-fields can be determined from the system of the Maxwell equations by means of $u^{1,2}(\mathbf{r})$:

$$E_{x_1}^g = \frac{j}{k\epsilon} \frac{\partial u^1}{\partial x_2}, \quad H_{x_1}^g = -\frac{j}{k\mu} \frac{\partial u^2}{\partial x_2}$$

$$E_{x_2}^g = -\frac{j}{k\epsilon} \frac{\partial u^1}{\partial x_1}, \quad H_{x_2}^g = \frac{j}{k\mu} \frac{\partial u^2}{\partial x_1}$$

where $\epsilon = \epsilon(\mathbf{r})$ and $\mu = \mu(\mathbf{r})$ are assumed to be piece-wise constant functions.

The conditions stated above provide for the uniqueness of the solutions u^i, $i = 1, 2$, if they exist. We assume in what follows that the indices $i = 1$ and $i = 2$ correspond to the problems E and H, respectively, and call $G^E = E_{x_3}^0 + u^1$ and $G^H = H_{x_3}^0 + u^1$ Green's functions of the problems E and H.

We look for the solutions of the problems E and H in the local co-ordinate systems (r_p, ϕ_p) as linear superpositions of the Fourier series

$$
u^n = \begin{cases}
\displaystyle\sum_{p=1}^{N_S} \delta_p \sum_{m=-\infty}^{\infty} x_m^{np} A_m^n(\hat{k}r_p, \hat{k}a_p) \exp(jm\phi_p) \\[2mm]
\displaystyle + \sum_{p=1}^{N_\epsilon} \sum_{m=-\infty}^{\infty} x_m^{nN_S+p} B_m^n(\hat{k}r_{p+N_S}, \hat{k}a_{p+N_S}) \exp(jm\phi_{p+N_S}) \\[2mm]
(r_p, \phi_p) \in R^2 \setminus (S \cup \Gamma_\epsilon) \\[2mm]
\displaystyle\sum_{p=1}^{N_S} \alpha_{pl} \sum_{m=-\infty}^{\infty} x_m^{np} A_m^n(kr_p\sqrt{\epsilon_i\mu_i}, ka_p\sqrt{\epsilon_i\mu_i}) \exp(jm\phi_p) \\[2mm]
\displaystyle + \sum_{m=-\infty}^{\infty} y_m^{nl} D_m^n(kr_{l+N_S}\sqrt{\epsilon_l\mu_l}, ka_{l+N_S}\sqrt{\epsilon_l\mu_l}) \exp(jm\phi_{l+N_S}) \\[2mm]
(r_p, \phi_p) \in \Omega_\epsilon
\end{cases}
\tag{2.9}
$$

where

$$
A_m^1(\mathbf{r}) = \begin{cases}
H_m^{(1)'}(x_2) J_m(x_1), & |x_1| < |x_2| \\
H_m^{(1)}(x_1) J_m'(x_2), & |x_1| > |x_2|
\end{cases}
$$

$$
A_m^2(\mathbf{r}) = \begin{cases}
J_m(x_1)[J_m(x_2)]^{-1}, & |x_1| < |x_2| \\
H_m^{(1)}(x_1)[H_m^{(1)}(x_2)]^{-1}, & |x_1| > |x_2|
\end{cases}
$$

$$
\begin{array}{ll}
B_m^1(\mathbf{r}) = H_m^{(1)}(x_1) J_m(x_2) & \qquad D_m^1(\mathbf{r}) = H_m^{(1)}(x_2) J_m(x_1) \\[2mm]
B_m^2(\mathbf{r}) = H_m^{(1)}(x_1)[H_m^{(1)}(x_2)]^{-1} & \qquad D_m^2(\mathbf{r}) = J_m(x_1)[J_m(x_2)]^{-1}
\end{array}
$$

$$
\delta_p = 1 - \sum_{l=1}^{N_S} \alpha_{pl}, \qquad \alpha_{pl} = \begin{cases} 1, & (r_p, \phi_p) \in \Omega_\epsilon \\ 0, & (r_p, \phi_p) \notin \Omega_\epsilon \end{cases}
$$

$(x_m^{np})_{m=-\infty}^{\infty}$, $(y_m^{nl})_{m=-\infty}^{\infty}$ $(p = 1, 2, \ldots, N_\epsilon)$ are the sought for diffraction harmonics. Using the large index asymptotics for the cylindrical functions (Watson, 1949) it is easy to show that u^n are infinitely differentiable functions everywhere where they satisfy eqn. 2.4 and the radiation conditions if it is assumed that the vectors of diffraction harmonics belong to the space l_2. As far as the limiting values of u^n on $\Gamma_{S\epsilon} = S \cup \Gamma_\epsilon$ are concerned, we formulate the following *a priori* assumption C which is equivalent to the assumptions A and B from Section 1.1:

The series in eqn. 2.9, as well as the series obtained after the term-wise differentiation of eqn. 2.9, converge uniformly on $\Gamma_{S\epsilon} \setminus S_\delta$ where S_δ denotes arbitrarily small vicinities of the endpoints of S ("edges"). We will show below that the assumption C provides the possibility to use the addition theorems for the cylindrical functions. We will limit ourselves in what follows to the problem E and to the case $\alpha_{pl} = 0$ (when the perfectly conducting screens S are situated out of the dielectric inclusions Ω_ϵ) because the problem H and other possible situations of the screens disposition are considered in the similar way.

In addition to the above assumptions, we demand for the problem E that the limiting values of u^2 and $\frac{\partial u^2}{\partial \mathbf{n}}$ on Γ_ϵ satisfy conditions given by eqn. 2.6. With this purpose we will represent these functions, using the addition theorems for cylindrical functions and the local co-ordinate systems (r_p, ϕ_p) in the vicinity of every circle from the set Ω_ϵ, in the form of the series similar to eqn. 2.9. After equivalent transformations we get the following relations:

$$y_p^{2l} = x_p^{2N_S+l} + \sum_{q \neq l}^{N_\epsilon} \sum_{m=-\infty}^{\infty} x_m^{2N_S+l} \frac{H_{m-p}^{(1)}(\hat{k}s_{N_S+q\,N_S+l})J_p(\hat{k}a_{N_S+l})}{H_{m-p}^{(1)}(\hat{k}s_{N_S+q\,N_S+l})}$$

$$\times \exp[j(m-p)\psi_{N_S+q\,N_S+l}]$$

$$+ \sum_{q=1}^{N_S} \sum_{m=-\infty}^{\infty} x_m^{2q} \frac{T_{m-p}^2(\hat{k}s_{q\,N_S+l})J_p(\hat{k}a_{N_S+l})}{R_m^2(\hat{k}a_q)} \exp[j(m-p)\psi_{q\,N_S+l}]$$

$$+ d_p^{N_S+l} J_p(\hat{k}a_{N_S+l})$$

$$y_p^{2l} \sqrt{\frac{\epsilon_l \mu_0}{\mu_l \epsilon_0}} \frac{J_p'(ka_{N_S+l}\sqrt{\epsilon_l \mu_l})}{J_p(ka_{N_S+l}\sqrt{\mu_l \epsilon_l})} = x_p^{2N_S+l} \frac{H_p^{(1)'}(\hat{k}a_{N_S+l})}{H_p^{(1)}(\hat{k}a_{N_S+l})}$$

$$+ \sum_{q \neq l}^{N_\epsilon} \sum_{m=-\infty}^{\infty} x_m^{2N_S+q} \exp[j(m-p)\psi_{N_S+q\,N_S+l}] \frac{H_{m-p}^{(1)}(\hat{k}s_{N_S+q\,N_S+j})J_p'(\hat{k}a_{N_S+j})}{H_m^{(1)}(\hat{k}a_{N_S+l})}$$

$$+ \sum_{q=1}^{N_S} \sum_{m=-\infty}^{\infty} x_m^{2q} \frac{T_{m-p}^2(\hat{k}s_{q\,N_S+l})J_p(\hat{k}a_{N_S+l})}{R_m^2(\hat{k}a_q)} \exp[j(m-p)\psi_{q\,N_S+l}]$$

$$+ d_p^{N_S+l} J_p'(\hat{k}a_{N_S+l}) \quad (2.10)$$

Here $l = 1, 2, \ldots, N_\epsilon$

$$R_m^2(\hat{k}a_q) = \begin{cases} H_m^{(1)}(\hat{k}a_q), & a_q < s_{qN_S+l} \\ J_m(\hat{k}a_q), & a_q > s_{qN_S+l} \end{cases}$$

$$T_m(\hat{k}s_{qN_S+l}) = \begin{cases} H_m^{(1)}(\hat{k}s_{qN_S+l}), & a_q < s_{qN_S+l} \\ J_m(\hat{k}s_{qN_S+l}), & a_q > s_{qN_S+l} \end{cases}$$

the values $d_p^{N_S+l}$ are determined by the incident field:

$$d_p^{N_S+l} = \begin{cases} (-j)^p \exp(-j\hat{k}s_{1N_S+l}c_l) \\ \frac{j}{4}H_p^{(1)}(\hat{k}r_0^{N_S+l})\exp(-jp\phi_0^{N_S+l}) \end{cases} \quad c_l = \cos(\psi_{1N_S+l} - \beta)$$

$r_0^{N_S+l}, \phi_0^{N_S+l}$ are the cylindrical wave source co-ordinates in the $N_S + l$th co-ordinate system.

Eqn. 2.10 yields the system of relations with respect to the unknowns x_p^{2m}:

$$x_p^{2N_S+l} = \sum_{q \neq l}^{N_\epsilon} \sum_{m=-\infty}^{\infty} x_m^{2N_S+q} b_{pm}^{N_S+q\,N_S+l} + \sum_{q=1}^{N_S} \sum_{m=-\infty}^{\infty} x_m^{2q} b_{pm}^{N_S+lq} + b_p^{N_S+l}$$

$$(l = 1, 2, \ldots, N_\epsilon) \quad (2.11)$$

where

$$b_{pm}^{N_S+lq} = \frac{T_{m-p}^2(\hat{k}s_{N_S+lq})J_p(ka_{N_S+l})}{R_m^2(\hat{k}a_q)} B_{pm}^{N_S+lq}$$

$$b_{pm}^{N_S+l\,N_S+q} = \frac{H_{m-p}^{(1)}(\hat{k}s_{N_S+l\,N_S+q})J_p(\hat{k}a_{N_S+l})}{H_m^{(1)}(\hat{k}a_{N_S+q})} B_{pm}^{N_S+lq} \quad (2.12)$$

$$b_p^{N_S+l} = d_p^{N_S+l}\Pi_p(\hat{k}a_{N_S+l}, z)J_p(\hat{k}a_{N_S+l})$$

$$B_{pm}^{N_S+lq} = \Pi_p(\hat{k}a_{N_S+l}, z)\exp[j(m-p)\psi_{q\,N_S+j}]$$

and

$$\Pi_p(\mathbf{r}) = \frac{x_2 J_p'(x_1 z) J_p(x_1) - J_p(x_1 z) J_p'(x_1)}{J_p(x_1 z) H_p^{(1)'}(x_1) - x_2 J_p'(x_1 z) H_p^{(1)'}(x_1)} \frac{J_p'(x_1)}{J_p(x_1)}$$

Here the diffraction harmonics y_p^{2l} describing the field inside the inclusions Ω_ϵ are obtained after extracting the harmonics x_p^{2l} and $x_p^{2N_S+q}$ ($q \neq l$):

$$y_p^{2l} = \frac{2j J_p(w\sqrt{\epsilon_l \mu_l}) x_p^{2N_S+l}}{\pi \hat{w} H_p^{(1)}(\hat{w})[J_p(\hat{w}) J_p'(w z_1) z_1 - J_p'(w) J_p(w\sqrt{\epsilon_l \mu_l}))]} \tag{2.13}$$

where

$$z = \sqrt{\frac{\epsilon_l \mu_l}{\mu_0 \epsilon_0}}, \quad z_1 = \sqrt{\frac{\epsilon_l \mu_0}{\mu_l \epsilon_0}}, \quad \hat{w} = \hat{k} a_{N_S+l}, \quad w = k a_{N_S+l}$$

The system of equations 2.11–2.13 is equivalent (under the assumption C) to the conjugation conditions given by eqn. 2.6 but this system is not well-posed and should be supplemented by the relations equivalent to the boundary conditions given by eqn. 2.5. With this purpose let us represent the solution u^2 in the local co-ordinate systems (r_q, ϕ_q) connected with an arc from the set S using the addition theorems for cylindrical functions

$$u^2 = \sum_{m=-\infty}^{\infty} x_m^{2q} A_m^2(\hat{k} r_q, \hat{k} a_q) \exp(jm\phi_q)$$

$$+ \sum_{\substack{l \neq \\ l}}^{N_S} \sum_{m=-\infty}^{\infty} x_m^{2l}$$

$$\times \frac{\sum_p \hat{Q}_p^2(\hat{k} r_q) \tilde{T}_{m-p}^2(\hat{k} l q l) \exp(jp\phi_q) \exp[j(m-p)\psi_{lq}]}{\hat{R}_m^2(\hat{k} a_l)}$$

$$+ \sum_{l=1}^{N_\epsilon} \sum_{m=-\infty}^{\infty} x_m^{2N_S+l}$$

$$\times \frac{\sum_p \tilde{R}_p(\hat{k} r_q) \tilde{T}_{m-p}(\hat{k} l q \, N_S + l) \exp(jp\phi_q) \exp[j(m-p)\psi_{N_S+l \, q}]}{H_m^{(1)}(\hat{k} a_{N_S+l})} \tag{2.14}$$

where

$$\tilde{T}_p(\hat{k} s_{qN_S+l}) = \begin{cases} H_p^{(1)}(\hat{k} s_{qN_S+l}), & a_q < s_{qN_S+l} \\ J_p(\hat{k} s_{qN_S+l}), & a_q > s_{qN_S+l} \end{cases}$$

$$\tilde{R}_p(\hat{k} r_q) = \begin{cases} J_p(\hat{k} r_q), & r_q < s_{qN_S+j} \\ H_p^{(1)}(\hat{k} r_q), & r_q > s_{qN_S+l} \end{cases}$$

and we assume that if $S \subset \Gamma_\epsilon$, then

$$\hat{R}_m^2(\cdot) = \hat{Q}_m(\cdot) = \hat{T}_m^2(\cdot) = J_m(\cdot)$$

and if $S \cap \Gamma_\epsilon = \emptyset$), then

$$\hat{R}_m^2(\cdot) = H_m^{(1)}(\cdot), \quad \hat{Q}_m^2(\cdot) = J_m(\cdot), \quad \hat{T}_m^2(\cdot) = H_m^{(1)}(\cdot)$$

for $a_i \leq s_{il}$, and

$$\hat{R}_m^2(\cdot) = H_m^{(1)}(\cdot), \quad \hat{Q}_m^2(\cdot) = H_m^{(1)}(\cdot), \quad \hat{T}_m^2(\cdot) = J_m(\cdot)$$

for $a_i > s_{il}$, where, in accordance with the notation introduced in Section 1.1, the indices i, l correspond to the local co-ordinate systems connected with the arcs from the set S and with the circles from Ω_ϵ, respectively. The properties of cylindrical functions and assumption C yield absolute convergence of the series in eqn. 2.14.

Now one can substitute eqn. 2.14 into eqn. 2.5:

$$\sum_p [x_p^{2q} + L_1 + L_2 d_p^q J_p(\hat{k}a_q)] \exp(jp\phi_q) = 0$$

$$\theta_q < |\phi_q - \psi_q| \quad (q = 1, 2, \ldots, N_S) \tag{2.15}$$

where

$$L_1 = \sum_{q \neq l}^{N_S} \sum_{m=-\infty}^{\infty} x_m^{2l} \frac{\hat{T}_{m-p}^2(\hat{k}s_{ql})\hat{Q}_p^2(\hat{k}a_q)}{R_m^2(\hat{k}a_l)} \exp[j(m-p)\psi_{lq}]$$

$$L_2 = \sum_{l=1}^{N_\epsilon} \sum_{m=-\infty}^{\infty} x_m^{2l} \frac{\tilde{T}_{m-p}(\hat{k}s_{N_S+lq})\tilde{R}_p(\hat{k}a_q)}{H_{m-p}^{(1)}(\hat{k}a_{N_S+l})} \exp[j(m-p)\psi_{N_S+lq}]$$

In addition to eqn. 2.5, we demand that $\frac{\partial u^2}{\partial \mathbf{n}}$ in eqn. 2.14 is continuous in the vicinities of the elements from the set S what yields

$$\sum_p \hat{E}_{pq} x_p^{2q} \exp(jq\phi_p) = 0, \quad \theta_q > |\phi_q - \psi_q| \quad (q = 1, 2, \ldots, N_S) \tag{2.16}$$

where

$$\hat{E}_{pq} = \frac{1}{J_p(\hat{k}a_q) H_p^{(1)}(\hat{k}a_q)}$$

Now assuming that the assumption C holds, it is not difficult to see that, due to the Sobolev theorem (Sobolev, 1966), the function $u^2(\mathbf{r})$ given by eqn. 2.9, where x_m^{2q} and y_m^{2l} satisfy eqns. 2.11, 2.15 and 2.16, is a solution of the problem given by eqns. 2.4–2.8. In other words, the coupled system of functional equations 2.11, 2.15 and 2.16 with trigonometric kernels is equivalent to the initial diffraction problem. There exists a left regulariser (a regularising operator) of this system which may be constructed by the method of the Riemann-Hilbert problem (Shestopalov, 1971). In order to prove its existence, let us introduce the new unknowns

$$z_p^q = \exp(jp\psi_p)[x_p^{2q} + L_1 + L_2] + d_p^q J_p(\hat{k}a_q)$$

and rewrite eqns. 2.15 and 2.16 in the form

$$
\begin{cases}
\displaystyle\sum_p z_p^q \exp(jpf) = 0, & \theta_q < |f| \\[2mm]
\displaystyle\sum_p \hat{E}_{pq} z_p^q \exp(jpf) = \sum_p \hat{E}_{pq} \gamma_{pq} \exp(jpf), & \theta_q > |f|
\end{cases}
\tag{2.17}
$$

$(q = 1, 2, \ldots, N_S)$ where

$$\gamma_{pq} = \hat{L}_1 + \hat{L}_2 + d_p^q \exp(jp\psi_q) J_p(\hat{k}a_q)$$

$$\hat{L}_1 = \sum_{q\neq l}^{N_S} \sum_{m=-\infty}^{\infty} \exp(jp\psi_q) x_m^{2l} \frac{\hat{T}_{m-p}^2(\hat{k}s_{ql})\hat{Q}_p^2(\hat{k}a_q)}{R_m^2(\hat{k}a_l)} \exp[j(m-p)\psi_{lq}]$$

$$\hat{L}_2 = \sum_{l=1}^{N_e} \sum_{m=-\infty}^{\infty} \exp(jp\psi_q) x_m^{2l} \frac{\hat{T}_{m-p}(\hat{k}s_{N_S+lq})\tilde{R}_p(\hat{k}a_q)}{H_{m-p}^{(1)}(\hat{k}a_{N_S+l})} \exp[j(m-p)\psi_{N_S+lq}]$$

The left-hand sides in eqn. 2.17 determine the operator acting in the space s_2 which corresponds to the problem of diffraction on a single screen (Shestopalov, 1971) when the set S consists of one element. The right-hand side of the second equation in eqn. 2.17 is an infinitely differentiable function if $(x_m^{2q})_{m=-\infty}^{\infty} \in s_2$. Then one can carry out the left regularisation of eqn. 2.17 by the method developed by Shestopalov (1983) and obtain an equivalent system

$$z_m^q = \sum_p z_p^q \epsilon_p(\hat{k}a_q) F_m^p(u_q) - \frac{j}{\pi} \sum_p \hat{E}_{pq} \gamma_{pq} F_m^p(u_q)$$

where

$$F_m^p(z) = \begin{cases} -\ln \frac{1+z}{2}, & m = p = 0 \\ \frac{1}{p} V_{p-1}^{-1}(z), & m = 0, p \neq 0 \\ \frac{1}{m} V_{m-1}^{p-1}(z), & m \neq 0 \end{cases}$$

$$\epsilon_p(z) = |p| + \frac{j}{\pi J_p(z) H_p^{(1)}(z)}, \quad u_q = \cos\theta_q$$

The system with respect to the unknowns x_m^{2q} takes the form

$$x_m^{2q} = \sum_{l=1}^{N_S} \sum_{p=-\infty}^{\infty} b_{mp}^{ql} x_m^{2l} + \sum_{l=1}^{N_\epsilon} \sum_{p=-\infty}^{\infty} b_{mp}^{qN_S+l} x_m^{2N_S+l} + b_m^q$$

$$(q = 1, 2, \ldots, N_S) \qquad (2.18)$$

where

$$b_{mp}^{qq} = \epsilon_p(\hat{k}a_q) F_m^p(u_q) \exp[j(m-p)\psi_q] \quad (q = 1, 2, \ldots, N_S)$$

$$b_{mp}^{ql} = \left[\sum_{\bar{q}\neq 0} \hat{Q}_{\bar{q}}^2(\hat{k}a_q) \hat{T}_{p-\bar{q}}^2(\hat{k}s_{ql}) |\bar{q}| F_m^{\bar{q}}(u_q) \exp[j(p-\bar{q})\psi_{lq}] \exp(j\bar{q}\psi_q) \right.$$

$$\left. - \frac{\hat{T}_{m-p}^2(\hat{k}s_{ql}) \hat{Q}_m^2(\hat{k}a_q)}{\hat{R}_p^2(\hat{k}a_l)} \exp[j(p-m)\psi_{lq}] \right] \frac{\exp(-jm\psi_q)}{\hat{R}^2(\hat{k}a_l)}$$

$$q \neq l \quad (q, l = 1, 2, \ldots, N_S)$$

$$b_{mp}^{qN_S+l} = \frac{\exp(-jm\psi_q)}{H_p^{(1)}(\hat{k}a_{N_S+l})} \left[\sum_{\bar{q}\neq 0} \tilde{R}_{\bar{q}}(\hat{k}a_q) \hat{T}_{p-\bar{q}}^2(\hat{k}s_{qN_S+l}) |\bar{q}| \right.$$

$$\times F_m^{\bar{q}}(u_q) \exp[j(p-\bar{q})\psi_{N_S+lq}] \exp(j\bar{q}\psi_q)$$

$$\left. - \frac{\tilde{T}_{m-p}(\hat{k}s_{qN_S+l}) \tilde{R}_m^2(\hat{k}a_q)}{H_p^{(1)}(\hat{k}a_{N_S+l})} \exp[j(p-m)\psi_{N_S+lq}] \right]$$

$$(q = 1, 2, \ldots, N_S, \; l = 1, 2, \ldots, N_\epsilon)$$

$$b_m^q = - d_m^q J_m(\hat{k}a_q) F_m^0(u_q) + \sum_{\bar{q}\neq 0} d_{\bar{q}}^q J_{\bar{q}}(\hat{k}a_q) |\bar{q}| F_m^{\bar{q}}(u_q)$$

$$(q = 1, 2, \ldots, N_S)$$

$$(2.19)$$

Using the representations for u^1 similar to eqn. 2.9 and the results of Section 1.1, we obtain the infinite system of linear algebraic equations corresponding to problem H

$$x_p^{1N_S+l} = \sum_{q \neq l}^{N_\epsilon} \sum_{m=-\infty}^{\infty} a_{pm}^{N_S+l\, N_S+q} x_m^{1N_S+q} + \sum_{q=1}^{N_S} \sum_{m=-\infty}^{\infty} a_{pm}^{N_S+lq} x_m^{1q} + a_p^{N_S+l}$$

$$(l = 1, 2, \ldots, N_S)$$

$$x_p^{1q} = \sum_{l=1}^{N_\epsilon} \sum_{m=-\infty}^{\infty} a_{pm}^{qN_S+l} x_m^{1N_S+l} + \sum_{l=l}^{N_S} \sum_{m=-\infty}^{\infty} a_{pm}^{ql} x_m^{l} + a_p^{q}$$

$$(q = 1, 2, \ldots, N_S)$$

$$(2.20)$$

Here the matrix elements $a_{pm}^{qN_S+l}$ and a_{pm}^{ql} are determined by eqn. 1.13 and

$$a_{pm}^{N_S+lq} = \frac{T_{m-p}(\hat{k}s_{N_S+lq}) R_m(\hat{k}a_q)}{H_p^{(1)}(\hat{k}a_{N_S+l})} \Pi_p(\hat{k}a_{N_S+l}, z_1)$$
$$\times \exp[j(m-p)\psi_{q\,N_S+l}] \quad (q = 1, 2, \ldots, N_S)$$

$$(2.21a)$$

$$a_{pm}^{N_S+l\, N_S+q} = \frac{H_{m-p}^{(1)}(\hat{k}s_{N_S+l\, N_S+q}) J_m(\hat{k}a_{N_S+q})}{H_p^{(1)}(\hat{k}a_{N_S+l})} \Pi_p(\hat{k}a_{N_S+l}, z_1)$$
$$\times \exp[j(m-p)\psi_{N_S+q\, N_S+l}] \quad (q \neq l)$$

$$(2.21b)$$

The functions $R(\cdot)$ and $T_m(\cdot)$ are defined in eqn. 1.8 and the free terms $a_p^{N_S+l}$, a_p^q $(l = 1, 2, \ldots, N_\epsilon, q = 1, 2, \ldots, N_S)$ are determined by the incident field and have the form

$$a_p^{N_S+l} = \frac{(-j)^p \exp\{-j[\hat{k}s_{1N_S+l} \cos(\psi_{1N_S+l} - \beta) + p\beta]\}}{H_p^{(1)}(\hat{k}a_{N_S+l})} \Pi_p(\hat{k}a_{N_S+l}, z_1)$$

$$(l = 1, 2, \ldots, N_\epsilon)$$

$$a_p^q = j\pi(\hat{k}a_q)^2(-1)^p \exp\{-j[\hat{k}s_{1q} \cos(\psi_{1q} - \beta) + p\psi_q]\}$$

$$\times \sum_{m=-\infty}^{\infty} j^m J_m'(\hat{k}a_q) F_p^m(u_q) \exp[jm(\psi_q - \beta)]$$

$$(q = 1, 2, \ldots, N_S),$$

$$(2.22a)$$

for the plane wave and

$$a_p^{N_S+l} = \frac{1}{4} \exp(-jp\phi_0^{N_S+l}) \frac{H_p^{(1)}(\hat{k}r_0^{N_S+l})}{H_p^{(1)}(\hat{k}a_{N_S+l})} \Pi_p(\hat{k}a_{N_S+l}, z_1)$$

$$(l = 1, 2, \ldots, N_\epsilon)$$

$$a_p^q = -\frac{1}{4}\pi(\hat{k}a_q)(-1)^p \exp(-jp\psi_q)$$

$$\times \sum_{m=-\infty}^{\infty} (-1)^m f_m(\hat{k}r_0^q, \hat{k}a_q) F_p^m(u_q) \exp[jm(\psi_q - \psi_0^q)]$$

$$(q = 1, 2, \ldots, N_S)$$

$$(2.22b)$$

for the cylindrical wave, where

$$f_m(\mathbf{r}) = \begin{cases} J_m(x_1) H_m^{(1)'}(x_2), & x_2 \geq x_1 \\ J_m'(x_2) H_m^{(1)'}(x_1), & x_2 < x_1 \end{cases}$$

(r_0^q, ϕ_0^q) are the source co-ordinates in the (r_q, ϕ_q) co-ordinate system. The free terms representations are obtained with the help of the incident waves expansions (Ivanov, 1968) in the form of the series of cylindrical functions.

Let us indicate the main stages of constructing the infinite systems given by eqns. 2.11, 2.18 and 2.20 which define operators and operator equations equivalent to the initial problems E and H. Assuming the existence of the solutions to eqns. 2.4–2.8, we represent the sought for functions u^1 and u^2 in the form given by eqn. 2.9, where every term in the finite sums coincides with the function determining the field of the key diffraction problems for a dielectric cylinder considered by Ivanov (1968) or for a unclosed circular cylindrical surface (Shestopalov, 1983). Then we make the assumptions formulated above concerning the limiting values of the solutions on the boundaries S and Γ_ϵ and construct the infinite systems of linear algebraic equations corresponding to the conjugation conditions and the systems of coupled functional equations with a trigonometric kernel corresponding to the boundary-value conditions.

Application of the left regularising operator constructed by the method of the Riemann-Hilbert problem yields the final infinite systems given by eqns. 2.18 and 2.20.

2.2 The existence and equivalence theorems

Let us prove the equivalence of the systems given by eqns. 2.11, 2.18 and 2.20 and the problems E and H, respectively, together with the existence and uniqueness of the solutions to the former in the space l_2. As a corollary, we establish existence of the Green functions G^E and G^H for the problems E and H and justify the variables separation in the local polar co-ordinate systems, as well as the possibility to apply the method of the Riemann-Hilbert problem for the solution to the problem stated by eqns. 2.4–2.8.

We will limit ourselves in what follows to problem H since problem E may be treated in a similar way.

Let us consider the infinite system given by 2.20. As has been pointed out above, the matrices $\|a_{pm}^{ql}\|$ $(q = 1, 2, \ldots, N_S, l = 1, 2, \ldots, M)$ coincide with the corresponding matrices in eqn. 1.12. Hence, in accordance with asymptotic estimates for $|a_{pm}^{ns}(k)|$, they define (kernel) Hilbert-Schmidt operators $A^{ql} : l_2 \to l_2$. The matrices $\|a_{pm}^{N_S+lq}\|$ $(q = 1, 2, \ldots, N_S, l = 1, 2, \ldots, M)$ differ from the matrices in eqn. 1.8 by the factor

$$\Pi_p(\hat{k}a_{N_S+l}, z_1) \sim \frac{\epsilon_l - \epsilon_0}{\epsilon_l + \epsilon_0} \qquad (2.23)$$

and one can show that they also define Hilbert-Schmidt operators A^{N_S+lq} : $l_2 \to l_2$.

To show that the free terms in eqn. 2.22 belong to l_2 (considering only the case of the plane wave excitation), it follows from eqn. 2.23 that

$$|a_{pm}^{N_S+lq}| < \text{const} \; \frac{|\hat{k}a_{N_S+l}|^{|p|}}{2^{|p|}|p|!} \frac{|\epsilon_l - \epsilon_0|}{|\epsilon_l + \epsilon_0|}, \quad |p| \to \infty \qquad (2.24)$$

The representation for $F_p^m(u_q)$ via the Legendre polynomials (Shestopalov, 1971) enables us to obtain from eqn. 2.22 the relations

$$\begin{aligned}
a_p^q =& \beta_p^q \{ J_p'(\hat{k}a_q) p^{-1} V_{p-1}^{p-1}(u_q) j^p \exp(-jp\beta) \\
&+ \frac{1}{2}(-1)^p \exp(jp\psi_q) \sum_{m \neq p} j^m J_p'(\hat{k}a_q) \exp[jm(\psi_q - \beta)] \\
&\times [P_{p-1}(u_q)P_m(u_q) - P_{m-1}(u_q)P_p(u_q)](p - m)^{-1} \}
\end{aligned}$$

where

$$\beta_p^q = j\pi(\hat{k}a_q)^2 \exp[-j\hat{k}l_{1q}\cos(\psi_{1q} - \beta)]$$

Now the asymptotic estimates for $J_q(\hat{k}a_q)$ and the Legendre polynomials
$(|P_q(u_q)| < \text{const } |q|^{-0.5})$ for $|q| \to \infty$ allow us to assert that the sequences
(a_p^q) and $a_p^{N_s+l}$ belong to the space l_2 because

$$a_p^q = (-1)^p [C_1^q P_p(u_q) + C_2^q P_{p-1}(u_q)] \exp(-jp\psi_q) + O(|p|^{-2.5}) \qquad (2.25)$$

where C_1^q and C_2^q are the constants which do not depend on p.

Finally we can rewrite eqn. 2.20 as a system of operator equations in l_2:

$$\mathbf{x}^q = \sum_{l=1}^{M} A^{ql} \mathbf{x}^l + \mathbf{a}^q \quad (q = 1, 2, \ldots, M)$$

where $\mathbf{x}^q = (x_p^{1q})$ is the vector of unknowns and $\mathbf{a}^q = (a_p^q)$ is the vector of
right-hand sides.

It has been established in Section 1.3 that every solution of the homogeneous system given by eqn. 1.34 admits the representation

$$x_p^{1q} = (-1)^p p^{-1} \exp(-jp\psi_q)[C_1^q P_{p-1}(u_q) + C_2^q P_p(u_q)] + O(|p|^{-2.5})$$
$$|x_p^{N_s+l}| < C_3^j |p| \gamma_l^{|p|} \frac{\epsilon_l - \epsilon_0}{\epsilon_l + \epsilon_0} \quad (0 < \gamma_l < 1) \qquad (2.27)$$

for $|p| \to \infty$. One can obtain such asymptotic estimates for every solution of
the homogeneous system given by eqn. 1.34 using eqns. 2.24 and 2.25 and
taking into account the fact that the matrices given by eqns. 2.21 and 1.8
are asymptotically "close". Assume now that the function u^1 in eqn. 2.9 is
constructed with the help of a solution x_p^{1q} $(q = 1, 2, \ldots, M)$ to eqn. 2.26
and y_p^{1l} in eqn. 2.20 is expressed by means of $x_p^{1N_s+l}$. Then, similarly to
the proof of lemma 1.4, one can show that u^1 satisfies the Meixner condition
given by eqn. 2.7 and the assumption C. The results of Podolsky (1965)
and the estimates given by eqns. 2.27, 2.24 and 2.52 enable us to justify the
application of the method of the Riemann-Hilbert problem for constructing
the system given by eqn. 2.26.

Let us formulate the equivalence theorem.

Theorem 2.1: Let $\mathbf{x}^q = (x_p^{1q}) \in l_2$ $(q = 1, 2, \ldots, M)$ be the solution to the system of operator equations given by eqn. 2.26. Then the function

$$
u^1 = \begin{cases}
\displaystyle\sum_{p=1}^{N_S} \sum_{m=-\infty}^{\infty} x_m^{1q} A_m^1(\hat{k}r_q, \hat{k}a_q)\exp(jm\phi_q) \\[2ex]
\displaystyle + \sum_{q=1}^{N_\epsilon} \sum_{m=-\infty}^{\infty} x_m^{1N_S+q} J_m^n(\hat{k}a_{N_S+q}) H_m^{(1)}(\hat{k}r_{q+N_S})\exp(jm\phi_{N_S+q}) \\[2ex]
\qquad (r_q, \phi_q) \in R^2 \setminus (S \cup \Gamma_\epsilon) \\[2ex]
\displaystyle \frac{2l}{\pi \hat{k} a_{N_S+l}} \sum_{m=-\infty}^{\infty} x_m^{1N_S+l} \Phi_m^1\left(\hat{k}a_{N_S+l}, \sqrt{\frac{\epsilon_l}{\epsilon_0}}, \sqrt{\frac{\mu_l}{\mu_0}}\right) \\[2ex]
\qquad \times J_m(kr_{N_S+l}\sqrt{\epsilon_l\mu_l})\exp(jm\phi_{N_S+l}) \\[2ex]
\qquad (r_{N_S+l}, \phi_{N_S+l}) \in \Omega_\epsilon
\end{cases}
$$

$$(2.28)$$

is a solution of the problem H. Here A_m^1 are determined by eqn. 2.9 and

$$
\Phi_m^1(x, y, z) = J_m(x)[J_m(x)J_m'(xyz)zy^{-1} - J_m(xyz)J_m'(x)]^{-1}
$$

Let $\mathbf{x}^q \in l_2$ satisfy eqn. 2.26 or eqn. 2.20. Let us construct u^1 by eqn. 2.28. Then this function satisfies eqn. 2.4 in a classical sense everywhere outside of the curves S and Γ_ϵ and the radiation conditions given by eqn. 2.8. It follows from eqn. 2.37 that the limiting values of u^1 and $\frac{\partial u^1}{\partial n}$ on Γ_ϵ satisfy the assumption C from Section 2.1 and, hence, the condition given by eqn. 2.7. Taking this into account and using the addition theorems for cylindrical functions we obtain from eqn. 2.20 that u^1 satisfies the conjugation conditions on Γ_ϵ (eqn. 2.6) and the boundary value conditions on S (eqn. 2.5). Thus if there exists a solution of eqn. 2.26 belonging to l_2, eqn. 2.28 gives the corresponding solution to the initial diffraction problem.

Let us prove the existence and uniqueness of the solution to the system given by eqn. 2.26.

Theorem 2.2: There exists a unique solution of the system given by eqn. 2.26.

Let us prove first that this system may be represented as a canonical Fredholm equation (Krein, 1971)

$$
\mathbf{x} = A\mathbf{x} + \mathbf{a}
$$

$$(2.29a)$$

in the Hilbert space l_2^M defined in eqn. 1.15. Here

$$A = \|A^{ns}\|_{n,s=1}^M \quad \mathbf{x} = \left\| \begin{array}{c} \mathbf{x}^1 \\ \vdots \\ \mathbf{x}^M \end{array} \right\| \quad \mathbf{a} = \left\| \begin{array}{c} \mathbf{a}^1 \\ \vdots \\ \mathbf{a}^M \end{array} \right\| \tag{2.29b}$$

and A defined by eqn. 1.36 is a bounded linear operator acting in l_2^M. Since A^{ql} are the Hilbert-Schmidt operators, for every $\epsilon > 0$ there exist finite-dimensional operators $T^{ql} = T^{ql}(\epsilon)$ such that $T = \|T^{ns}\|_{n,s=1}^M$ is a finite-dimensional operator in l_2^M and

$$\|A^{ql} - T^{ql}\| < \frac{\epsilon}{M} \quad (q, l = 1, 2, \ldots, M) \tag{2.30a}$$

Following lemma 1.7 one can prove the estimate

$$\|A\| \leq \left(\sum_{q,l=1}^M \|A^{ql}\|^2 \right)^{\frac{1}{2}} \tag{2.30b}$$

which implies

$$\|A - T\| \leq \left(\sum_{q,l=1}^M \|A^{ql} - T^{ql}\|^2 \right)^{\frac{1}{2}} < \epsilon \tag{2.30c}$$

and due to Riesz and Nagy (1952) one can conclude that A is a compact operator in l_2^M.

Let us show that the homogeneous equation $\mathbf{x} = A\mathbf{x}$ has only a trivial solution $\mathbf{x} = 0$, or, in other notation $x_p^{1q} = 0$ where $\mathbf{x} = (\mathbf{x}^q)$, $\mathbf{x}^q = (x_p^{1q})$ $(q = 1, 2, \ldots, M, \ p = \pm 1, \pm 2, \ldots)$. Let $\mathbf{x}$ be a solution of the homogeneous equation. Then the function u^1 constructed by eqn. 2.28 is a solution of the problem H with $H_z^0 \equiv 0$. Due to the uniqueness theorem (Hönl et al., 1962) $u^1 \equiv 0$ in $\mathbb{R}^2 \setminus (S \cup \Omega_\epsilon)$ for $\text{Im } \epsilon_0 > 0$, $\text{Im } \mu_0 > 0$ and the cylindrical functions completeness (Vekua, 1953) yields the required result. It follows from this theorem and theorem 1.4 that $G^H = H_{x_3}^0 + u^1$ is Green's function for the considered problem if u^1 is given by eqn. 2.28 and (x_p^{1q}) is a solution of eqn. 2.20.

Theorems 2.1 and 2.2 are proved for the problem E in a similar way with the help of asymptotic estimates given by eqns. 2.24, 2.25 and 2.27 for the free terms in eqns. 2.12 and 2.10 and for the solutions of eqns. 2.11 and 2.18, respectively.

Let us prove the existence and uniqueness of the solution to the system given by eqn. 2.26.

Theorem 2.3: Assume that there exists a unique solution (x_p^{2q}) of the system given by eqns. 2.11 and 2.18. Then

$$
u^2 = \begin{cases}
\displaystyle\sum_{p=1}^{N_S}\sum_{m=-\infty}^{\infty} x_m^{2q} A_m^2(\hat{k}r_q, \hat{k}a_q)\exp(jm\phi_q) \\[2ex]
\displaystyle + \sum_{q=1}^{N_\epsilon}\sum_{m=-\infty}^{\infty} x_m^{2N_S+q}\frac{H_m^{(1)}(\hat{k}r_{q+N_S})}{H_m^{(1)}(\hat{k}a_{q+N_S})}\exp(jm\phi_{N_S+q}), \\[2ex]
\qquad (r_q, \phi_q) \in R^2 \setminus (S \cup \Gamma_\epsilon) \\[2ex]
\displaystyle\frac{2l}{\pi \hat{k}a_{N_S+l}}\sum_{m=-\infty}^{\infty} x_m^{2N_S+l}\Phi_m^2\left(\hat{k}a_{N_S+l}, \sqrt{\frac{\epsilon_l}{\epsilon_0}}, \sqrt{\frac{\mu_l}{\mu_0}}\right) J_m(kr_{N_S+l}\sqrt{\epsilon_l\mu_l}) \\[2ex]
\qquad \times \exp(jm\phi_{N_S+l}), \\[2ex]
\qquad (r_{N_S+l}, \phi_{N_S+l}) \in \Omega_\epsilon
\end{cases}
$$

$$\tag{2.31}$$

where

$$
\Phi_m^2(x, y, z) = \{H_m^{(1)}(x)[J_m(x)J_m'(xyz)yz^{-1} - J_m(xyz)J_m'(x)]\}^{-1}
$$

is a corresponding solution of problem E and $G^E = E_{x_3}^0 + u^2$ is its Green's function.

The problem given by eqns. 2.4–2.8 which enables one to determine the Green functions $G^{H,E}$ describes forced monochromatic oscillations and is tightly connected with the corresponding E and H spectral problems considered in Chapter 1.

Let us limit ourselves here to the analysis of problem H given by eqns. 2.4–2.8. It has been shown above that the Fredholm alternative is valid for the system given by eqn. 2.20, i.e. for Im $\epsilon_0 \geq 0$, Im $\mu_0 \geq 0$ there exists the bounded operator $(I - A)^{-1}$ defined in the whole space l_2^M, where A is given by eqns. 2.29 and 1.13. $(I - A)^{-1}$ is continuous as a function of the parameter $\hat{k} = k\epsilon_0\mu_0$ for Im $\hat{k} = 0$ and is an analytical function for Im $\hat{k} > 0$ ($k = \omega c^{-1}$ where ω is the incident field frequency). Hence the resolvent set $\rho(A)$ is not empty which enables us to apply the Fredhom theorem and to prove that the fundamental frequencies (eigenvalues) of the problem given by eqns. 1.1–1.5 form a discrete set on the complex plane $\hat{k}$. An analytical continuation of $(I - A)^{-1}$ through the axis $\text{Im}\,\hat{k} = 0$ to $\text{Im}\,\hat{k} > 0$ may have pole-type singularities. At the same time, applying the kernel property of operators A^{ql} and the technique based on the calculation of the Fredholm determinants (Hönl *et al.*, 1962) one can show that G^H admits a meromorphic continuation to $\text{Im}\,\hat{k} = 0$ as a function of the parameter $\hat{k}$ and its singularities (poles) coincide with the values $\hat{k}$ where eqn. 2.29 or eqn. 2.20 with $a = 0$ have nontrivial solutions and which, in turn, coincide with eigenvalues of problem H given by eqns. 1.1–1.5 due to theorem 1.1. The residues in these poles determine the eigenfunctions satisfying the general radiation condition given by eqn.1.5 and describing the H-type eigenoscillations [known as "damped resonances" by Reichardt (1960) and Wineberg (1972)].

Similar conclusions are valid for the problem E and the corresponding spectral problem.

2.3 Numerical solution to systems of second-kind operator equations

As shown in Sections 2.1 and 2.2, the study of forced electromagnetic oscillations is reduced to the solution of the infinite systems given by eqns. 2.11, 2.18 and 2.20. The Green functions G^E and G^H are calculated (by summing infinite series) with the help of eqns. 2.28 and 2.31. It is of special interest to investigate Green's functions in the resonance frequency range $kL = O(1)$ (L is the characteristic size of a structure) where the high Q-factor resonances may exist. But in this frequency range it is not possible to obtain the required solutions in a closed (analytical) form for arbitrary values of the geometrical and electrodynamical parameters. Therefore, various numerical methods are widely applied here and among them are the reduction and successive approximation methods. We will justify in what follows the applicability of the latter for the solution of the infinite systems given by eqns. 2.11, 2.18 and 2.20 and verify the error estimates of the method.

Let us consider eqns. 2.11, 2.18 and 2.20 as one operator equation

$$\mathbf{x} = C\mathbf{x} + \mathbf{d} \qquad\qquad (2.32)$$

where

$$C = \|C^{il}\|_{i,l=1}^{M} \quad x = \left\| \begin{matrix} \mathbf{x}^1 \\ \vdots \\ \mathbf{x}^M \end{matrix} \right\| \quad d = \left\| \begin{matrix} \mathbf{d}^1 \\ \vdots \\ \mathbf{d}^M \end{matrix} \right\|$$

$$C_{mp}^{il} = \begin{cases} a_{mp}^{il} & \text{(problem H)} \\ b_{mp}^{il} & \text{(problem E)} \end{cases} \qquad d_p^i = \begin{cases} a_p^i & \text{(problem H)} \\ b_p^i & \text{(problem E)} \end{cases}$$

The matrices $\|a_{mp}^{il}\|$ and $\|b_{mp}^{il}\|$ are given by eqns. 1.13, 2.21, 2.19 and the free terms d_p^i by eqns. 2.22 and 2.19.

Following the general scheme of the reduction method presented in Section 1.11 we introduce the sequence of projectors

$$P_Q : l_2^M \to l_2^M, \quad Q = 0, 1, 2, \ldots$$

$$P_Q = \|\delta_i^l d_Q\|_{i,l=1}^M, \quad d_Q : l_2 \to l_2$$

$$d_L x = y, \quad x = (x_m)_{m=-\infty}^{\infty}, \quad y = (y_m)_{m=-\infty}^{\infty}, \quad y_m = \begin{cases} 0, & |m| > Q \\ 1, & |m| \le Q \end{cases}$$

and construct the sequence of equations

$$x_Q = P_Q d + P_Q C P_Q x_Q, \quad Q = 0, 1, 2, \ldots \tag{2.33}$$

in the finite-dimensional spaces $P_Q(l_2^M)$ such that each of them is a finite system of the second kind of linear algebraic equations, with block-diagonal matrices consisting of the elements $\|C_{mp}^{il}\|_{p,m=-Q}^{Q}$. The solutions of such systems will be called approximate solutions of eqn. 2.32 or of the infinite systems given by eqns. 2.11, 2.18 and 2.20.

In order to justify this reduction method we will use the eigenconvergence technique developed by Vainikko and Karma (1974) and prove that

(i) there exists such a positive number Q_0 that for all $Q \ge Q_0$ eqn. 2.33 has a unique solution;

(ii) $\|x_e - x_Q\| < \delta_Q, \delta_Q \to 0, Q \to \infty$ (with respect to the l_2^M-norm) where x_e is an exact solution of eqn. 2.32 and estimate δ_Q, i.e. determine the rate of convergence.

Since all the results are based on asymptotic estimates of a_{pm} and b_{pm}^{il} for $|p|, |m| \to \infty$ it is sufficient to consider problem H because problem E is treated in a similar way.

As shown in Section 2.2, eqn. 2.32 is a canonical Fredholm equation, i.e. C is a compact operator in l_2^M. Using eqn. 2.30 and the definition of P_Q, we get

$$\|C - P_Q C P_Q\|^2 \leq \sum_{i,l=1}^{M} \sum_{|p|,|m|>Q} \|C^{il}\|^2, \quad \|C - P_Q C\|^2 \leq \sum_{p} \sum_{m} \sum_{i,l=1}^{M} |C_{pm}^{il}|^2$$
$$(2.34)$$

Since C^{il} are the kernel operators (see lemma 1.1), eqn. 2.34 and the estimates given by eqns. 2.24 and 2.25 yield

$$\|C - P_Q C P_Q\| \to 0, \qquad \|C - P_Q C\| \to 0, \quad Q \to \infty$$

and

$$\|d - P_Q d\|^2 = \sum_{i=1}^{M} \sum_{|p|>Q} |d_p^i|^2 \to 0, \quad Q \to \infty \tag{2.35}$$

Then it follows from theorems 5 and 3 proved in the paper by Vainikko and Karma (1974) that the assertions (i) and (ii) hold and, in particular, there exists the inverse operator $(C - P_Q C P_Q)^{-1}$ for all $Q \geq Q_0 > 0$ and the following rough estimate is valid for the rate of convergence

$$\|\mathbf{x}_e - \mathbf{x}_Q\| < \delta_Q, \quad \delta_Q = \hat{Q}_1(\|\mathbf{x}_e\|\|P_Q - P_Q C P_Q\|) + \|\mathbf{d} - P_Q \mathbf{d}\| \tag{2.36}$$

where $\hat{Q}_1$ does not depend on Q for $Q \geq Q_0$.

Let us specify the last estimate, limiting ourselves to the case when the curves S lie out of the inclusions Ω_ϵ ($N_{\epsilon S} = 0, N_S = N_{eS}$). We will use the notations introduced in Section 1.1 and put $P_i = B_{O_i}^{a_i}$:

$$w_{il} = \max_{|\beta| < \pi - \theta_i} \sqrt{a_i^2 + l_{il}^2 - 2a_i l_{il} \cos \psi_i + \psi_{il} + \beta} \quad (i = 1, 2, \ldots, N_S)$$

$$\hat{w}_{il} = \min_{|\beta| < \pi - \theta_i} \sqrt{a_i^2 + l_{il}^2 - 2a_i l_{il} \cos \psi_i + \psi_{il} + \beta} \quad (l = 1, 2, \ldots, N_\epsilon)$$

It is obvious that

$$w_{il} < a_l, \quad S_i \subset S_l, \quad \hat{w}_{il} > a_l, \quad S_i \cap P_i = \emptyset \quad (i, l = 1, 2, \ldots, N_S)$$

$$w_{iN_S+l} < a_i, \quad \Omega_{\epsilon N_S+l} \subset P_i, \quad \hat{w}_{il} > a_i, \quad \Omega_{\epsilon N_S+l} \cap P_i \neq \emptyset$$
$$(i = 1, 2, \ldots, N_S, \ l = 1, 2, \ldots, N_\epsilon)$$

Lemma 2.1: The following estimates hold for $\|C^{il} - d_Q C^{il} d_Q\|$ when $Q \to \infty$:

$$\|C^{il} - d_Q C^{il} d_Q\| < B^{ii} Q^{-\frac{3}{2}}, \quad \|C^{il} - d_Q C^{il} d_Q\| < B^{il} \gamma_{il}^{Q+1}, \quad i \neq l \quad (2.37a)$$

where

$$\gamma_{il} = \begin{cases} w_{il} a_i^{-1}, & S_i \subset S_l \\ a_i w_{il}^{-1}, & S_i \cap S_l = \emptyset \end{cases}$$

if $i, l = 1, 2, \ldots, N_S$;

$$\|C^{iN_S+l} - d_Q C^{iN_S+l} d_Q\| < B^{iN_S+l} \alpha_{iN_S+l}^{Q+1} \quad (2.37b)$$

where

$$\alpha_{iN_S+l}^{Q+1} = \begin{cases} w_{iN_S+l} a_i^{-1}, & S_{N_S+l} \subset P_l \\ a_i \hat{w}_{il}^{-1}, & S_{N_S+l} \cap P_i = \emptyset \end{cases} \quad (2.37c)$$

if $i = 1, 2, \ldots, N_S, \; l = 1, 2, \ldots, N_\epsilon$;

$$\|C^{N_S+il} - d_Q C^{N_S+il} d_Q\| < B^{N_S+il} \frac{|\epsilon_i - \epsilon_0|}{|\epsilon_i + \epsilon_0|} \left(\frac{a_{N_S+i}}{l_{N_S+il} - a_i} \right)^{Q+1} \quad (2.37d)$$

if $i = 1, 2, \ldots, N_\epsilon, \; l = 1, 2, \ldots, M$ and $l_{N_S+il} > a_l$ for $i, l \leq N_S$;

$$\|C^{N_S+l} - d_Q C^{N_S+l} d_Q\| < (Q+1) B^{N_S+il} \frac{|\epsilon_i - \epsilon_0|}{|\epsilon_i + \epsilon_0|} \left(\frac{a_{N_S+i}}{a_l - l_{N_S+il}} \right)^Q \quad (2.37e)$$

if $i = 1, 2, \ldots, N_\epsilon, \; l = 1, 2, \ldots, N_S$ and $l_{N_S+il} < a_l$.

Here B^{il} are the constant values depending on the geometrical parameters $(a_i, \theta_i, l_{il}, \ldots)$, the structure and the frequency of the incident field.

Lemma 2.1 may be proved with the help of the estimates obtained in Chapter 1. Only the proof of the first estimate in eqns. 2.37 is demonstrated.

In fact, one can show that for $i, l = 1, 2, \ldots, N_S$ (and $m > Q + 1$)

$$\|C^{ii} - d_Q C^{ii} d_Q\| < \text{const} \sum_{|p|>Q} p^{-4} + \sum_{|m|>Q, m \neq p} (p - m)^{-2} |p|^{-1} |m|^{-3}$$

where

$$\sum_{\substack{|m|>Q, \\ m\neq p}} (p-m)^{-2}|p|^{-1} = \sum_{p=Q+1}^{m-1} (m-p)^{-2}|p|^{-1} + \sum_{p=1}^{\infty} (p+m)^{-1}p^{-2}$$

$$+ \sum_{p=Q+1}^{\infty} (p+m)^{-2}p^{-1} \tag{2.38}$$

and

$$\sum_{p=Q+1}^{m-1} (m-p)^{-2}p^{-1} = m^{-2}\sum_{k=1}^{\infty} km^{1-k} \sum_{p=Q+1}^{m-1} p^{k-2} \tag{2.39}$$

The last factor in the right-hand side of eqn. 2.39 is calculated explicitly and after a series of transformations of eqns. 2.38 and 2.39 we have

$$\sum_{p=1}^{\infty} p^{-2}(p+m)^{-1} < 2m^{-1}, \quad \sum_{p=Q+1}^{\infty} (m+p)^{-2}p^{-1} < \frac{1}{2}m^{-1}(Q+1)^{-1} \tag{2.40}$$

and, finally, the inequality

$$\sum_{|p|>Q, p\neq m} (p-m)^{-2}|p|^{-1} \leq 12m^{-1} \tag{2.41}$$

Eqn. 2.41 enables one to get the required estimate in eqns. 2.37.

The estimates given by eqns. 2.25 and 2.37 yield

$$\|C - P_Q C P_Q\| < \text{const}\, Q^{-\frac{3}{2}}, \quad \|\mathbf{d} - P_Q \mathbf{d}\| < \text{const}\, Q^{-1} \tag{2.42}$$

Then it follows from 2.36 that

$$|\mathbf{x}_e - \mathbf{x}_Q| < \hat{Q}_1 \|\mathbf{x}_e\| Q^{-\frac{3}{2}} + \hat{Q}_2 Q^{-1} \|\mathbf{d} - P_Q \mathbf{d}\| < \text{const}\, Q^{-1} \tag{2.42}$$

where $\hat{Q}_1$ and $\hat{Q}_2$ are constant values depending on the geometrical parameters, the type and frequency of the incident field. One can improve this estimate of the rate of convergence by the method proposed by Kantorovich and Akilov (1977) if eqn. 2.32 is rewritten in the form

$$\mathbf{z} = C\mathbf{z} + C\mathbf{d} \tag{2.44}$$

with respect to the unknowns $\mathbf{x} = \mathbf{z} + \mathbf{d}$ and the corresponding approximate equation is considered

$$\mathbf{z}_Q = P_Q C\mathbf{d} + P_Q C P_Q \mathbf{z}_Q, \quad Q = 0, 1, 2, \ldots \tag{2.45}$$

Then it follows from eqns. 2.44 and 2.45 that

$$\|\mathbf{z}_Q - \mathbf{z}\| < \|(I - P_Q C P_Q)^{-1}\|(\|C - P_Q C P_Q\|\|\mathbf{z}\| + \|C - P_Q C\|\|\mathbf{d}\|)$$

Using the estimates given by eqns. 2.37 one can show that

$$\|C - P_Q C P_Q\| < \text{const } Q^{-\frac{3}{2}}$$

and hence

$$\|\mathbf{z}_Q - \mathbf{z}\| < \hat{Q}_3 Q^{-\frac{3}{2}} \tag{2.46}$$

Eqns. 2.43 and 2.46 determine the average rate of convergence for the reduction method with respect to the l_2^M-norm. The corresponding co-ordinate estimates for $|x_{em}^i - x_{em}^{iQ}| \to 0$, $Q \to \infty$ turn out to be of one order greater with the constants $\hat{Q}_{1,2,3} \sim \omega L c^{-1}$ where L is a characteristic size of a resonator (as a rule, $L = \max a_i$). Such estimates for the rate of convergence ensure applicability of the reduction method in the resonance frequency range when the wavelength is commensurable with resonators dimensions.

Let us determine the limits of applicability for the successive approximations method. With this purpose we will represent the operators defined by eqns. 2.11, 2.18 and 2.20 as an operator sum

$$C = R_1 + R_2 + R_3 + R_4, \qquad R_n = \|R_n^{il}\|_{i,l=1}^M \quad (n = 1, 2, 3, 4)$$

where $R_n^{il} : l_2 \to l_2$ are the so-called interaction operators describing interaction of the perfectly conducting screens $S = \cup_{i=1}^{N_S} S_i$ ($n = 2$, $i = 1, 2, \ldots, N_S$), dielectric inclusions Ω_ϵ ($n = 3$, $i = N_S + 1, N_S + 2, \ldots, M$) and the interaction of S and Ω_ϵ ($n = 4$, $i = 1, 2, \ldots, M$):

$$R_1^{il} = \begin{cases} C^{ii}, & i = j \\ I_{\emptyset}, & i \neq j \end{cases} \qquad R_2^{il} = \begin{cases} C^{il}, & i \neq l \text{ and } i, l \leq N_S \\ I_{\emptyset}, & i = l \text{ or } i > N_S \text{ or } l > N_S \end{cases}$$

$$R_3^{il} = \begin{cases} C^{il}, & i \neq l \text{ and } i, l > N_S \\ I_{\emptyset}, & i = l \text{ or } i, l \leq N_S \end{cases}$$

$$R_4^{il} = \begin{cases} C^{il}, & i \leq N_S, \ l > N_S \text{ or } i > N_S, \ l \leq N_S \\ I_{\emptyset}, & i, l \geq N_S \text{ or } i, l \leq N_S \end{cases}$$

here $I_{\emptyset}$ denotes a zero operator in l_2. It is easy to see that $R_3^{i+N_S(l+1)}$ ($i, l = 1, 2, \ldots, N_\epsilon$) coincide with the interaction operators introduced by

Ivanov (1968) and $\|R_3^{i+N_S(l+1)}\| \to 0$ when $\epsilon_i \to \epsilon_0$, $\mu_i \to \mu_0$ (for "weak" dielectric inclusions). It follows from the asymptotic behaviour of $|a_{pm}^{ns}(k)|$ and estimates for V_{m-1}^{n-1} obtained by Shestopalov (1971) that $\|R_n il\| \to 0$ when $a_{i,l} l_{il}^{-1} \to 0$ $(n = 2, 3, 4)$ or $u_{i,l} \to -1$ $(n = 2)$ where $u_i = -\cos\theta_i$ (i.e. when the screens are sufficiently small or the distance between them are sufficiently big).

Let us consider the operator R_1. If to formally annul R_n $(n = 2, 3, 4)$, eqn. 2.32 splits into M independent equations in l_2

$$\mathbf{x}^i = R_1^{ii}\mathbf{x}^i + \mathbf{d} \quad (i = 1, 2, \ldots, N_S) \tag{2.47}$$

$$\mathbf{x}^{N_S+l} = \mathbf{d}^{N_S+l} \quad (l = 1, 2, \ldots, N_\epsilon) \tag{2.48}$$

where each of the equations corresponds to the diffraction problem of the $\mathbf{H}^0, \mathbf{E}^0$ field on the screen S_i (eqn. 2.47) or on the dielectric inclusion $\Omega_{\epsilon l}$ (eqn. 2.48) and the solutions of eqns. 2.47 and 2. 48 may be thus interpreted as a "zero" approximation which does not take into account interaction.

Accordingly, condition $\|C\| < 1$ which ensures the successive iterations convergence (Kantorovich and Akilov, 1977) holds in the following cases:

(i) "weak" dielectric inclusions are placed in small screens: $u_i \to -1$ $(i = 1, 2, \ldots, N_S)$ and $\epsilon_l \to \epsilon_0$, $\mu_l \to \mu_0$ $(l = 1, 2, \ldots, N_\epsilon)$;

(ii) dielectric inclusions are far from each other and from the screens and screens are small: $u_i \to -1$ $(i = 1, 2, \ldots, N_S)$ and $aL^{-1} \ll 1$ where $a = \max_{1 \le i \le N_\epsilon} a_{N_S+i}$, $L = \min_{1 \le i,j \le M} l_{il}$;

(iii) dielectric inclusions and screens are far from each other and relative resonator's dimensions are small: $aL^{-1} \ll 1$ and $ka \ll 1$ where $a = \max_{1 \le i \le M} a_{N_S+i}$, $L = \min_{1 \le i,j \le N_S+i} l_{il}$.

The successive approximations method may be also applied when $ka \sim 1$ if eqn. 2.32 is regularised and rewritten in the form

$$\mathbf{x} = (I - R_1)^{-1}(R_2 + R_3 + R_4)\mathbf{x} + (I - R_1)^{-1}\mathbf{d} \tag{2.49}$$

where $\|R_2 + R_3 + R_4\| \ll 1$ and hence

$$\|(I - R_1)^{-1}(R_2 + R_3 + R_4)\| < 1 \quad \text{for} \quad al_{il}^{-1} \ll 1 \quad (i, l = 1, 2, \ldots, M)$$

A zero approximation is obtained by the reduction method as a solution of eqns. 2.47 and 2.48, as well as all the successive approximations (with different right-hand sides). An analogous approach was applied by Veliev *et al.* (1981). The solution in a closed form may be obtained if $(I - R_1)^{-1}$ is known, i.e. when the solution of eqns. 2.47 and 2.48 may be constructed explicitly.

2.4 Spectral problems and excitation of arbitrary cylindrical screens: compact boundaries

The method for solving spectral and excitation problems applied in the previous Sections for cylindrical structures with the cross-sections formed by circular domains from the family CB_c^ε is developed here (in the case of the Dirichlet boundary conditions on the screens) for arbitrary cylindrical structures whose cross-sections belong to the family CB^e. Following Tuchkin (1985) and Shestopalov *et al.* (1994) the initial boundary-value problem is reduced to coupled summation-type equations which admit equivalent regularisation and reduction to a canonical Fredholm operator equation.

Let us formulate the spectral problem. Consider an open resonator formed by a finite number N_S of infinitely thin, perfectly conducting, unclosed circular cylindrical surfaces placed in a homogeneous medium with complex permittivity and permeability, ε_0, μ_0 (Im $\varepsilon_0 \geq 0$, Im $\mu_0 \geq 0$). The resonator's cross-section belongs to the family CB^e and its boundary S consists of N_S finite closed and unclosed piece-wise smooth contours (see Section 1.1).

We will limit ourselves in what follows to the case $N_S = 1$ assuming that all the results may be obtained in a similar way for an arbitrary whole number $N_S > 1$.

Let H_0 be the Riemann surface of the analytical continuation for the fundamental solution $E_0(k; \mathbf{r} - \mathbf{q}) = -\frac{i}{4} H_0^{(1)}(k|\mathbf{r} - \mathbf{q}|)$ (or, equally, for the function $\ln k$) of the two-dimensional Helmholtz equation $\mathcal{L}(k)u \equiv \Delta u + k^2 u = 0$ with respect to the spectral parameter k ($\mathbf{r}, \mathbf{q} \in \mathbb{R}^2$).

It is necessary to determine the values $k \in H_0$ for which there exist non-trivial solutions $u(p) \in C(\mathbb{R}^2) \cap C^2 \mathbb{R}^2 \setminus S$ of the homogeneous Helmholtz equation

$$\Delta u(\mathbf{r}) + k^2 u(\mathbf{r}) = 0, \quad \mathbf{r} \in R^2 \setminus S$$

satisfying homogeneous boundary conditions $u(\mathbf{r}) = 0$ on S, the Meixner conditions in the form

$$\nabla u(\mathbf{r}) = \{|\mathbf{r} - \mathbf{r}_1^e|, |\mathbf{r} - \mathbf{r}_2^e|\}^{-\frac{1}{2}} h(\mathbf{r})$$

(which imply the fulfillment of eqn. 1.4) where $\mathbf{r}_1^e$ and $\mathbf{r}_2^e$ are the the endpoints of the contour S, the function $h(\mathbf{r})$ satisfies the Hölder condition in the vicinity of S, and the Reichardt radiation condition (see eqn. 1.5); there exists an $R_0 > 0$ such that for all $\mathbf{r} = (r, \phi)$: $r = |\mathbf{r}| \geq R_0$ the following representation holds

$$u(\mathbf{r}) = \sum_{m=-\infty}^{\infty} C_m H_m^{(1)}(kr) \exp(jm\phi)$$

We will call $u(\mathbf{r})$ the generalised eigenfunction of the Laplace operator in $\mathbb{R}^2 \setminus S$ corresponding to the eigenvalue (fundamental frequency) k.

The difference between internal and external boundary-value diffraction problems is in the fact that the generalised eigenfunctions are not bounded when $|\mathbf{r}| \to \infty$ and $\mathrm{Im}\, k < 0$.

The method for solving such a spectral problem that was used in the previous Sections for the case when S is a circular arc was based on separating variables in polar co-ordinates and the method of the Riemann-Hilbert problem for the equivalent reduction to a canonical Fredholm operator equation. This technique cannot be applied when S is an arbitrary unclosed contour and here the coupled second kind summation-type equations are obtained by means of the solution representations in the form of potentials and the special regularisation procedure applied for the boundary integral equation.

For real $k > 0$ every solution of the inhomogeneous (excitation) boundary-value problem for the Helmholtz equation with the boundary condition $u(\mathbf{r}) = f(\mathbf{r})$, $\mathbf{r} \in S$ satisfying the Sommerfeld radiation condition admits representation in the form of a (generalised) single-layer potential

$$u(\mathbf{r}) = \int_S E_0(k; \mathbf{r} - \mathbf{q}) Z(\mathbf{q}) \, d\, s_{\mathbf{q}} \tag{2.50}$$

where $Z(\mathbf{q})$ is a certain function defined on S such that

$$Z(\mathbf{q}) = \sqrt{s(\mathbf{q}, \mathbf{q}_1^e) \cdot s(\mathbf{q}, \mathbf{q}_2^e)} H(\mathbf{q}), \quad \mathbf{q} \in S \tag{2.51}$$

when S is an unclosed contour. Here $s(\mathbf{r}, \mathbf{q})$ is the length of the arc measured from $\mathbf{r}$ to $\mathbf{q}$ and $H(\mathbf{q})$ belongs to the Hölder class with respect to the $s(\mathbf{r}, \mathbf{q})$

norm. If S is a closed contour, the condition given by eqn. 2.51 is ignored and $Z(\mathbf{q})$ belongs to the Hölder class on S.

The Sommerfeld radiation condition and the Reichardt condition are equivalent for real $k > 0$. It may be shown similarly to the proof given in Section 1.5 that the representation given by eqn. 2.50 is valid when $k \neq 0$ is an arbitrary complex number and $u(\mathbf{r})$ satisfies the Reichardt condition.

Lemma 2.2: A solution to the boundary-value problem $\mathcal{L}(k)u = 0$, $u(\mathbf{r}) = 0$, $\mathbf{r} \in S$ satisfying the Meixner condition on S admits the representation given by eqns. 2.50 and 2.51 if and only if $u(\mathbf{r})$ satisfies the Reichardt condition.

Applying the boundary condition $u(\mathbf{r}) = 0$, $\mathbf{r} \in S$ for eqn. 2.50 we get the integral equation

$$\int_S E_0(k; \mathbf{r} - \mathbf{q}) Z(\mathbf{q}) \, d\, s_\mathbf{q} = 0, \quad \mathbf{r} \in S \tag{2.52}$$

with respect to $Z(\mathbf{q})$ in the class defined by eqn. 2.51.

Eqn. 2.52 formally coincides with the homogeneous integral equation of the diffraction problem $\mathcal{L}(k)u^s = 0$, $u^s(\mathbf{r}) = -u_0(\mathbf{r})$, $\mathbf{r} \in S$. Here u^s is the scattered field which admits representation in the form of a (generalised) single-layer potential

$$u^s(\mathbf{r}) = \int_S E_0(k; \mathbf{r} - \mathbf{q}) Z(\mathbf{q}) \, d\, s_\mathbf{q} \tag{2.53}$$

where the unknown density $Z(\mathbf{q})$ satisfies the integral equation

$$\int_S E_0(k; \mathbf{r} - \mathbf{q}) Z(\mathbf{q}) \, d\, s_\mathbf{q} = -u_0(\mathbf{r}), \quad \mathbf{r} \in S \tag{2.54}$$

Assume that S is a simply-connected contour which may be complemented to a closed contour S' without points of self-intersection. Then S' is a topological equivalent of the unit circle

$$Q[-\pi, \pi] = \{\exp(j\theta), \ \theta \in [-\pi, \pi]\}$$

and there exist continuous functions $x = x(\theta)$, $y = y(\theta)$ determining the one-to-one correspondence between Q and S': $[x(\theta), y(\theta)] \in S'$, $\theta \in [-\pi, \pi]\}$.

Assume that S' is sufficiently smooth, namely, $x(\theta)$ and $y(\theta)$ admit the 2π-periodical continuations on the whole axis which are once continuously differentiable and their second and third derivatives are piece-wise continuous functions.

Let the prototype of S correspond to the closed interval $\theta \in [-d, d]$ for a certain $d \in [0, \pi]$ and let θ_1 and θ_2 be the prototypes of the points $\mathbf{r}, \mathbf{q} \in S$. Then eqn. 2.54 may be rewritten as

$$\int_{-d}^{d} \mathrm{E}_0(k; R(\theta_1, \theta_2)) Z(\theta_2) \, \mathrm{d}\theta_2 = G(\theta_1), \quad \theta_1 \in (-d, d) \qquad (2.55)$$

where $G(\theta_1)$ is the known function and $R(\theta_1, \theta_2)$ is the distance between the images θ_1, θ_2 on the xy-plane.

Let us define the function $Z^0(\theta)$, $\theta \in [-\pi, \pi]$ as

$$Z^0(\theta) = \begin{cases} Z(\theta), & \theta \in (-d, d) \\ 0, & \theta \in [-\pi, \pi] \setminus (-d, d) \end{cases} \qquad (2.56)$$

It follows from eqns. 2.55 and 2.56 that $Z^0(\theta)$ satisfies the equation

$$\int_{-\pi}^{\pi} \mathrm{E}_0(k; R(\theta_1, \theta_2)) Z^0(\theta_2) \, \mathrm{d}\theta_2 = G(\theta_1), \quad \theta_1 \in (-d, d) \qquad (2.57)$$

Here

$$\mathrm{E}_0(k; R(\theta_1, \theta_2)) = j\frac{2}{\pi}[\ln|2\sin\Theta_{12}| + H(\theta_1, \theta_2)], \quad \Theta_{12} = \frac{\theta_1 - \theta_2}{2} \qquad (2.58)$$

and we will assume that the Fourier coefficients h_{mn} of the double Fourier series

$$H(\theta, \tau) = \sum_{m=-\infty}^{\infty} \sum_{n=-\infty}^{\infty} h_{mn} \exp[j(m\theta + n\tau)], \quad \theta, \tau \in [-\pi, \pi] \qquad (2.59)$$

are known. In fact,

$$\ln|2\sin\Theta_{12}| = -\frac{1}{2} \sum_{n=-\infty, n\neq 0}^{\infty} \frac{1}{|n|} \exp(jn\theta_1) \exp(-jn\theta_2), \quad \theta_1, \theta_2 \in [-\pi, \pi] \tag{2.60}$$

We will look for the solution $Z^0(\tau)$ in the form of the Fourier series

$$Z^0(\tau) = \sum_{n=-\infty}^{\infty} z_n \exp(jn\tau), \quad \tau \in [-\pi, \pi] \qquad (2.61)$$

Substituting eqns. 2.58–2.61 into eqn. 2.57 and taking into account the orthogonality of $\{\exp(nj\tau)\}_{n=-\infty}^{\infty}$ we have

$$\sum_{\substack{n=-\infty \\ n\neq 0}}^{\infty} \frac{1}{|n|} z_n \exp(jn\theta) - 2 \sum_{n=-\infty}^{\infty} z_n \sum_{p=-\infty}^{\infty} h_{p,-n} \exp(jp\theta) = \sum_{n=-\infty}^{\infty} g_n \exp(jn\theta),$$

$$\theta \in [-d, d]$$

(2.62)

where $\{g_n\}_{n=-\infty}^{\infty}$ are the Fourier coefficients of the function $\frac{i}{2}G(\theta)$ smoothly continued on $[-\pi, \pi]$. Eqns. 2.61 and 2.56 yield

$$\sum_{n=-\infty}^{\infty} z_n \exp(jn\tau) = 0, \quad \tau \in [-\pi, \pi] \setminus (-d, d) \qquad (2.63)$$

Eqns. 2.62 and 2.63 form coupled summation-type equations with respect to the unknown infinite vector $\{z_n\}_{n=-\infty}^{\infty}$ of the Fourier coefficients satisfying the conditions

$$\sum_{n=-\infty}^{\infty} (|n| + 1)^{-1} |z_n|^2 < \infty \qquad (2.64)$$

where

$$\sum_{n=-\infty}^{\infty} (|n| + 1) |g_n|^2 < \infty$$

The following theorem gives the rate of decrease for the coefficients h_{mn} and, hence guarantees applicability of the regularisation procedure.

Theorem 2.4: Assume that the functions $x(\theta)$ and $y(\theta)$ satisfy the smoothness conditions formulated above. Then

$$\sum_{n=-\infty}^{\infty} \sum_{m=-\infty}^{\infty} (|n| + 1)^2 < (|m| + 1)^2 |h_{nm}|^2 < \infty \qquad (2.65)$$

Generalisation of the approach proposed by Tuchkin and Shestopalov (1985) allows us to regularise the system defined by eqns. 2.62 and 2.63 using the technique different from the method of the Riemann-Hilbert problem and to obtain the equivalent infinite algebraic system of the second kind

$$(I + D + BH)\mathbf{x} = B\mathbf{g} \qquad (2.66)$$

Here

$$D = \begin{pmatrix} D^{(+)} & 0 \\ 0 & 0 \end{pmatrix} \quad B = \begin{pmatrix} B^{(+)} & 0 \\ 0 & B^{(-)} \end{pmatrix} \quad H = \begin{pmatrix} H^{(+,+)} & H^{(+,-)} \\ H^{(-,+)} & H^{(-,-)} \end{pmatrix}$$

$$(2.67)$$

are the matrix operators such that

$$D^{(+)} = \{d_{sn}^{(+)}\}_{s,n=0}^{\infty} \quad B^{(+)} = \{b_{sn}^{(\pm)}\}_{s,n=0}^{\infty} \quad H^{(\pm,\pm)} = \{h_{sn}^{(\pm,\pm)}\}_{s,n=0}^{\infty} \quad (2.68)$$

where

$$d_{00}^{(+)} = -1 - \sqrt{2}\ln\frac{1-t_0}{2}, \quad d_{s0}^{(+)} = -\sqrt{2}p_s \ (s \geq 1), \quad d_{sn}^{(+)} = 0 \ (n \geq 1)$$

$$b_{00}^{(+)} = 0, \quad b_{0n}^{(+)} = p_n \ (n \geq 1), \quad b_{s0}^{(+)} = 0 \ (s \geq 1)$$

$$b_{sn}^{(+)} = \delta_{sn} - Q_{s-1,n-1}^{(1,0)}(t_0), \quad b_{sn}^{(-)} = \delta_{sn} - Q_{s-1,n-1}^{(0,1)}(t_0) \ (s,n \geq 1)$$

$$h_{sn}^{(+,\pm)} = -\frac{\tau_s\tau_n}{1+\delta_{s0}}[(h_{s,-n} \pm h_{sn}) + (h_{-s,-n} \pm h_{-s,n})] \ (s \geq 0, \ n \geq 1)$$

$$h_{sn}^{(-,\pm)} = -\tau_s\tau_n[(h_{s,-n} \pm h_{sn}) - (h_{-s,-n} \pm h_{-s,n})] \ (s \geq 1, \ n \geq 1)$$

$$(2.69)$$

Here

$$p_n = \frac{1}{n}(1+t_0)\hat{P}_{n-1}^{(0,1)}(t_0), \quad t_0 = \cos d, \quad \tau_n = \sqrt{n} \ (n \geq 1)$$

and $\hat{P}_n^{(\alpha,\beta)}(x)$ are the Jacobi polynomials (for the definition see Tuchkin and Shestopalov, 1985). The vectors of right-hand sides and unknowns in eqn. 2.66 have the form

$$\mathbf{g} = (\mathbf{g}^{(+)}, \mathbf{g}^{(-)})^T, \qquad\qquad \mathbf{y} = (\mathbf{y}^{(+)}, \mathbf{y}^{(-)})^T$$

$$\mathbf{g}^{(+)} = \{g_n^{(+)}\}_{n=0}^{\infty}, \qquad\qquad \mathbf{g}^{(-)} = \{g_n^{(-)}\}_{n=1}^{\infty}$$

$$\mathbf{y}^{(+)} = \{y_n^{(+)}\}_{n=0}^{\infty}, \qquad\qquad \mathbf{y}^{(-)} = \{y_n^{(-)}\}_{n=1}^{\infty}$$

$$y_n^{(+)} = \frac{z_n + z_{-n}}{\tau_n(1+\delta_{n0})}, \qquad\qquad y_n^{(-)} = \frac{z_n - z_{-n}}{\tau_n}$$

$$g_n^{(+)} = \frac{\tau_n(g_n + g_{-n})}{1+\delta_{n0}}, \qquad\qquad g_n^{(-)} = \tau_n(g_n - g_{-n})$$

Let us introduce the space of vector-columns $l_2^{(0)} = l_2^{(+)} \otimes l_2^{(-)}$ where

$$l_2^{(+)} = \left\{ \{x_n\}_{n=0}^{\infty} : \sum_{n=0}^{\infty} |x_n|^2 < \infty \right\}$$

$$l_2^{(-)} = \left\{ \{x_n\}_{n=1}^{\infty} : \sum_{n=1}^{\infty} |x_n|^2 < \infty \right\}$$

Theorem 2.5: Assume that the conditions given by eqn. 2.65 hold. Then the system given by eqn. 2.66 is the second kind algebraic system in $l_2^{(0)}$, i.e. it has the form

$$(I + K)\mathbf{y} = \mathbf{b}$$

where $\mathbf{y}, \mathbf{b} \in l_2^{(0)}$ and K is a compact operator in $l_2^{(0)}$.

We have carried out the regularisation of eqn. 2.52 for real $k > 0$. This procedure may be literally repeated for every complex $k \neq 0$ and as a result one obtains the operator equation in l_2

$$[I + D + BH(k)]\mathbf{x} = \mathbf{o}, \quad x \in l_2 \tag{2.71}$$

identical to eqn. 2.66 (with $\mathbf{g} = \mathbf{o}$) where D is a compact operator, B is a bounded operator in l_2 and the properties of $H(k)$ are stated below.

Lemma 2.3: The compact operator (operator-valued function) $H(k)$: $l_2 \to R(H)$, $R(H) \in l_2$ is defined in the domain $D(H) = H_0 \setminus \{0, \infty\}$.

Lemma 2.4: $H(k)$ is an analytical operator-valued function in $D(H)$.

Let us denote by F the linear transformation described above which establishes a correspondence between a solution $u(\mathbf{r})$ of the initial spectral problem and a (nontrivial) solution x of eqn. 2.71. Then the following equivalence theorem is valid.

Theorem 2.6: 1. Every solution $u(\mathbf{r})$ of the initial spectral problem if exists defines a solution $\mathbf{x} = F(u) \in l_2$ of eqn. 2.71. 2. There exists the inverse transformation such that for every solution $\mathbf{x}$ of eqn. 2.71 the function $u = F^{-1}\mathbf{x}$ is a solution of the spectral problem.

Eqn. 2.71 is a problem on characteristic numbers for a Fredholm analytical operator-valued function $I + D + BH(k)$. The initial spectral problem has only a trivial solution for all $k \in \{k : 0 < \arg k < \pi\}$ and hence for such k there exists a bounded operator $[I + D + BH(k)]^{-1}$ which is a meromorphic operator-valued function on $D(H)$ with not more than a countable set of poles. These assertions together with the results of theorem 2.6 enable us to prove the following principle results.

Theorem 2.7: Fundamental frequencies (eigenvalues) of the initial spectral problem form not more than a countable set of (isolated) points in $D(H)$.

Theorem 2.8: Green's function of the initial boundary-value problem is a meromorphic function of k on the Riemann surface of $\ln k$.

In conclusion let us note that it is easy to construct and justify an algorithm for calculating the fundamental frequencies using the approach proposed in Section 1.11. Let us introduce the sequences of projectors $P_N : l_2 \to U_N$, $U_N \subset l_2$, $\dim U_N = N$, $U_1 \subset U_2 \subset \cdots \subset U_n \subset \ldots$ and finite-dimensional operator-valued functions $B_N(k) = P_N[I+D+BH(k)]P_N$. Then the following assertion is valid.

Theorem 2.9: Let k_0 be the fundametal frequency (eigenvalue) of the initial spectral problem. Then there exists a sequence $k_N \in \sigma_N$ of the characteristic numbers of B_N such that $|k_N - k_0| \to 0$ $(N \to \infty)$. On the other hand, if $k_N \in \sigma_N$ and $|k_N - k_0| \to 0$, $k_0 \in D(H)$ $(N \to \infty)$, then k_0 is a fundametal frequency of the initial spectral problem.

In accordance with the results of Section 1.11, the approximate characteric numbers $k_N \in \sigma_N$ may be determined as the roots of the equations $\det B_N(k) = 0$.

2.5 Spectral properties and excitation of arbitrary cylindrical screens: noncompact boundaries. Solution of problems in the generalised statement

Following the scheme presented in Sections 1.2, 1.3 and 1.7 we will consider the problems on the E- and H-type fundamental frequencies of open resonators with noncompact boundaries whose cross-sections are formed by the domains from the family NB_c (introduced in Section 1.1) in terms of determination of the nontrivial solutions to homogeneous boundary-value problems given by eqns. 1.1–1.5. We will begin with the H-case corresponding to the Neumann boundary-value conditions for the sought for function $u = u(\mathbf{r})$ (coinciding with longitudinal component of the H-field).

Let λ be a fixed value of a spectral parameter and $R > R_0 \geq 0$ a positive number which is not a root of the equation $H_n^{(1)}(\sqrt{\epsilon_1}\lambda r) = 0$ $(n = 0, 1, 2, \ldots)$ for this given λ. It is easy to see that this condition holds for all $R > R_0$ if $4k\pi \leq \arg \lambda \leq 2(2k+1)\pi$, $k = 0, \pm 1, \pm 2, \ldots$.

Assume that $r > R$ is fixed, the functions u and v belong to the class M and u is an eigenfunction of the problem given by eqns. 1.1–1.5. Then

$$u(r, \phi) = \sum_{n=0}^{\infty} C_n H_n^{(1)}(\sqrt{\epsilon_1}\lambda r) \cos n\phi \tag{2.72}$$

is the Fourier series of the twice continuously differentiable function u. The equality of its left and right limiting values on the half-circle $C_R = \{r : |r| = R\} \cap \{r : x_2 > 0\}$ yields

$$C_n = \frac{2}{\pi} \int_0^{\pi} u(R, \phi) \cos n\phi \, d\phi / H_n^{(1)}(\sqrt{\epsilon_1}\lambda R), \quad n \geq 1$$

$$C_0 = \frac{1}{\pi} \int_0^{\pi} u(R, \phi) \, d\phi / H_0^{(1)}(\sqrt{\epsilon_1}\lambda R) \tag{2.73}$$

The expression for the normal derivative of u on C_R has the form

$$b_\lambda^R(u) = \frac{\partial u}{\partial r}\Big|_R$$

$$= \frac{1}{\pi} \int_0^{\pi} u(R, \phi) \, d\phi \frac{\partial H_0^{(1)}(\sqrt{\epsilon_1}\lambda R)/\partial r}{H_0^{(1)}(\sqrt{\epsilon_1}\lambda R)}$$

$$+ \sum_{n=1}^{\infty} \frac{2}{\pi} \int_0^{\pi} u(R, \psi) \cos n\psi \, d\psi \frac{\partial H_n^{(1)}(\sqrt{\epsilon_1}\lambda R)/\partial r}{H_n^{(1)}(\sqrt{\epsilon_1}\lambda R)} \cos n\phi \tag{2.74}$$

Dividing eqn. 1.1 by $\epsilon > 0$, multiplying it by $\bar{v}$ and integrating by parts in the domain $\Omega_R = \Omega \cap U_R$, we obtain, taking account of eqn. 1.2, the integral identity

$$\iint_{\Omega_R} \left(\frac{1}{\epsilon} \nabla u \nabla \bar{v} - \lambda u \bar{v} \right) d\mathbf{r} - \int_{C_R} \frac{1}{\epsilon} b_\lambda^R(u) \bar{v} \, dl = 0 \tag{2.75}$$

Let us note that the boundary of Ω_R satisfies the cone condition.

We will consider the problem in the generalised statement and formulate the definition of the solution.

Definition 2.1: A nontrivial element $u \in W_2^1(\Omega_R)$ satisfying eqn. 2.75 for every function $v \in W_2^1(\Omega_R)$ and admitting representation in the form of a series given by eqn. 1.5 in the domain $\Omega \setminus \overline{\Omega}_R$ with the coefficients given by eqn. 2.73 is said to be the generalised eigenfunction of the problem given by eqns. 1.1–1.5.

One can show that every classical eigenfunction of the problem given by eqns. 1.1–1.5 is its generalised eigenfunction.

The definition of the generalised eigenfunction does not depend on R. Let u_{R_1} and u_{R_2} be the generalised eigenfunctions defined for $R = R_1$ and $R = R_2$ with $R_2 > R_1 > R_0$. We prove then that u_{R_1} satisfies the conditions formulated in the definition for $R = R_2$ while u_{R_2} satisfies these conditions for $R = R_1$. The identity

$$\iint\limits_{\Omega_{R_1}^{R_2}} \left(\frac{1}{\epsilon} \nabla u_{R_i} \nabla \bar{v} - \lambda u_{R_i} \bar{v} \right) \, \mathrm{d}\mathbf{r}$$

$$- \int\limits_{C_{R_1}} \frac{1}{\epsilon_1} \frac{\partial u_{R_i}}{\partial r} \bar{v} \, \mathrm{d}l + \int\limits_{C_{R_2}} \frac{1}{\epsilon_1} \frac{\partial u_{R_i}}{\partial r} \bar{v} \, \mathrm{d}l = 0 \quad (i = 1, 2)$$

$$(2.76)$$

obtained by means of variables separation follows from the representations given by eqn. 1.5 for $r > R_0$ and the theorems on the internal smoothness of generalised solutions. Here $\Omega_{R_1}^{R_2} = \{\mathbf{r} : R_1 < |\mathbf{r}| < R_2\}$. Adding eqns. 2.75 and 2.76 for $R = R_1$, $i = 1$ and subtracting eqn. 2.76 from eqn. 2.75 for $R = R_2$ and $i = 2$, we come to the required assertion.

Let D_λ^R be a simply-connected domain in the Riemann surface of the function $H_n^{(1)}(\sqrt{\epsilon_1 \lambda} R)$ such that $0 \notin \bar{D}_\lambda^R$. Then $H_n^{(1)}(\sqrt{\epsilon_1 \lambda} R)$ has no zeros in D_λ^R [due to the asymptotics of the Hankel functions $H_n^{(1)}(z)$ for big values of $|z|$] and the following estimate holds

$$\left| \left[\frac{\partial H_n^{(1)}(\sqrt{\epsilon_1 \lambda} R)}{\partial r} \right] \Big/ H_n^{(1)}(\sqrt{\epsilon_1 \lambda} R) \right| \leq C_2(\lambda) = \sqrt{\epsilon_1 \lambda}(1 + C_1)$$

$$(2.77)$$

$$(n = 0, 1, 2, \ldots)$$

for all $\lambda \in D_\lambda^R$ and sufficiently large $R > 0$ where C_1 is a small positive number.

Application of the compact embedding theorems (Adams, 1975) enables us to prove that the identity given by eqn. 2.75 is equivalent for every $\lambda \in D_\lambda^R$

to the operator equation

$$u - (\lambda + 1)A^H u - B^H(\lambda)u = 0 \qquad (2.78)$$

in $W_2^1(\Omega_R)$ where A^H is a compact operator in $W_2^1(\Omega_R)$.

Let us state the properties of the operator-valued function $B^H(\lambda)$ defined by the relation

$$\iint\limits_{\Omega_R} \left(\frac{1}{\epsilon} \nabla B^H u \nabla \bar{v} + B^H u \bar{v} \right) d\mathbf{r} = \int\limits_{C_R} \frac{1}{\epsilon} b_\lambda^R(u) \bar{v} \, dl \qquad (2.79)$$

It follows from eqns. 2.74 and 2.77, the Parseval equality and the trace properties of the functions from $W_2^1(\Omega_R)$, that the one-and-a-half-linear form given by eqn. 2.79 is bounded in $W_2^1(\Omega_R)$, which yields the compactness of $B^H(\lambda)$, $\lambda \in D_\lambda^R$ [caused by the compactness of the embedding of the traces of bounded functions from $W_2^1(\Omega_R)$ to $L_2(C_R)$].

The choice of R enables us to assert using the results of Kato (1972) that the operator-valued function $B^H(\lambda)$ is analytical in D_λ^R.

Hence we have proved the following lemma.

Lemma 2.5: The problem given by eqns. 1.1–1.5 on determination of the H-type fundamental frequencies of open resonators whose cross-sections are formed by the domains from the family NB_c is equivalent in the generalised statement to the problem on characteristic numbers

$$K^H(\lambda)u \equiv [1 - (\lambda + 1)A^H - B^H(\lambda)]u = 0, \quad \lambda \in D_\lambda^R \qquad (2.80)$$

where $K^H(\lambda) : W_2^1(\Omega_R) \to W_2^1(\Omega_R)$ is an analytical Fredholm operator-valued function in D_λ^R.

In order to prove this lemma it is sufficient to check that a nontrivial solution of eqn. 2.80 continued to Ω for a fixed $\lambda \in D_\lambda^R$ by means of eqn. 1.5 with the coefficients given by eqn. 2.73 is a generalised eigenfunction of the problem given by eqns. 1.1–1.5 and every its eigenfunction satisfies eqn. 2.80.

Let us consider the E-case. Here the coefficients in eqn. 1.5 has the form

$$C_n = \frac{2}{\pi} \int\limits_0^\pi u(R, \phi) \sin n\phi \, d\phi / H_n^{(1)}(\sqrt{\epsilon_1 \lambda} R) \quad (n = 1, 2, \dots) \qquad (2.81)$$

and

$$b_\lambda^R(u) = \sum_{n=1}^\infty \frac{2}{\pi} \int_0^\pi u(R,\psi) \sin n\psi \ d\psi \frac{\partial H_n^{(1)}(\sqrt{\epsilon_1\lambda}R)/\partial r}{H_n^{(1)}(\sqrt{\epsilon_1\lambda}R)} \sin n\phi$$

The classical eigenfunctions satisfy the integral identity

$$\iint\limits_{\Omega_R} [\nabla u \nabla \bar{v} - \lambda\epsilon(\mathbf{r})u\bar{v}] \ d\mathbf{r} - \int\limits_{C_R} b_\lambda^H(u)\bar{v} \ dl = 0 \qquad (2.82)$$

for every function v continuously differentiable in Ω_R such that $\text{supp}\, v \subset \Omega$.

Definition 2.2: A nontrivial element $u \in \tilde{W}_2^1(\Omega_R)$ satisfying eqn. 2.82 for every function $v \in \tilde{W}_2^1(\Omega_R)$ and admitting representation in the form of a series given by eqn. 1.5 in the domain $\Omega \setminus \overline{\Omega}_R$ with the coefficients given by eqn. 2.81 is said to be the generalised eigenfunction of the problem given by eqns. 1.1–1.5.

Here $\tilde{W}_2^1(\Omega_R)$ is the subspace of $W_2^1(\Omega_R)$ consisting of functions having zero trace on $\partial\Omega$.

The proof of the following assertion is similar to that of Lemma 2.5.

Lemma 2.6: The problem given by eqns. 1.1–1.5 on determination of the E-type fundamental frequencies of open resonators whose cross-sections are formed by the domains from the family NB_c is equivalent in the generalised statement to the problem on characteristic numbers

$$K^E(\lambda)w \equiv [1 - (\lambda + 1)A^E - B^E(\lambda)]w = 0, \quad \lambda \in D_\lambda^R$$

where $K^E(\lambda) : \tilde{W}_2^1(\Omega_R) \to \tilde{W}_2^1(\Omega_R)$ is an analytical Fredholm operator-valued function in D_λ^R.

Let us specify the conditions concerning the structure of the set A of the gap lines of the function $\epsilon(\mathbf{r})$. Assume that for every point $\mathbf{r} = \mathbf{r}_0 \in \Omega$ there exists a smooth unclosed curve Γ_0 starting in $\mathbf{r}_0$ and ending in the domain $\Omega \cap \{\mathbf{r} : |\mathbf{r}| > R_0\}$ such that each point $\mathbf{r}_A = \Gamma_0 \cap A$ (of the intersection of Γ_0 with a curve from A) has a vicinity where the corresponding element of A is an infinitely smooth curve.

Let us denote by E the set of such functions $\epsilon(\mathbf{r})$ taking a fixed constant value for $|\mathbf{r}| > R_0$.

Lemma 2.7: Let $\epsilon(\mathbf{r}) \in E$. Then the set $\{\lambda| : 2k\pi \leq \arg \lambda \leq 2(2k+1)\pi, \; k = 0, \pm 1, \pm 2, \dots\}$ does not contain the eigenvalues of the problem given by eqns. 1.1–1.5.

We will limit ourselves to the proof of this lemma for the E-case assuming for the sake of simplicity that $\epsilon(\mathbf{r})$ takes two constant values ϵ_1 and ϵ_2, i.e. $\Omega = \Omega_1 \cup \Omega_2 \cup A$ where $\Omega_1 = \{\mathbf{r} : \epsilon(\mathbf{r}) = \epsilon_1\}$ is an unbounded subdomain of Ω.

Let us fix a certain $R > R_0$. The generalised eigenfunctions of the problem given by eqns. 1.1–1.5 satisfy eqn. 2.82 for all $v \in \tilde{W}_2^1(\Omega_R)$ and eqn. 1.1 in the domain $\Omega \setminus \Omega_R$. Integration by parts of eqn. 1.1 multiplied by $\bar{u}$ over the domain $\Omega_R^{R_1}$ ($R_1 > R$) yields

$$\iint\limits_{\Omega_R^{R_1}} (|\nabla u|^2 - \lambda \epsilon_1 |u|^2) \, \mathrm{d}\mathbf{r} + \int\limits_{C_R} b_\lambda^R(u)\bar{u} \, \mathrm{d}l - \int\limits_{C_R} \frac{\partial u}{\partial r}\bar{u} \, \mathrm{d}l = 0 \qquad (2.83)$$

Adding eqn. 2.82 with $v = u$ and eqn. 2.83 we get the equality

$$\iint\limits_{\Omega_{R_1}} (|\nabla u|^2 - \lambda \epsilon(\mathbf{r})|u|^2) \, \mathrm{d}\mathbf{r} = \int\limits_{C_{R_1}} \frac{\partial u}{\partial r}\bar{u} \, \mathrm{d}l \qquad (2.84)$$

Calculating the limit in eqn. 2.84 for $R_1 \to \infty$ and $\lambda \in \{\lambda| : 4k\pi < \arg \lambda < 2(2k + 1)\pi\}$ (with an account of eqn. 1.5 and the asymptotics of the Hankel functions) we finally obtain that $u = 0$ almost everywhere in Ω.

If $\arg \lambda = 2k\pi$, i.e. $\sqrt{\lambda}$ is real, the following identity holds for all $v \in \tilde{W}_2^1(\Omega_R)$

$$\iint\limits_{\Omega_R} [\nabla \bar{u} \nabla v - \lambda \epsilon(\mathbf{r})\bar{u}v] \, \mathrm{d}\mathbf{r} = \int\limits_{C_R} \overline{b_\lambda^R(u)}v \, \mathrm{d}l \qquad (2.85)$$

Term-wise subtraction of eqns. 2.82 and 2.85 for $v = u$ yields

$$\int\limits_{C_R} [b_\lambda^R(u)\bar{u} - \overline{b_\lambda^R(u)}u] \, \mathrm{d}l = 0$$

and the chain of equalities which hold for real values of $\sqrt{\lambda}$

$$0 = \int\limits_{\dot{C}_R} (b_\lambda^R(u)\bar{u} - \overline{b_\lambda^R(u)}u) \, \mathrm{d}\, l$$

$$= \sum_{n=1}^{\infty} \int\limits_0^{\pi} \sin n\phi |C_n|^2 W[H_n^{(1)}(\sqrt{\epsilon_1\lambda}R), H_n^{(1)}(\sqrt{\epsilon_1\lambda}R)]R \, \mathrm{d}\, \phi$$

$$+ \sum_{n=1}^{\infty} \sum_{s=1,s\neq n-s}^{\infty} \int\limits_0^{\pi} \sin s\phi \sin(n-s)\phi (\overline{C}_{n-s}C_s H_{n-s}^{(1)}(\sqrt{\epsilon_1\lambda}R)H_n^{(1)'}(\sqrt{\epsilon_1\lambda}R)$$

$$- C_{n-s}\overline{C}_s H_{n-s}^{(1)}(\sqrt{\epsilon_1\lambda}R)H_s^{(1)'}(\sqrt{\epsilon_1\lambda}R))R \, \mathrm{d}\, \phi$$

$$= \frac{\pi}{2} \sum_{n=1}^{\infty} |C_n|^2$$

where

$$C = RW[H_n^{(1)}(\sqrt{\epsilon_1\lambda}R), \quad H_n^{(1)}(\sqrt{\epsilon_1\lambda}R)]$$

Hence all C_n are equal to zero and, consequently, $u = 0$ for $r > R_0$ due to the independence of the definition of the generalised eigenfunctions on R. Further application of the embedding and internal smoothness theorems, integral representations of the solution to the problem given by eqns. 1.1–1.5 and independence of its definition on R enable us to state that u is analytic in the domains Ω_1 and Ω_2, which implies $u = 0$ in Ω_1. Since $u \in W_2^2(G)$ for every domain G such that $\overline{G} \subset \Omega_R$ and u is a piece-wise analytic in Ω, it follows from the theorems concerning boundary values of analytic functions (Hörmander, 1986) applied for infinitely smooth parts of the boundary of G that $u = 0$ in Ω which completes the proof.

Now we can state the discreteness of the fundamental frequencies spectrum for the considered family of open resonators.

Theorem 2.10: The problem given by eqns. 1.1–1.5 on the determination of E- and H-type fundamental frequencies of open resonators whose cross-sections are formed by the domains from the family NB_c has the discrete spectrum formed by the finite-multiplicity eigenvalues which are situated out of the set $\{\lambda : \mathrm{Im}\sqrt{\lambda} \geq 0\}$ and has no finite accumulation points in $C_\lambda \setminus \{0\}$ where C_λ is the complex λ-plane.

The principle corollaries of this theorem are that there may exist only complex E- and H-type fundamental frequencies of the open resonators considered with strictly negative imaginary parts and, consequently, the closed upper half-planes of the complex planes λ and $\omega = \sqrt{\lambda}$ consist of only the regular points where the resolvent operators are bounded (except for $\lambda = 0$). The latter is in complete correspondence with the results of asymptotic analysis obtained in Section 1.9 for the image slot resonator. Of course, this property holds for all other open slotted resonators (in particular, with rectilinear slots which have been studied in Sections 1.7–1.9) whose cross-sections are formed by the domains from the families NB_p, NB_b, NB_u, NB_b^r and NB_u^r.

The excitation problems for open resonators with cross-sections from the family NB_c are naturally stated as inhomogeneous boundary-value problems for the Helmholtz equation, eqn. 1.1 (for a fixed regular value of λ, e.g. such that $\mathrm{Im}\,\lambda = 0$) satisfying the edge and radiation conditions given by eqns. 1.4 and 1.5 (the latter are equivalent to the Sommerfeld radiation condition at real frequencies) with two types of inhomogeneities in the boundary-value conditions

(i) given on $\partial\Omega$:

$$u = f^1 \quad \text{(E-case)} \qquad \text{or} \qquad \frac{\partial u}{\partial \mathbf{n}} = f^2 \quad \text{(H-case)}$$

($\partial/\partial \mathbf{n}$ is the normal derivative defined on the regular parts of $\partial\Omega$) where $f^l = f^l(\mathbf{x})$, $l = 1,2$ are the given continuous compactly supported functions defined on the regular parts of $\partial\Omega$;

(ii) given on $A = \Gamma_\epsilon$ as inhomogeneous transmission conditions and considered only for open resonators with rectilinear slots whose cross-sections are formed by the domains from the families NB_p:

$$[u] = 0, \quad \left[\frac{\partial u}{\partial \mathbf{n}}\right] = \psi^1 \quad \text{(E-case)}$$

$$[u] = \psi^2, \quad \left[\frac{1}{\epsilon}\frac{\partial u}{\partial n}\right] = 0 \quad \text{(H-case)}$$

where $\psi^l = \psi^l(x_1)$, $l = 1,2$ are the given continuous compactly supported functions defined on A.

Case (i) models excitation by the sourses situated on the boundary and case (ii) corresponds to the problem of the plane E- or H- polarised wave diffraction on a "plane" slotted scatterer.

These excitation problems are equivalently reduced in each of the cases (i) and (ii) in the manner described previously to the Fredholm equations

$$K^l(\lambda)u \equiv [I - (\lambda + 1)A^l - B^l(\lambda)]u = F^l \quad (l = 1, 2)$$

with the Fredholm operator-valued pencils $K^l(\lambda)$ acting in the corresponding Sobolev spaces where I is the identical operator, A^l is a completely continuous operator, $B^l(\lambda)$ is holomorphic completely continuous operator-valued function $(\lambda \in \mathcal{H} \setminus \{0\})$ and F^l is determined by excitation conditions.

Properties of the spectrum of the problem given by eqns. 1.1–1.5 enable us to state the uniqueness theorem for the excitation problems.

Theorem 2.11: There exists a unique solution to each of the excitation problems for open resonators whose cross-sections are formed by the domains from the families NB_c [case (i)] and NB_p [case (ii)] for every $\lambda \neq 0$, $\text{Im}\sqrt{\lambda} \geq 0$ except for not more than a finite number of isolated values of $\lambda : \text{Im}\sqrt{\lambda} < 0$ which exhaust all eigenvalues of the corresponding boundary eigenvalue problems.

2.6 Fundamental frequencies and resonance oscillations for types of open coaxial slot resonators

In the previous Sections the physical characteristics (general properties of the complex eigenfrequencies and the eigenoscillations classification) were established for the following specific types of open cylindrical resonators (see Figure 2.1): coaxial slot resonators formed by two perfectly conducting unclosed circular screens, symmetric confocal and non-confocal resonators formed by two perfectly conducting strip screens with lossy circular dielectric pivots.

It was shown that the eigenoscillations spectra of open cylindrical resonators are discrete and were effectively described in terms of characteristic numbers of (analytical or meromorphic) Fredholm matrix operator-valued functions of the frequency spectral parameter having the form

$$\begin{pmatrix} A^{11}(ka_1) - I & A^{12}(ka_1) \\ A^{21}(ka_2) & A^{122}(ka_2) - I \end{pmatrix} : l_2^{(2)} \to l_2^{(2)}$$

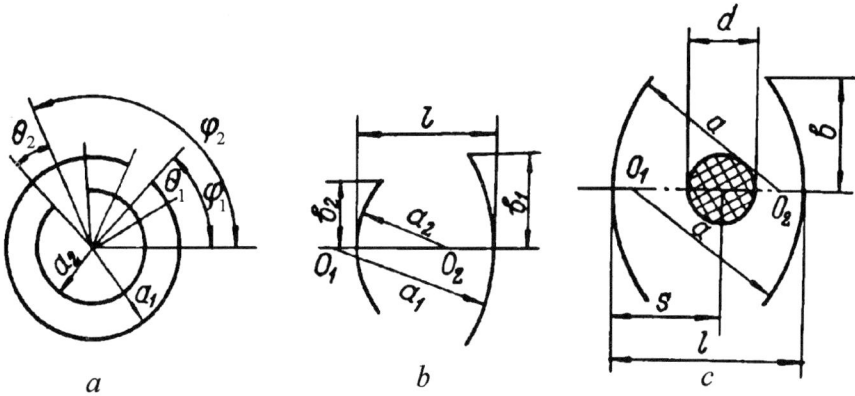

Figure 2.1 *Different types of two-dimensional open resonators*

for resonators formed by perfectly conducting strip screens and

$$\begin{pmatrix} A^{11}(ka_1) - I & A^{12}(ka_1) & A^{13}(ka_1) \\ A^{21}(ka_2) & A^{22}(ka_2) - I & A^{23}(ka_2) \\ A^{31}(ka_3) & A^{32}(ka_3) & A^{33}(ka_3) - I \end{pmatrix} : l_2^{(3)} \to l_2^{(3)}$$

for resonators formed by strip screens with inclusions in the form of circular dielectric pivots. Here $A^{im}(ka_i) : l_2 \to l_2$ are the kernel operator-valued functions determined by eqns. 1.8 and 1.13. The characteristic numbers of $A^{ii}(ka_i) - I$ coincide with the eigenfrequencies of the resonator formed by the ith mirror ($i = 1, 2$) or by the inclusion ($i = 3$). The operator-valued functions $A^{im}(ka_i)$ ($i \neq m$, $i, m = 1, 2$) describe interaction between the mirrors and $A^{3i}(ka_3)$, $A^{i3}(ka_{3i})$ ($i = 1, 2$) describe interaction between the mirrors and inclusions.

The software constructed on the basis of numerical algorithms presented in Section 1.11 enables one to calculate complex fundamental frequencies with five significant digits taking the reduction order P_i of the operator-valued functions $A^{im}(ka_i)$ equal to $[|ka_i|] + 10$ (here $[\cdot]$ is the entire part of $|ka_i|$).

We will pay special attention in what follows to the study of spectral characteristics (fundamental frequencies and Q-factors) of the considered open resonators as functions of various non-spectral parameters (curvature radii of the mirrors, distances between the mirrors, permittivities and diameters of the inclusions) when $L \sim \lambda$ where L is the characteristic size of a resonator and λ is the resonance wavelength.

Let us consider the H-type eigenoscillations ($E_{x_3} = 0$) of open coaxial slot resonators (see Figure 2.1) in the long-wave ($|\kappa| < 1$) and short-wave ($|\kappa| \sim 1$) ranges ($\kappa = ka_1$, a_1 is the curvature radius of the external mirror).

In the long-wave range a circular slotted cylinder has one fundamental frequency κ_0 corresponding to a quasi-static eigenoscillation which is called the slot resonance, while an open coaxial slot resonator has two fundamental frequencies $\kappa_1 = 0.35626 - j\,1.9634 \cdot 10^{-2}$ (the first eigenoscillation) and $\kappa_2 = 0.60097 - j\,1.1554 \cdot 10^{-5}$ (the second eigenoscillation) calculated for $\theta_1 = \theta_2 = 5°$, $a_2 = 0.7a_1$. Analysis of the isolines $|H_{x_3}| = $ const and $\arg H_{x_3} = $ const of the only non-zero component of the field excited in the open coaxial slot resonator by a plane unit-amplitude magnetically polarised wave (at the frequencies $\mathrm{Re}\,\kappa_1$ and $\mathrm{Re}\,\kappa_2$) allows us to assert that the first eigenoscillation is similar to the slot resonance (the magnetic field is practically homogeneous, the electric field is concentrated close to the slot and $\arg H_{x_3}$ is constant inside the resonator). Besides, with broadening the slot ($\theta_1 \to 180°$) the fundamental frequency κ_1 gradually transforms to that of a slot resonator (see Table 2.1 where the data for $a_2 = 0.7a_1$ and $\theta_1 = 5°$ are presented) and with narrowing the slot ($\theta_1 \to 0°$) $\mathrm{Re}\,\kappa_1 \to 0$. The Q-factors of this oscillation (for $\psi_2 = 0$) and of the quasi-static oscillation in a slotted circular cylinder have the same order. The presence of the internal screen enables one to optimise some characteristics of the slot resonance: variations of ψ_2 and $a_2 a_1^{-1}$ may cause the first-order growth of the Q-factor accompanied by a fall of $\mathrm{Re}\,\kappa_1$.

Table 2.1

ka_1	θ_2°			
	0	1	3	5
Re	0.2446	0.3376	0.3507	0.3563
$-$ Im	$4.2 \cdot 10^{-3}$	$1.6 \cdot 10^{-2}$	$1.8 \cdot 10^{-2}$	$1.9 \cdot 10^{-2}$
Re	0.4757	0.5652	0.6052	0.6309
$-$ Im	$1.4 \cdot 10^{-2}$	$1.7 \cdot 10^{-3}$	$1.7 \cdot 10^{-4}$	$1.1 \cdot 10^{-5}$

Thus we may conclude that the physical nature of the first eigenoscillation corresponding to the fundamental frequency κ_1 is similar to the one of the slot resonance.

But this conclusion is not valid for the second eigenoscillation in an open coaxial slot resonator corresponding to the fundamental frequency κ_2. Its Q-factor reaches the level $Q \sim 2.81 \cdot 10^4$ (for $\theta_1 = \theta_2 = 5°$, $a_2 = 0.7a_1$, $\psi_2 = 0$) and the resonance wavelength $\lambda = \frac{2\pi a_1}{0.631}$ is almost ten times as high as the curvature radius of the external screen. The isolines $\arg H_{x_3} = \text{const}$ have two strictly anti-phase domains bounded by the internal screen and situated between the internal and external screens except for vicinities of the edges. Maximum values of the magnetic field are concentrated in the parts of the screens facing the slots and the electric field is mainly concentrated close to the internal slot. To make an acoustic analogy (Koshparenok *et al.*, 1980), one can note that the anti-phase character of the compression and rarefaction of the medium (which plays the role of the magnetic field component) directs the acceleration of the medium's mass (or the electric field component) from one domain to another and only a small part of the energy goes out to free space. This explains the physical nature of the high Q-factor characteristic of the second eigenoscillation.

Let us analyse the dependence of the fundamental frequency κ_2 on the geometrical parameters $(\theta_1, \theta_2, a_1, a_2, \psi_2)$. With broadening the internal (external) slot (having the angle width θ_2 or θ_1) the fundamental frequency κ_2 gradually transforms to the fundamental frequency κ_{11}^c of the H_{11}^c oscillation of a slot resonator [in accordance with the classification given by Shestopalov (1983)]. If $\theta_2 \to 180°$, the Q-factor of this oscillation monotonically decreases and if $\theta_1 \to 180°$, it has a local maximum for $\theta_1 = 9°$ (when the anti-phase domains have equal areas and the communication with external space is minimal). In the last case $\operatorname{Re} \kappa_2$ is practically constant with respect to θ_1 varying in the interval $(0°, 30°)$.

An analogous physical mechanism lies in the basis of the Q-factor dependence on the rotation angle ψ_2 of the internal screen and the ratio of the curvature radii a_2 and a_1. If the slotwidths (or the parameters $\theta_1 = \theta_2 = 5°$, $a_2 = 0.7a_1$) are fixed, then $a_2 a_1^{-1} = 0.82$ or $\psi_2 = \psi_2^0$ are optimal values for the Q-factor. When ψ_2 varies from $0°$ to $180°$, the second resonance oscillation corresponding to the fundamental frequency κ_2 gradually transforms to a low Q-factor oscillation ($Q \sim 10$) for which two "piston" waves emerge between the screens (Borzenkov and Sologub, 1975).

The indicated properties of the second resonance oscillation are good pre-conditions for practical applications of open coaxial slot resonators. The presence of two close long-wave fundamental frequencies enables one to broaden the resonance scattering band because of the values of ψ_2 when the Q-factors of these resonance oscillations coincide.

Let us consider the average wavelength resonances $(|\kappa| \sim 1)$. Since the fundamental frequencies spectrum of the considered family of open resonators is discrete, the eigenoscillations may be numbered with respect to the growing fundamental frequencies moduli (see Table 2.2 where $\theta_1 = \theta_2 = 5°$, $a_2 = 0.7a_1$, $\psi_1 = \psi_2 = 0$). Systematic calculations allow us to assert that in the long-wave and average wavelength ranges the spectrum is simple.

Table 2.2

κ	m			
	1	2	3	4
Re	0.35626	0.63097	1.1884	1.5998
Im	$1.9634 \cdot 10^{-2}$	$8.4754 \cdot 10^{-4}$	$8.4754 \cdot 10^{-4}$	$2.9487 \cdot 10^{-5}$

κ	m			
	5	6	7	8
Re	2.3748	2.5850	2.6324	3.5573
Im	$2.9487 \cdot 10^{-5}$	$4.2407 \cdot 10^{-2}$	$2.1043 \cdot 10^{-7}$	$1.8831 \cdot 10^{-4}$

In order to clear up the structure of these resonance oscillations, one should consider the magnetically polarised oscillations $(E_{x_3} = 0)$. It is known (Weinstein, 1951) that the higher order eigenoscillations $\mathrm{H}_{mn}^{s(c)}$ ($m > 0$ and the signs c or s mean that the oscillations are even or odd with respect to the axis $\psi_2 = 0$) in a closed coaxial cylindrical resonator $(\theta_1 = \theta_2 = 0)$ having the double degeneration caused by resonator symmetry. Irregular perturbations of the walls in the form of longitudinal slots destroy this degeneration. The third $(n = 3)$ and the fourth $(n = 4)$ resonance oscillations corresponding to the fundamental frequencies κ_3 and κ_4 may be considered as the perturbed H_{11}^s and H_{11}^c eigenoscillations of a closed coaxial cylindrical resonator. The slots turn out to be in the knot of the magnetic field of the H_{11}^s oscillation ($n = 3$) and in the antinode of the magnetic field of the H_{11}^c oscillation ($n = 4$). In the last case one observes two characteristic maxima of the magnetic field close to the slots (the external slot "breaks" the antinode). In the same manner we define the eigenoscillations corresponding to the fundamental frequencies κ_n ($n = 5, 6, 7$) as the perturbed H_{21}^s, H_{21}^c and H_{31}^s eigenoscillations. However, the seventh oscillation has no analogue in a closed coaxial cylindrical resonator. Its field is concentrated in the domain bounded by the internal screen and the isolines $|H_{x_3}| = \text{const}$ and $\arg H_{x_3} = \text{const}$ are identical to the ones of the H_{11}^s eigenoscillations of a circular slotted cylinder (Veliev *et al.*, 1981). Further, it is easy to see that $\text{Re}(ka_2) = \text{Re}\,\kappa_7$ ($a_2 = 1.8427a_1$) and the difference between $\text{Re}\,\kappa_7$ and the real part of the H_{11}^s eigenoscillations does not exceed 0.08%. Hence, one can classify the seventh oscillations as the H_{11}^s eigenoscillation of the resonator bounded by the internal screen. This oscillation has the smallest diffraction losses in the average wavelength range. We may conclude that in this range open coaxial cylindrical resonators have a denser spectrum compared with the corresponding closed coaxial resonator.

Let us determine how the eigenoscillations of open coaxial cylindrical resonators depend on the angle slotwidth θ_2 and the angle ψ_2. It follows from the numerical analysis that for fixed $a_2 = 0.7a_1$, $\psi_2 = 0°$ and $\theta_1 = 5°$ the fundamental frequencies κ_n transform to the ones $\kappa_{mn}^{s(c)}$ of a circular slotted cylinder:

$$\kappa_4 \rightarrow \kappa_{21}^c, \quad \kappa_5 \rightarrow \kappa_{21}^s, \quad \kappa_6 \rightarrow \kappa_{01}^s$$

$$\kappa_7 \rightarrow \kappa_{31}^s, \quad \kappa_8 \rightarrow \kappa_{31}^c$$

The whole range of the variation of θ_2 may be separated into three characterisic intervals $5° - 45°$, $45° - 150°$ and $150° - 180°$. In the first interval the fundamental frequencies and Q-factors practically do not depend on θ_2 and the eigenoscillations of an open resonator may be defined here as the perturbed eigenoscillations of a closed coaxial cylindrical resonator (except for the seventh oscillation corresponding to κ_7). In the second interval the eigenoscillations of an open coaxial slot resonator are destroyed and those of an open circular slotted cylinder are formed. And in the third interval the eigenoscillations may be identified as the perturbed eigenoscillations of an open circular slotted cylinder. An anomalous behaviour which is analysed below is characteristic for the fundamental frequency κ_6 of the sixth oscillation and its Q-factor as functions of $\theta_2 \in (150°, 180°)$ (the Q-factor reaches the level $5.402 \cdot 10^5$ when $a_2 = 0.7a_1$, $\theta_2 = 160°$ and $\theta_1 = 5°$)

The real parts of the fundamental frequencies are different for all $\theta_2 \in (5°, 180°)$ except for one value when $\mathrm{Re}\,\kappa_5 = \mathrm{Re}\,\kappa_7$. But in this intersection point the Q-factors are essentially different and hence the eigenoscillation spectrum remains simple during the deformation process considered ($\theta_2 \to 180°$). Analogous properties are characteristic for the cases when fundamental frequencies are considered as functions of ψ_2. Their real parts are practically constant with respect to ψ_2 except for κ_3 and κ_4 and the Q-factors have a distinct oscillating (resonance) behaviour. This is connected with the fact that the internal slot may be either in the antinode (where the Q-factor is maximal) or in the knot (where the Q-factor is minimal) of the magnetic field concentrated between the screens.

Let us consider in detail the anomalous behaviour of the Q-factor of the sixth eigenoscillation as a function of the internal slotwidth. The real part of its fundamental frequency and the Q-factor have a distinct oscillating (resonance) behaviour in the interval $\theta_2 \in 120° - 180°$. The sixth eigenoscillation continuously transforms to the H_{01} oscillation of an open circular slotted cylinder when $\theta_2 \to 180°$ but the order of the Q-factor of the former is four times as high as that of the latter for $a_2 = 0.7a_1$, $\theta_2 = 160°$ and $\theta_1 = 5°$. If $\theta_2 \neq 160°$, there exist two maxima of the Q-factor considered as a function of the rotation angle ψ_2. The internal screen position does not influence the real part of fundamental frequencies while the Q-factor has a second-order growth for certain values of ψ_2.

The $|H|_{x_3}$ isolines of the eigenoscillation considered calculated at the resonance value θ_2 show that the magnetic field amplitude in resonator's centre is 70 times as high as that of the H_{01} oscillation of an open circular slotted cylinder.

A sharp growth of the Q-factor becomes clear after the analysis of the surface current distributions of the H_{01} oscillation in the corresponding closed resonator containing an internal screen ($\theta_1 = 0$, $\theta_2 \neq 180°$). The spectral problem in this case has only isolated real eigenvalues (see Section 1.8) which may be determined by the approximate formula

$$\kappa_{01} = \nu_{01} \left[1 - \gamma^2 (1 - \gamma) \frac{\nu_{01}^2}{4} \tan^2 \Delta \right] + O(\Delta^4) \qquad (2.86)$$

for the H_{01} fundamental frequencies perturbed by a "narrow" internal screen $(\theta_2 \sim 180°)$. Here $\gamma = a_2 a_1^{-1}$, $\Delta = \pi(1 - \theta_2 / 180°$ and ν_{01} is the first root of the equation $J_0'(x) = 0$. The surface current on the external screen are given by the approximate representation

$$j(\phi) = 2\kappa_{01}^2 \left(\frac{1}{\nu_{01}^2 - \kappa_{01}^2} - \frac{2\cos\phi}{\nu_{21}^2 - \kappa_{01}^2} \right) - \frac{(\gamma + 2\cos\phi)(1 - \gamma)}{1 + 2\gamma\cos\phi + \gamma^2} \qquad (2.87)$$

One can see from eqn. 2.87 that the surface current in a cylinder with a "narrow" internal screen may vanish or have a minimum for $\phi = \phi_0$ and hence a slot cut in the vicinity of this angle co-ordinate will not cause cinsiderable diffractional losses. These considerations allow us to conclude that a sharp growth of the Q-factor of the sixth eigenoscillation when $\theta_2 \in [120° - 180°]$ is connected with "assymetric" behaviour of the surface current in the corresponding closed resonator.

2.7 Fundamental frequencies of strip mirror open resonators

Physical characteristics of open cylindrical resonators formed by two perfectly conducting circular strips are studied in detail by Poyedinchuk (1983a), Koshparenok *et al.* (1985) and Shestopalov (1985) who pay the main attention to the study of the magnetically polarised oscillations in the case when the resonance wavelength and resonators' dimensions are commensurable ($1 \leq |\kappa| \leq 20$, $b \sim a$ where $\kappa = ka$). An example of such a resonator is presented in Figure 2.1. The curvature centres of the mirrors may coincide with each other or with the mirrors' centres or lie inside (outside) the resonator. There are no limitations on the curvature radii a_i, the aperture dimensions b_i and the distance between the mirrors l.

Let us consider first a symmetrical confocal resonator ($a_2 = l = a$, $b_1 = b_2 = b$). It is known that a symmetrical confocal resonator in the frequency range $|\kappa| \gg 1$ ($\kappa = ka$) for $b \ll a$ has fundamental frequencies corresponding to the high Q-factor "bouncing ball" oscillations. Systematic calculations carried out on the basis of the rigorous approach described in Sections 1.8 and 1.11 show that such resonators have distinct resonance properties in the frequency range $1 \leq |\kappa| \leq 20$. Several fundamental frequencies corresponding to the highest Q-factor eigenoscillations (for $b = 0.63a$) are presented in order of increasing moduli in Table 2.3.

Table 2.3

H_{mn}	κ	
	Re	Im
H_{00}	1.4947	0.25712
H_{01}	3.7178	$2.0279 \cdot 10^{-2}$
H_{11}	5.4903	0.10531
H_{02}	6.9948	$3.7385 \cdot 10^{-2}$
H_{12}	8.6431	0.23391
H_{03}	10.142	$6.0803 \cdot 10^{-4}$
H_{22}	10.201	0.56659
H_{31}	11.231	0.81146
H_{13}	11.757	$4.1188 \cdot 10^{-3}$
H_{51}	12.837	0.85143
H_{04}	13.294	$7.9296 \cdot 10^{-4}$
H_{14}	16.894	$8.5723 \cdot 10^{-2}$
H_{05}	16.450	$5.3688 \cdot 10^{-4}$
H_{34}	16.615	$1.9441 \cdot 10^{-2}$

The classification of eigenoscillations is a result of solving the problem of resonator excitation by a plane unit-amplitude magnetically polarised wave (at the frequencies Re κ where κ is a resonance frequency) and of investigating the isolines $|H_{x_3}| = $ const and arg $H_{x_3} = $ const of the only non-zero component of the excited field. A lot of such data has been obtained by Poyedinchuk (1983a), Koshparenok *et al.* (1985) and Shestopalov (1985). It was shown that as well as in asymptotic models (Weinstein, 1966) one could identify the eigenoscillations by the numbers of zeros of the magnetic field along two perpendicular axes of symmetry, denoting them by the symbol H_{mn} where n corresponds to resonator's axis ($n \gg m$ for the "bouncing ball" oscillations). It is clear that, for example, the H_{41} and H_{51} oscillations presented in Table 2.3 cannot be described within the frame of asymptotic models. Under certain conditions these oscillations may interact with the H_{03} and H_{13} oscillations (see Section 2.8) which causes sharp growth of diffractional losses.

Let us consider the influence of the aperture dimensions on the eigenoscillations spectra. The results presented by Shestopalov (1985) show that $\mathrm{Re}\,\kappa$ are practically constant with respect to the ratio ba^{-1} for $n \geq 4$ (when the number of resonance wavelengths along mirror's curvature radius is greater than two). In fact, with growing index n the eigenfields are concentrated close to resonator's axis and, therefore, small vicinities of the edges do not take part in forming the resonance fields. But the eigenoscillations with $n < 4$ essentially depend on ba^{-1} (as well as for $m \geq n$) because here edge diffraction plays a crucial role. With growing m the dependence on ba^{-1} becomes stronger due to the fact that internal resonance fields are mainly concentrated close to resonator's mirrors which causes the "aperture" resonances. The field amplitudes have local maxima close to the edges and hence diffraction fields make the greatest contribution to the resonance fields. That is why even the slightest variations of ba^{-1} leads to sharp changes of fundamental frequencies. The fundamental frequencies of the H_{mn} oscillations for $m \neq 0$ as functions of ba^{-1} have behaviour similar to that of the H_{0n} oscillations for $n > m$.

The Q-factors of all oscillations (except for H_{03} and H_{13}) monotonically grow with respect to ba^{-1} (for $b = \sin 60° a$, the resonator becomes a closed one).

From the theoretical and practical viewpoints it is important to compare the results obtained for the H_{mn} oscillations ($m < 0$) in the frequency range $1 < |\kappa| < 20$ with the data calculated by the asymptotic formula

$$\mathrm{Re}\,\kappa = \pi\left(n + \frac{1}{2}m + \frac{1}{4}\right) + O(n^{1-\epsilon}) + O(n^{-1}) \qquad (2.88)$$

which does not take into account the dependence on ba^{-1} but sufficiently well describes the real parts of fundamental frequencies beginning from $n = 4$. Eqn. 2.88 verified within the frame of asymptotic models is uniform with respect to $m \in [0, Cn^{1-\epsilon}]$ ($0 < \epsilon < 1$, $C = \mathrm{const}$). With growing n the relative error rapidly decreases (it equals to 0.3% for the H_{05} oscillation). It is interesting to point out that eqn. 2.88 gives an upper estimate for the real parts of the H_{0n} fundamental frequencies. Such a comparison enables one to establish the boundaries of the approximate models applicability.

The normalised diffraction far field patterns of the fields scattered by a resonator (excited by a magnetically polarised plane wave at frequencies equal to the real parts of its fundamental frequencies) allow us to study the diffraction eigenfields. Such far field patterns have a complicated structure except for the H_{00} oscillation whose pattern is isotropic (due to its long-wave character). Other oscillations have multi-petal far field patterns. The shadow petal of the H_{0n} oscillations narrows with increasing index n (for the H_{05} oscillation it is almost three times as narrow as that of the H_{01} oscillation) since the moduli of their fundamental frequencies grow and the optical properties of the scattered field begin to play a more noticeable role.

Let us consider symmetrical non-confocal resonators with mirror aberrations, such as the forward mirror displacement, warps, different curvature radii ($a_1 \neq a_2$) and apertures ($b_1 \neq b_2$). Averbakh *et al.* (1965), Boitzov and Fradkin (1968), Gloge (1966), Lazutkin (1968), Sanderson (1969), Weinstein (1966) and others studied deformations of the "bouncing ball" fundamental frequencies caused by various aberrations in the frequency range $|\kappa| \gg 1$ ($b_i \ll l, i = 1, 2$).

We will consider the influence of aberrations in the frequency range $1 \ll |\kappa| \ll 20$ ($b_i \sim l, i = 1, 2$). Let us clarify first how the fundamental frequencies of a symmetrical resonator ($b_1 = b_2 = b$, $a_1 = a_2 = a$) depend on the relative distance between the mirrors $\hat{l} = la^{-1}$. The real parts of the fundamental frequencies $\operatorname{Re} \kappa$ and Q-factors Q monotonically decrease when $\hat{l} \in [1, 2]$ (the boundary points of this interval correspond to confocal and concentric resonators, respectively). For $\hat{l} \in [1.5, 1.8]$ the Q-factors of the H_{02l} oscillations decay less intensively than that of the H_{02l+1} oscillations. A quite different picture is observed for $\hat{l} < 1$ where $\operatorname{Re} \kappa$ monotonically grows with decreasing $\hat{l}$. For certain values of $\hat{l}$ the real parts of the H_{03}, H_{22} and H_{41} fundamental frequencies coincide but their Q-factors have essential differences (of about three orders) and, hence the spectrum has only a partial degeneration. The Q-factors of the H_{mn} oscillations ($m < n$) have local maxima in the interval $\hat{l} \in [0.8, 1]$ while the Q-factors of the H_{mn} oscillations ($m \geq n$) monotonically grow. Further decrease of $\hat{l}$ leads to the monotonic growth of the Q-factors for all types of oscillations (for $\hat{l} = 2(1 - \sqrt{1 - b^2 a^{-2}})$ the resonator becomes a closed one).

The asymptotic formula

$$\operatorname{Re} \kappa = \begin{cases} \pi \left[n + m + \tfrac{1}{2} - (m + \tfrac{1}{2}) \arccos \hat{l} - 1 \right] \hat{l}^{-1}, & 1 \leq \hat{l} < 2 \\ \left[\pi n + (m + \tfrac{1}{2}) \arccos 1 - \hat{l} \right] \hat{l}^{-1}, & \hat{l} \leq 1 \end{cases}$$

which is valid for the H_{mn} oscillations $(n \geq m)$ gives a 1.5% error already for the H_{22} oscillation but with increasing n the relative error rapidly falls and becomes equal to 0.4% for the H_{04} oscillation in the whole interval $\hat{l} \in [1,2)$.

Thus we may conclude that the real parts of the fundamental frequencies and Q-factors of the H_{nm} oscillations $(n > m, n \geq 3)$ in the frequency range $1 \ll |\kappa| \ll 20$ behave (as functions of the relative distance between the mirrors) similarly to the H_{mn} "bouncing ball" oscillations $(n \gg m)$ in the frequency range $|\kappa| \gg 1$.

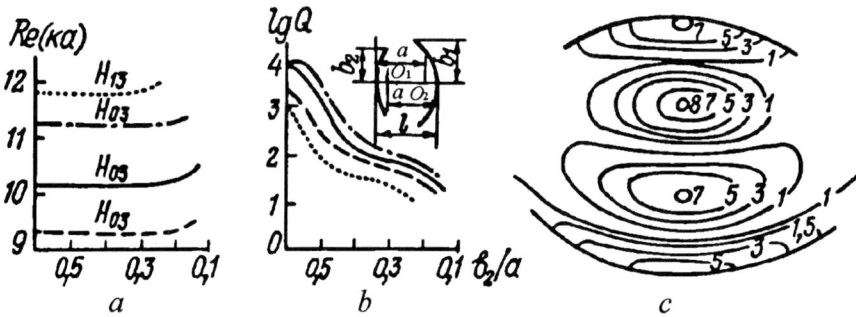

Figure 2.2 *Influence of mirrors' aberration on the characterisitcs of an open resonator*

Let us analyse the fundamental frequencies deformation with respect to the decreasing aperture b_2 and curvature radius a_2 taking the H_{03} and H_{13} oscillations as examples for consideration. The data presented in Figure 2.2 (where the values $\hat{l} = 0.8, 0.9, 1.0, 1.1$ correspond to the dot curves, dash-dot curves, bold curves and dash curves, respectively) are evidence of weak dependence of the real parts of the H_{03} fundamental frequencies on the ratio $b_2 a^{-1}$ in the interval $0.41 \leq \frac{b_2}{b_1} \leq 1$ for the under-confocal $(\hat{l} < 1)$, confocal $(\hat{l} = 1)$ and over-confocal $(\hat{l} > 1)$ resonators. This interval becomes much smaller $(0.7 \leq \frac{b_2}{b_1} \leq 1)$ for the H_{13} oscillation and here the Q-factors of both oscillations essentially fall (if $\frac{b_2}{b_1} = 0.41$, the Q-factor of the H_{13} oscillation in a confocal resonator is equal to 100 while $\mathrm{Re}\,\kappa$ remains practically constant). The field is "compressed" towards the smaller mirror (see Figure 2.2 where the isolines $H_{x_3} = \mathrm{const}$ are presented for the H_{03} oscillation of the confocal resonator, $ka = 10.153$, $\frac{b_1}{a} = 0.60$, $\frac{b_2}{a} = 0.48$). If $\frac{b_2}{b_1} \to 0$, resonance oscillations of the open resonator are destroyed and the low Q-factor eigenoscillations of a cylindrical strip appear. In fact, when $\frac{b_2}{b_1} \to 0$, the operator-valued functions $A^{21}(ka)$ describing interaction between the mirrors and $A^{22}(ka) - I$ describing resonance properties of the mirror with the aperture b_2 tends to zero operator and unit operator, respectively (in the operator norm).

The real parts of the H_{03} and H_{13} fundamental frequencies of a confocal resonator weakly depend on the warp angle $\delta \in (0° - 20°)$ (they change in the third significant digit) while the Q-factors Q decrease almost in two orders ($Q = 185$ for $\delta = 20°$). Resonance fields shift to the mirrors' edges and yield a fall in the Q-factors since the greater field amplitudes on the edges correspond to greater diffraction losses. However, Re κ essentially depend on δ in non-confocal resonators although the structure of their eigenoscillations are similar to that in confocal resonators.

Systematical calculations show that the real parts of the H_{mn} fundamental frequencies ($m < n$) of under-confocal resonators (as well as of over-confocal resonators) monotonically grow with decreasing $\frac{a_2}{a_1}$ (in the case of fixed distance l between the mirrors) but the rates of decrease are essentially different for the H_{0n} and H_{mn} oscillations ($m \neq 0$). The Q-factor of the H_{03} oscillation in a under-confocal resonator as a function of $\frac{a_2}{a_1}$ decreases much more rapidly than the same Q-factor in an over-confocal resonator. Actually, in the first case the 0.1 variation of $\frac{a_2}{a_1}$ causes a first-order reduction of the Q-factor while in the second case the same variation of $\frac{a_2}{a_1}$ lowers the Q-factor by 1.8 times. One can see after considering the isolines $H_{x_3} = $ const that the resonance fields in both cases have an identical phase structure but in the second case the field amplitude reduces in five times. The isolines $H_{x_3} = $ const get narrow towards the resonator's axis near the small mirror and broaden in the vicinity of the opposite mirror where the field amplitude grows as well as diffraction losses. Similar features are characteristic for over-confocal resonators but the indicated growth of the field amplitude close to mirrors' edges (with decreasing $\frac{a_2}{a_1}$) is much weaker. One cannot describe these phenomena within the frames of the known heuristic models of open resonators when the mirrors are assumed to be infinite and the edge diffraction is not taken into account. In fact, if the distance l between the mirrors is equal to one of the curvature radii (or, in our notation $l = a_2$, $l < a_1$), the size of field's spot is infinite on one of the mirrors and is zero on the other.

2.8 Intertype oscillations in open cylindrical resonators

Shteinshleiger (1955) was the first to work out an approximate mathematical model of the interaction of oscillations in volume (closed) electromagnetic resonators. This model is based on the classical perturbation theory of non-self-adjoint operators and cannot be applied to open resonators because the boundary-value problems describing their spectral charactersitics are non-selfadjoint due to the presence of radiation conditions in infinity (see Section 1.2), the fundamental frequencies are complex quantities and eigenoscillations (eigenfunctions of the boundary eigenvalue problems) do not form the basis (Wineberg, 1972). Existing asymptotic models characterised in Section 2.7 allow one to study only a part of the spectrum, namely short-wavelength "bouncing ball" or "whispering gallery" oscillations. But this information is not enough to construct a mathematical model of the interaction phenomena in open resonators. A rigorous statement of corresponding problems was proposed by Koshparenok *et al.* (1979a, 1979b) who considered interaction of oscillations in a perfectly conducting circular cylinder with a longitudinal slot using the method of the Riemann-Hilbert problem. It was shown that there existed such values of slotwidth ("the points of coincidence") where certain fundamental frequencies of the open resonator considered (a slotted cylinder) were close on the complex plane and the corresponding eigenoscillations began to interact. After the points of coincidence these eigenoscillations are transformed to each other.

Here we will investigate the intertype interaction of oscillations in the open resonator formed by two perfectly conducting infinite cylindrical strips (whose transversal cross-section is formed by the domain from the family CB_c^e) considering the H-type oscillations in the corresponding symmetrical confocal resonator ($a_1 = a_2 = l = a$, $b_1 = b_2 = b$).

As shown in Section 1.3, the spectral problem for such open resonators is equivalent to the problem on characteristic numbers for Fredholm matrix operator-valued functions which is reduced to the dispersion equation. The latter in the case considered splits into four independent equations for the H_{mp} oscillations

$$\det \|\delta_p^m - [(a_{pm}^{11}(ka) + (-1)^p a_{pm}^{12}(ka)) + (-1)^m (a_{p,-m}^{11}(ka) + (-1)^p a_{p,-m}^{12}(ka))]\|$$

here δ_p^m is the Kronecker delta.

Hence the spectrum of the H-type oscillations of a symmetrical confocal resonator consists of the H_{2m2p}, H_{2m2p+1} H_{2m+12p} and $H_{2m+12p+1}$ oscillations. It is evident that the intertype interaction is possible only for the oscillations belonging to one symmetry class.

Let us denote by $\hat{b} = \frac{b}{a}$ the ratio of mirror's aperture to its curvature radius and by κ_{mp} (Q_{mp}) the fundamental frequencies (Q-factors) of the H_{mp} oscillations. Numerical calculations show that when $\hat{b}$ varies the H_{03} and H_{13} oscillations from the symmetry class H_{2m2p+1} interact with the H_{41} and H_{51} oscillations from the symmetry class $H_{2m+12p+1}$.

The high-Q-factor H_{03} oscillation is characterised by three field's variations along resonator's axis. Its Q-factor Q_{03} has a local maximum $Q_{03} \sim 10^4$ in the point $\hat{b} = 0,65$ and is comparable with Q_{41} in the vicinity of the point of coincidence $\hat{b} = 0.775$ where $|\kappa_{03} - \kappa_{41}|$ is minimal and $Q_{03} = Q_{41}$. The curves for Re κ_{mp} as functions of $\hat{b} \in [0.73, 0.83]$ are similar to the Wien graphs (see Figure 2.3 where the bold lines and the dashed lines correspond to the H_{03} and H_{41} oscillations, respectively).

Analysis of the isolines $|H_{x_3}| = $ const of the resonance fields (obtained a result of the solution to the problem of resonator's excitation by a plane unit-amplitude magnetically polarised wave at the frequency Re κ_{03}) shows that for $\hat{b} = 0.69$, $ka = 10.139$ the excited field corresponds to the H_{03} oscillation, for $\hat{b} = 0.775$, $ka = 10.130$ it has a hybrid character (the H_{03} and H_{14} oscillations are "mixed") and for $\hat{b} = 0.82$, $ka = 10.001$ the excited field corresponds to the H_{41} oscillation. The same situation occurs for the excitation frequency Re κ_{41}: if $\hat{b} < 0.775$ ($\hat{b} > 0.775$), the excited field is connected with the H_{41} (H_{03}) oscillation and it has a hybrid character for $\hat{b} = 0.775$ (see Figure 2.4). The interaction dynamics $H_{03} \Leftrightarrow H_{41}$ in the vicinity of the point of coincidence $\hat{b} = 0.775$ is clearly shown by the current distributions on the mirrors presented in Figure 2.5 with respect to the varying distance between them.

We may conclude that the H_{03} and H_{41} oscillations interact intensively close to the point $\hat{b} = 0.775$ and their mutual transformation occurs after this point. In addition to this, the diffraction losses of the high-Q-factor H_{03} oscillation grow sharply and reach the level of the losses characteristic for the low-Q-factor H_{41} oscillation.

A similar interaction takes place for the H_{13} and H_{51} oscillations in the

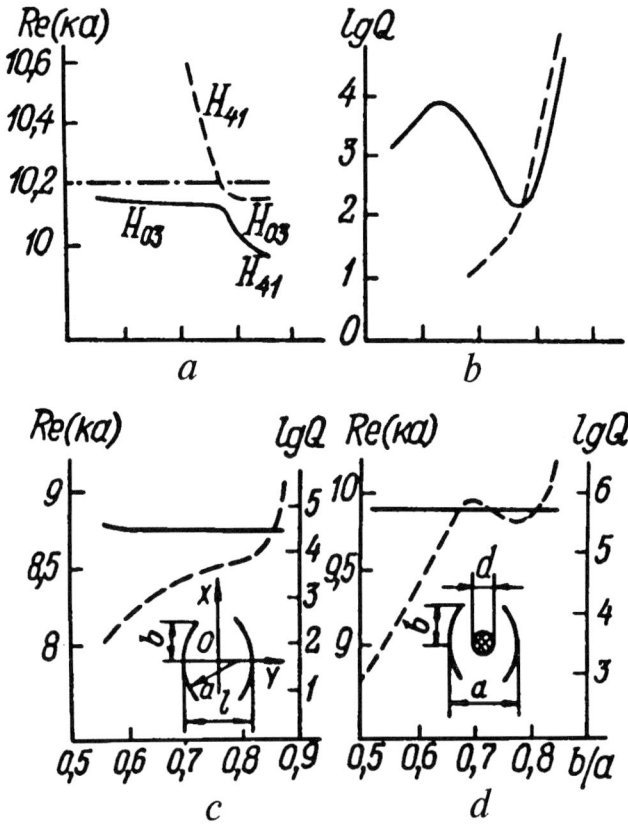

Figure 2.3 *Characterisitcs of an open resonator at the H_{03}- and H_{41}-oscillations*

point $\hat{b} = 0.790$.

Let us investigate how the distance between the mirrors and location of dielectric bodies inside a resonator influence the interaction processes.

As shown in Section 2.6, the increasing relative distance $\hat{l}$ between the mirrors yields growth of diffraction losses for all types of oscillations. The losses increase much faster for the H_{mn} oscillations $(m > n)$ compared with the H_{0n} oscillations and it suggests the idea to control the interaction of oscillations by appropriate variations of $\hat{l}$. One can see in Figure 2.3 that the real parts of fundamental frequencies (bold lines) are practically constant and the Q-factors (dashed lines) monotonically increase with respect to growing $\hat{b}$. These data are presented for $\hat{l} = 1.17$ and hence the 0.17 increment of $\hat{l}$

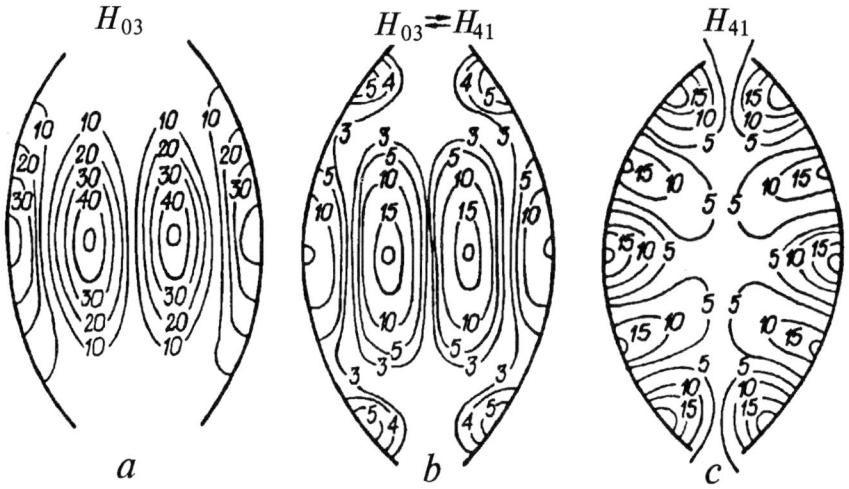

Figure 2.4   *Fields of H_{03}- and H_{41}-oscillations near the Morse critical point*

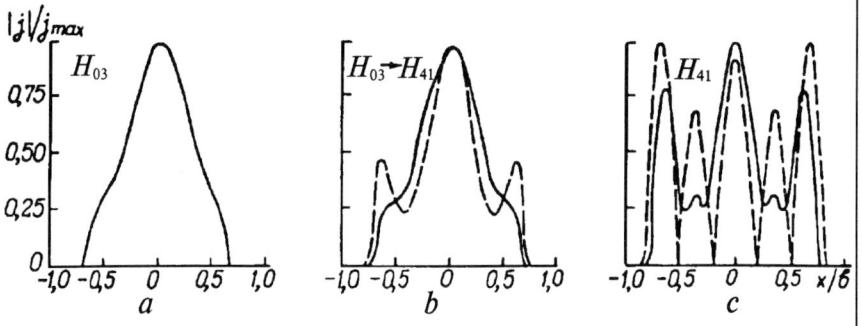

Figure 2.5   *Distribution of surface current on the mirrors of an open resonator for H_{03}- and H_{41}-oscillations near the Morse critical point*

yields a complete suppression of the $H_{03} \Leftrightarrow H_{41}$ interaction. The same effect is achieved if a dielectric body is placed inside a resonator. As an example, let us consider the case when a dielectric pivot with $\text{Re}\,\epsilon = 7.1$, $\text{Im}\,\epsilon = 0$, $d = 0.05a$ is placed inside a symmetric confocal resonator (see Figure 2.3). Then the real parts of fundamental frequencies are practically constant as functions of $\hat{b}$ and the Q-factor of the H_{03} oscillation has a certain fall in the vicinity of

$\hat{b} = 0.775$ where the $H_{03} \Leftrightarrow H_{41}$ interaction takes place in the corresponding empty resonator.

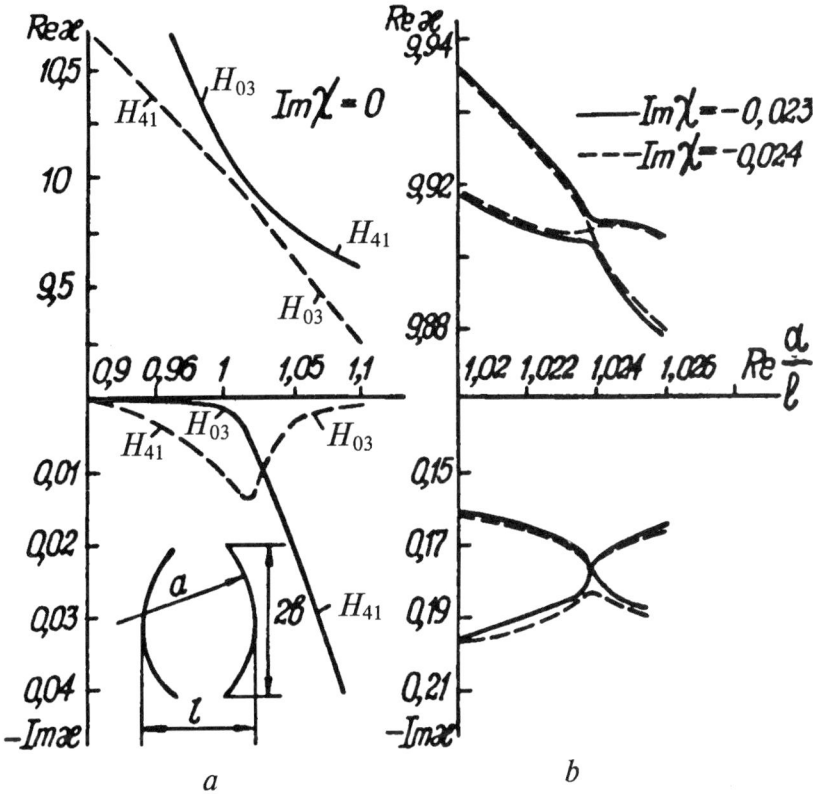

Figure 2.6 *Dispersion curves of a symmetric open resonator*

The dispersion curves of the H_{03} and H_{41} oscillations in an empty resonator calculated as the roots of eqn. 1.34 close to the Morse critical points are presented in Figure 2.6 (the curves 1 and 2 correspond to the values $\operatorname{Im}\chi = -0.023$ and $\operatorname{Im}\chi = -0.024$, respectively, where $\chi = \hat{b}$ is considered as a spectral parameter). The diffraction losses of the H_{03} oscillation essentially grow when $\hat{l}^{-1}$ increases from 1 to 1.05 due to its interaction with the H_{41} oscillations.

In conclusion we point out that it is not possible to study the intertype interaction using only asymptotic models. In fact, none of the H_{m1} oscillations ($m > 1$) in confocal resonators (in particular, H_{41} and H_{51} oscillations) may be described within the frame of these models. Eqn. 3.3 gives acceptable results for $\text{Re}\,\kappa_{mp}$ $(p > m)$ if this asymptotic relation is applied at the points situated sufficiently far from the point of coincidence (the relative errors are 0.6% for the H_{03} oscillation and 0.12% for the H_{13} oscillation). The Q-factors of these oscillations calculated by the asymptotic formulae proposed by Weinstein (1966) have nothing in common with the data presented in Figure 2.3.

2.9 Spectral characteristics of circular resonators with dielectric inclusions

Let us consider the spectral characteristics of open cylindrical resonators formed by two perfectly conducting circular strip mirrors containing inclusions in the form of lossy dielectric pivots with complex permittivity $\epsilon = \text{Re}\,\epsilon + j\,\text{Im}\,\epsilon$, $\text{Im}\,\epsilon \geq 0$ (see Figure 2.1). The main attention will be paid to a symmetrical confocal resonator with an inclusion placed along its axis. We have chosen such a resonator because corresponding empty confocal resonators are fully described in previous Sections for $1 \leq |\kappa| \leq 20$, $b \sim 2a$ and for $|\kappa| > 20$, $b \gg 2a$ they are studied in detail by Weinstein (1966). Also, the structure of the eigenoscillations in symmetrical confocal resonators is similar to that in nonconfocal resonators. At the same time, there is practically no information on the spectral properties of open resonators with dielectric inclusions.

As shown in Chapter 1 and in Section 2.7 the determination of fundamental frequencies is reduced to the problem of characteristic numbers of a meromorphic Fredholm matrix operator-valued functions of the frequency spectral parameter

$$\begin{pmatrix} A^{11}(ka_1) - I & A^{12}(ka_1) & A^{13}(ka_1) \\ A^{21}(ka_2) & A^{22}(ka_2) - I & A^{23}(ka_2) \\ A^{31}(ka_3) & A^{32}(ka_3) & A^{33}(ka_3) - I \end{pmatrix} : l_2^{(3)} \to l_2^{(3)} \qquad (2.89)$$

here $A^{ij}(ka_i) : l_2 \to l_2$ are the kernel operator-valued functions determined by eqns. 1.8 and 1.13 where one has to put $N = 2$, $M = 1$, $a_1 = a_2 = a$, $2a_3 = d$, $l_{21} = l_{12} = a$, $l_{31} = l_{13} = s$, $l_{23} = l_{32} = a - s$, $\psi_2 = \psi_{12} = \psi_{13} = \psi_{32} = 0$, $\psi_1 = \psi_{21} = \psi_{31} = \psi_{23} = \pi$, $u_1 = u_2 = \sqrt{1 - b^2 a^{-2}}$.

An eigenoscillation of the open resonator considered will be called the H_{mn} oscillation if for $\epsilon \to 1$, $\mu \to 1$ it continuously transforms to that of the corresponding closed resonator. The problem of characteristic numbers for the operator-valued function of the spectral parameter ka defined by eqn. 2.89 may be rewritten as a countable set of explicit equations

$$ka = F_{mn}(\epsilon, \mu, \hat{d}, \hat{s}, \hat{b}) \tag{2.90}$$

determining the required fundamental frequencies. We will study the functions given by eqn. 2.90 (calculated with five significant digits) for the following sets of non-spectral parameters $\hat{d} = \frac{d}{a}$, $\hat{s} = \frac{s}{a}$, $\hat{b} = \frac{b}{a}$, ϵ and μ:

(1) $1 \le \epsilon \le 10$, $\mu = 1$, $\hat{d} = $ const, $\hat{s} = 0.5$, $\hat{b} = $ const;
(2) $\epsilon = $ const, $\mu = 1$, $0 \le \hat{d} \le 1$, $\hat{s} = 0.5$, $\hat{b} = $ const;
(3) $\epsilon = $ const, $\mu = 1$, $\hat{d} = $ const, $\hat{d} \le 2\hat{s} \le 2 - \hat{d}$, $\hat{b} = $ const.

Let us consider the first case assuming that the origin of the inclusion coincide with resonator's centre O_R ($\hat{s} = 0.5$) and the following conditions hold: $\hat{d}\,\mathrm{Re}(ka)\sqrt{\mathrm{Re}\,\epsilon} \ll 1$, $\hat{d}\sqrt{\mathrm{Re}\,\epsilon}\lambda^{-1} \ll 1$ where $\lambda = 2\pi[\mathrm{Re}(ka)]^{-1}$. We will separate two possible positions of the inclusion: (A) the pivot is in the antinode and (K) the pivot is in the knot of the electric eigenfield. Numerical calculations are evidence of essential dependence of the functional relation $ka = F_{mn}(\mathrm{Re}\,\epsilon, 1, \mathrm{const}, 0.5, \mathrm{const})$ on the pivot's position (A or K). For the H_{m2n} oscillations (the electric eigenfield is minimal in O_R) the fundamental frequencies and Q-factors are practically constant with respect to $\mathrm{Re}\,\epsilon \in (1, \epsilon_0)$ where $|ka - k_0 a| < C10^{-3}$, $\mathrm{Re}(ka) < \mathrm{Re}(k_0 a)$. Here $k_0 a$ is the H_{m2n} fundamental frequency of the empty resonator, $C = C(ka)$ and ϵ_0) depends on the value $\hat{d}\,\mathrm{Re}(ka)\sqrt{\mathrm{Re}\,\epsilon}$. In the same manner one can show that the real parts of the H_{m2n} fundamental frequencies (the electric eigenfield is maximal in O_R) monotonically decrease and tend to the real parts of the fundamental frequencies of an open resonator containing a perfectly conducting inclusion (with the radius equal to that of the dielectric inclusion) when $\mathrm{Re}\,\epsilon \to \infty$.

Let us note that the properties of eigenoscillations described above are not observed when pivot's diameter is commensurable with the resonance wavelength, i.e. when $\hat{d}\,\mathrm{Re}(ka)\sqrt{\mathrm{Re}\,\epsilon} \sim 1$.

In the second case $\mathrm{Re}(ka)$ monotonically decrease as a function of $\hat{d}$ and the Q-factors of the H_{0n} oscillations have a distinct resonance character. For sufficiently low losses in the dielectric ($\mathrm{Im}\,\epsilon \sim 10^{-4}$) the Q-factor of an open resonator with an inclusion may be almost one order as high as that of the empty resonator. Let us analyse this phenomenon considering the H_{03} oscillation. Its Q-factor calculated as a function of $\hat{d}$ in a resonator with a dielectric inclusion has two distinct maxima as well as the Q-factor of the same but empty resonator (the first local maximum is in the point $\hat{d} = 0.18$ for $\mathrm{Im}\,\epsilon = 2 \cdot 10^{-4}$) and in the first local maximum the Q-factor of the former is one order as high as that of the latter. In the "resonance" values of $\hat{d}$, the Q-factors essentially depend on $\mathrm{Im}\,\epsilon$: they are maximal when $\mathrm{Im}\,\epsilon = 0$ and decrease when $\mathrm{Im}\,\epsilon$ grows. Q-factor's dependence on $\hat{d}$ has a monotonous character when $\mathrm{Im}\,\epsilon > 10^{-2}$ and $\mathrm{Re}(ka)$ are practically constant with respect to $\mathrm{Im}\,\epsilon$ (the difference is in the third significant digit for $\mathrm{Im}\,\epsilon = 0$ and $\mathrm{Im}\,\epsilon = 10^{-2}$).

To explain such behaviour of the Q-factors let us compare the resonance fields excited in the resonators with and without inclusions by a plane magnetically polarised wave at the frequencies $\mathrm{Re}\,\kappa$ (where κ are resonance frequencies) limiting ourselves to the case $\mathrm{Im}\,\epsilon = 0$ when anomalous phenomena become apparent in the clearest way and considering the isolines $|E_{x_2}| = \mathrm{const}$ of the normalised transversal component of the electric field. For the resonance values of $\hat{d}$ the E_{x_2} and H_{x_3} amplitudes grow sharply inside dielectric inclusions where the isolines $|H_{x_3}| = \mathrm{const}$ are identical to that of pivot's magnetic eigenfields. Hence, the resonance behaviour of the Q-factors with respect to $\hat{d}$ is caused by excitation of eigenoscillations of an inclusion itself. For non-resonance values of $\hat{d}$ the field excited inside an inclusion does not correspond to any of the pivot's eigenoscillations and the Q-factors vary due to the scattering of the resonator's eigenfield on the inclusion. The local maxima of the Q-factors as functions of $\hat{d}$ are removed for sufficiently high losses because the Q-factors of the pivot's eigenoscillations are small even for $\mathrm{Im}\,\epsilon = 0$ (except for the "whispering gallery" oscillations with $kd \sim 1$ and the "locked oscillations" which are absent in the frequency range considered and may be excited when $\mathrm{Re}\,\epsilon \gg 1$, $\mathrm{Im}\,\epsilon \ll 1$, $kd \sim 1$) and, consequently, its resonance properties are not observed in the frequency range $\omega d\sqrt{\mathrm{Re}\,\epsilon}\,c^{-1} = O(1)$ (c is the light velocity in vacuum).

The real parts of the fundamental frequencies in resonators with inclusions are practically constant with respect to $\mathrm{Im}\,\epsilon$ and monotonically decrease with growing $\hat{d}$, while the imaginary parts $-\text{Im}(ka)$ increase sharply due to the growth of heat losses in dielectric pivots and diffraction losses caused by the scattering of the resonator's eigenfield on inclusions. Therefore, the Q-factors of resonators with inclusions will also monotonically decrease with growing $\hat{d}$.

The anomalous growth of the Q-factors is characteristic not only for the H_{0n} oscillations in symmetrical confocal resonators but for the H_{22} and H_{13} oscillations in non-confocal resonators. Here the resonance values of $\hat{d}$ decrease with increasing $\text{Re}\,\epsilon$: e.g. if $\text{Im}\,\epsilon = 2 \cdot 10^{-4}$, the first local maxima for the H_{03} oscillation are observed when $\text{Re}\,\epsilon = 2.08$, $\hat{d} = 0.18$ and $\text{Re}\,\epsilon 3.8$, $\hat{d} = 0.12$. A similar situation occurs for the H_{0n} oscillations with increasing index n (when $\text{Re}\,\epsilon 2.08$): the first resonance values $\hat{d} = 0.06$ and $\hat{d} = 0.30$ are observed for the H_{05} and H_{02} oscillations, respectively.

Let us consider the third case when $ka = F_{mn}(\text{const}, 1, \text{const}, \hat{s}, \text{const})$ are treated as functions of the pivot's position. We will limit ourselves to the study of the H_{0n} oscillations in symmetrical confocal resonators containing dielectric circular pivots (lossless or with large losses, $\text{Im}\,\epsilon \sim 10^{-1}$) and perfectly conducting pivots ($\mu = \frac{1}{\epsilon}$, $\epsilon \to \infty$).

$\text{Re}(ka)$ and $\lg Q$ have a distinct resonance character as functions of $\hat{s}$ and the maximal deviations from $\text{Re}(k_0a)$ and $\lg Q_0$ of the corresponding empty resonators are observed when an inclusion is in the antinode (knot) of the electric (magnetic) field. With growing $\hat{s}$ the amplitudes of the $\text{Re}(ka)$ oscillations increase and the Q-factors maxima fall. With decreasing $\hat{s}$ there are practically no oscillations since an inclusion weakly perturbs the eigenfield of an empty resonator.

In the case of an ideal dielectric inclusion ($\text{Im}\,\epsilon = 0$) the real parts of the H_{0n} ($n = 1, 3$) fundamental frequencies and Q-factors oscillate. The behaviour of $\text{Re}(ka)$ is similar to that of a resonator with a lossy inclusion. The Q-factor of the H_{03} oscillation in the case of a lossy inclusion (for $\hat{s} = 0.5$) is much less than Q_0 (in the empty resonator) while in the case of a lossless inclusion the Q-factor is greater than Q_0. Such a behaviour may be explained if to take into account the excitation of pivot's eigenoscillations. If an inclusion is removed from resonator's centre ($\hat{s} \neq 0.5$) the Q-factor falls sharply and reaches its minimum when a pivot finds itself in the antinode of the electric field ($\hat{s} = 0.82$ for the H_{03} oscillation) because in this position an inclusion essentially perturbs the resonance eigenfield of the empty resonator.

In the case of a perfectly conducting inclusion three situations are possible: $\text{Re}(ka) > \text{Re}(k_0a)$ when an inclusion is in the knot (antinode) of the electric (magnetic) field; $\text{Re}(ka) < \text{Re}(k_0a)$ when an inclusion is in the antinode (knot) of the electric (magnetic) field; $\text{Re}(ka) = \text{Re}(k_0a)$ and here the amplitudes of the electric and magnetic fields may be different. Let us note that in the case of dielectric inclusions $\text{Re}(ka)$ is not greater than $\text{Re}(k_0a)$. $\text{Re}(ka)$ is a periodic function of $\hat{s}$ for $\hat{d} = 0.02$ whose period is approximately equal to the resonance wavelength of the empty resonator.

The Q-factors weakly oscillate for small inclusions. With increasing diameter of the pivot they are no longer periodic functions of $\hat{d}$ and sharply fall when an inclusion goes out of the resonator's centre (because a violation of symmetry may cause intensive scattering of resonator's eigenfield on the inclusion).

Spectral properties of open resonators with layered, anisotropic and plasma inclusions are studied in detail by Brovenko (1988) and Brovenko *et al.* (1993). Creation of a mathematical model is especially difficult for resonators with plasma layers diffused directly on metal mirrors when one should essentially modify the method of solution to the boundary-value problems (the method of the Riemann-Hilbert problem).

Chapter 3

Eigenwaves in open waveguides

There has been increasing interest during the last decade in the processes of electromagnetic wave propagation in complex guiding systems with inhomogeneous filling. Although different types of them have been widely used in various practical applications for more than 30 years and a lot of their physical properties have been established, interest in developing rigorous mathematical technique in this field of electrodynamics has appeared only recently. Modern guiding systems such as microstrip transmission lines and other types of open waveguides with non-compact boundaries produce new problems to be solved and require working out special methods for their studies.

Typical here are nonselfadjoint boundary eigenvalue problems for the systems of Helmholtz equations with piecewise-constant coefficients and transmission conditions where the spectral parameter enters in a nonlinear way. The methods developed in this Chapter for solving such problems are based on the spectral theory of operator-valued functions and operator pencils.

We proceed from the model assumptions that propagation and oscillations of electromagnetic fields are described in a magnetically homogeneous and isotropic medium by the system of Maxwell's equations

$$\begin{cases} \text{curl } \mathbf{E} = -j\omega\mu\mathbf{H} \\ \text{curl } \mathbf{H} = j\omega\epsilon\mathbf{E} \end{cases} \tag{3.1}$$

where $\mathbf{E}$, $\mathbf{H}(\mathbf{r}; x_3)$, $\mathbf{r} = (x_1, x_2)$, $\omega = $ const is the frequency of electromagnetic

171

field, $\mu = $ const (and here we take $\mu = 1$), $\epsilon = \epsilon(\mathbf{r})$ is a dielectric permeability of the medium.

Considering cylindrical resonators and waveguides (adding to the system of Maxwell equations the boundary-value conditions for $\mathbf{E}$ and $\mathbf{H}$ on perfectly conducting surfaces, on the boundaries between different media and the conditions at infinity in open domains) and looking for solutions of eqn. 3.1 in the form

$$\mathbf{E}, \mathbf{H} = \mathbf{E}, \mathbf{H}(\mathbf{r}) \ \exp(j\gamma x_3)$$

we get the problem of normal waves propagating along a cylindrical guiding surface and corresponding to certain complex values of the longitudinal wavenumber γ, which is regarded as a spectral parameter. For $\gamma = 0$ we have the specific problem of determining the solutions of eqn. 3.1 which do not depend on x_3 and correspond to certain complex values of the spectral parameter ω, the fundamental frequencies of electromagnetic field.

A great number of papers and works are devoted to the investigation of fundamental frequencies and normal waves in different types of resonators and waveguides [the detailed list of references is presented, for example, by Ilyinsky and Shestopalov (1989)] developing physical principles and calculating resonance properties such as scattered and accumulated power, far field patterns etc. We shall not review this literature here because it is a very specialised and complicated matter and we only turn readers' attention to several recent works based on the methods of operator-valued functions and considering nonlinear eigenvalue problems in electrodynamics.

Let us note that most of the results that have been obtained until recently, are related to shielded waveguides (transmission lines), open waveguides with compact boundaries or with noncompact boundaries having a given (coordinate) shape. In this Chapter we will study rather general families of both shielded waveguides and structures with noncompact boundaries of the cross-sections which have not been treated before, thus generalising the classical results of Tikhonov and Samarsky (1977) (see also Ilyinsky *et al.*, 1993) concerning the existence of normal waves in empty tubes.

In fact, among the main features of the boundary eigenvalue problems arising in the theory of guided waves in nonhomogeneous medium is the following: it is rather difficult to find eigenvalues using standard approaches of the spectral theory of differential operators due to the complicated character of the boundary "transmission" operators. In this way one has to refer to the recent studies on the method of boundary integral equations with strongly elliptic Fredholm operators for different types of boundary-value problems for the Helmholtz equation in Sobolev spaces carried out by von Petersdorff (1989) and some other works [see the references presented by von Petersdorff (1989)]. We will solve such problems in Sections 3.1 and 3.2 using the elements of this technique and the method of the integral operator-valued functions developed here for cylindrical resonators.

It will also be shown that the method of abstract polynomial operator pencils can be effectively and naturally used for solving the problems of normal waves. In particular, the known (Adams, 1975) connections between the spectral properties of operator pencils and certain non-self-adjoint operators which give an opportunity to use the fundamental results of operator theory in spaces with indefinite metric. The theory of operator pencils is sufficiently well elaborated but these results have found applications in the mathematical models of the electrodynamics of guidance systems only several years ago, beginning with the fundamental works of Smirnov (1987, 1990, 1991).

We will use this approach in Section 3.3 to consider the problem of wave propagation in arbitrary shielded waveguides in the generalised statement. We will construct the Fredholm operator pencils of the problems and prove the discreteness and existence of the normal waves spectra together with the completeness of the system of normal waves. One has to mention that it is not generally possible to obtain the same results using the methods of potential theory and boundary integral equations developed in Chapter 1, Sections 3.1 and 3.2.

The results of Section 3.3 generalise the classical assertions of Samarsky and Tikhonov (1947) concerning the completeness of normal waves in empty waveguides (tubes).

The constructive method described in this Chapter could be useful for both the determination of the general properties of the spectrum of normal waves, such as existence and discreteness, and for direct calculations of eigenvalues (complex longitudinal wavenumbers). The problem of fundamental frequencies will be naturally used as the first step towards the study of normal waves and simultaneously as a part of the more complicated two-parameter eigenvalue problem.

3.1 Reduction to the boundary eigenvalue problems for the system of Helmholtz equation

As was mentioned in the introduction to this Chapter, the problem of fundamental frequencies is a particular case of the problem of normal waves considered as a two-parameter eigenvalue problem. We will show here that the connection between these problems can be obtained from the formal investigation of their differential and integral operators.

First let us state the problem of normal waves in an open waveguide (a slot line) whose transversal cross-section is formed by an arbitrary domain either from the family BD_c or NB_c. It is convenient to put

$$u(\mathbf{r}) = gH_{x_3}(\mathbf{r}), \quad v(\mathbf{r}) = gE_{x_3}(\mathbf{r})$$

where

$$g = g(\mathbf{r}) = \lambda[\epsilon(\mathbf{r}) - \eta], \quad \eta = (\gamma/k)^2$$

and g is the squared value of the transversal wavenumber. It is easy to see that all other field components are expressed by means of u, v from the system of Maxwell's equations assuming that $\partial/\partial x_3 = j\gamma$. For these functions one gets, following Ilyinsky and Shestopalov (1989), the boundary eigenvalue problem

$$\Delta u + gu = 0 \qquad \Delta v + gv = 0, \quad \mathbf{r} \in \Omega \setminus \Gamma_\epsilon \tag{3.2}$$

$$v = 0, \quad \frac{\partial u}{\partial \mathbf{n}} = 0, \quad x \in \Gamma_S = \Gamma \cup S \cup S_a \tag{3.3}$$

$$[u] = 0, \qquad [v] = 0, \quad \mathbf{r} \in \Gamma_\epsilon \tag{3.4}$$

$$\gamma\left[\frac{\partial v}{\partial \tau}\right] - \left[\frac{\partial u}{\partial \mathbf{n}}\right] = 0, \quad \gamma\left[\frac{\partial u}{\partial \tau}\right] + \left[g\frac{\partial v}{\partial \mathbf{n}}\right] = 0, \quad \mathbf{r} \in \Gamma_\epsilon \tag{3.5}$$

with the edge conditions given by eqn. 1.4 [the edges coincide with the (irregular) points of Γ_S and Γ_ϵ where the (unit) normal and tangential vectors $\mathbf{n}$ and τ do not exist; in particular, all the endpoints of S and S_a are the edges] and the radiation conditions at infinity

$$
\begin{aligned}
v(\mathbf{r}) &= \sum_{m=1}^{\infty} a_m^1 H_m^{(1)}(\hat{k}r)\sin m\phi \\
u(\mathbf{r}) &= \sum_{m=0}^{\infty} a_m^2 H_m^{(1)}(\hat{k}r)\cos m\phi
\end{aligned}
\tag{3.6}
$$

where $\mathbf{r} = (r, \phi)$ and the series admit double term-wise differentiation with respect to $\hat{k}r \geq R_0$ and $\phi \in (0, \pi)$; Γ_ϵ denotes, in accordance with the notation introduced in Section 1.1, the gap lines of the piece-wise constant function $\epsilon(\mathbf{r})$ and it is assumed that $\mu(\mathbf{r}) = \text{const} = 1$. In the case of shielded slot lines when the cross-section is formed by a (bounded) domain from the family BD_c, Γ is a closed contour and Γ_S is a (multi-conneceted) compact contour, the radiation conditions are ignored.

When the cross-sections are formed by the domains from the families NB_p, NB_b^r, NB_u^r and BD_u^r [i.e. when Γ_ϵ consists of a finite number of rectilinear intervals ("slots") L] the transmission conditions given by eqn. 3.5 take the form

$$\gamma \left[\frac{\partial v}{\partial x_1} \right] - \left[\frac{\partial u}{\partial x_2} \right] = 0, \quad \gamma \left[\frac{\partial u}{\partial x_1} \right] + \left[g \frac{\partial v}{\partial x_2} \right] = 0, \quad \mathbf{r} \in \Gamma_\epsilon \qquad (3.5a)$$

We will call the problem given by eqns. 3.2-3.6 (for the domains from the families NB_c) or by eqns. 3.2-3.5 (for the domains from the families BD_c) the problem V of normal waves considered with respect to the spectral parameters g, γ or η where γ is the longitudinal wavenumber of the normal wave. The symbol σ_V will denote the (point) spectrum of the problem V assuming that σ_V is in the complex γ- or η-plane. The resolvent set ρ_V complements σ_V to the whole complex plane (or the complex manifold) where the problem V is considered.

Let us establish the connection between the problems on fundamental frequencies and normal waves. If to put $\gamma = 0$ in eqns. 3.2-3.6, one gets two independent boundary eigenvalue problems, H and E, respectively, for the functions u and v and they will be the problems on the H-type and E-type fundamental frequencies of a cylindrical slot resonator (with respect to the spectral parameter λ) treated in Chapters 1 and 2.

Let us consider, for the sake of simplicity, only the case of shielded slot lines. Then, if, for example $\lambda = \lambda_0$ is the eigenvalue of the problem H given by eqns. 1.1-1.4, we may formally put

$$\lambda = h(\gamma), \quad h(0) = \lambda_0$$

and take γ in the problem V as a parameter. The differential operator of the problem V is continuous with respect to γ in the neighbourhood of the origin and the function $\lambda = h(\gamma)$ is continuous too and may be calculated, at least for sufficiently small values of γ.

Eigenvalues of the problem V can be treated as the inverse functions of λ: $\eta = h^{-1}(\lambda)$ (if they exist) and may be calculated for the small values of λ (or for γ).

On the other hand, if we want to know whether $\gamma = 0$ (or $\eta = 0$) is the element of spectrum of the problem V, we may formally put $\gamma = 0$ in eqns. 3.2-3.6, thus obtaining two independent eigenvalue problems with respect to the spectral parameter λ. Let us consider two cases:

(i) if the fixed value $\lambda = \lambda_0$ is not an eigenvalue of at least one of these problems then $0 \in \rho_V$ together with a neighbourhood of the origin Ω_0 of the complex η-plane;

(ii) if, e.g. λ_0 is a H-type fundamental frequency, then $\gamma = 0$ is an element of the spectrum of the problem V (i.e. for this particular value of λ $\eta_0 = \eta(\lambda_0) = 0$), hence $\lambda_0 = \lambda(0) \in \sigma_H$, $\lambda = \lambda(\eta) \in \sigma_V$ and $\lambda(\eta)$ is continuous with respect to η for sufficiently small values of $|\eta|$.

In the first case we have $\Omega_0 \subset \rho_V \neq \emptyset$. In the second case we obtained the proof of the existence of a non-empty subset of the spectrum of the problem V, but certainly $\lambda_0 + \Delta\lambda \notin \sigma_H$ for sufficiently small values of $\Delta\lambda$. Therefore, there exist regular points of the problem V in the vicinity of the origin of the complex plane C_η.

Thus we have proved the following assertion:

Assertion: There always exists a deleted neighbourhood of the origin in the complex η-plane which belongs to the resolvent set of the problem V.

The existence of eigenvalues $\lambda(0)$ has been proved in Sections 1.8 and 1.9 for particular cases of shielded and open slot resonators. To show the existence of normal waves and, consequently, the existence of inverse functions $\eta = h^{-1}(\lambda)$ one can study the integral operator-valued function $K_V(z)$, $z = (\lambda, \eta)$ of the problem V, which will be constructed in the following section. An operator equation determining implicit functions $h(\eta)$ or $h^{-1}(\lambda)$ with initial values $\lambda(0) = \lambda_0$ may be obtained with the help of K_V.

3.2 Fredholm operator-valued functions and the discreteness of the eigenwave spectra

Let us consider waveguides whose cross-sections $\Omega = \Omega^1 \cup \Omega^2$ are formed by the domains from the families NB_p, NB_b^r, NB_u^r (open slot lines) or BD_b^r, BD_u^r (shielded slot lines) where $\Omega^2 = \Omega_1^2$ is assumed to be a (one) simply-connected bounded or unbounded $S_\Pi(L)$-domain (i.e. in accordance with the notation of Section 1.1, $N_2 = 1$ for the families NB_p, NB_b^r, BD_b^r, BD_u^r); $L = \Gamma_\epsilon$ denotes the "slots" (coinciding in general case with rectilinear intervals $\Gamma_S^I = \cup_{i=1}^{M_S} \Gamma_{Si}^I$, $\Gamma_{Si}^I \subset \{\mathbf{r} : x_2 = \text{const}\}$, $i = 1, 2, \ldots, M_S$). As in Section 1.7, we will limit ourselves to the case $M_S = 1$, $L = (-a, a) \subset \mathbf{r} : \{x_2 = 0\}$ corresponding to the families NB_p BD_b^r, BD_u^r assuming that the results may be generalised for an arbitrary $S_\Pi^c(L)$-domain and natural M_S (for the families NB_b^r, NB_u^r, BD_u^r and BD_b^r).

In the case considered the operator-valued function $K_V(\mathbf{z})$ may be constructed with the help of the methods developed in Sections 1.5 and 1.6, similar to the one of problem H (see Section 1.7). Let us represent the solution's components u and v of the problem given by eqns. 3.2-3.5 in each of the domains Ω^m as the generalised single-layer and double-layer Green's potentials given by eqns. 1.24 and 1.25, using the results of theorems 1.4 and 1.5. The kernels are determined with the help of Green's functions $G_m^i = G_m^i(\mathbf{r}, \mathbf{q}; g)$ of the ith boundary-value problem of the Helmholtz equation 3.2 of the domain Ω^m $(m = 1, 2)$ and the densities are given by eqn. 1.47. All the results and assumptions concerning analytical continuations of Green's functions formulated in Section 1.7 remain valid if to replace the complex parameter λ by g.

Applying the transmission conditions given by eqns. 3.4 and 3.5 to the integral representations given by eqns. 1.24 and 1.25 and using the theorems about the traces of the derivatives of the potentials (see Section 1.5) we finally obtain the system of two boundary integral equations which is similar to eqns. 1.48 and 1.49

$$K_V(\mathbf{z})\phi \equiv \int_L \mathcal{K}_V(t.s; \mathbf{z})\phi(s)\,\mathrm{d}s = \mathbf{o}, \quad t \in L \qquad (3.7)$$

$$\mathcal{K}_V(t,s;\mathbf{z}) = \sum_{m=1}^{2} \begin{pmatrix} (\dfrac{\epsilon_m}{\eta}-1)\overset{0}{G_m^2} & \overset{0}{G_m^2} \\ \overset{0}{G_m^2} & \dfrac{\eta}{\epsilon_m-\eta}\overset{0}{G_m^1} + \dfrac{\epsilon_m}{\epsilon_m-\eta}\overset{0}{F_m} \end{pmatrix}$$

where

$$\overset{0}{G_m^2} = G_m^2(\mathbf{r}^0,\mathbf{q}^0;g), \quad \mathbf{r}^0 = (t,0), \quad \mathbf{q}^0 = (s,0)$$

$$\overset{0}{F_m} = F_m(t,s;\mathbf{z}) = \int_{c_1}^{t} \frac{\partial}{\partial x_2}\int_{c_2}^{s}\frac{\partial}{\partial y_2}G_m^1(\mathbf{r}_1^0,\mathbf{q}_1^0;g)|_{x_2=y_2=0}\,\mathrm{d}s_1\,\mathrm{d}t_1$$

$$\mathbf{r}_1^0 = (t_1,x_2), \quad \mathbf{q}_1^0 = (s_1,y_2), \quad \phi(s) = (\phi_1(s),\phi_2(s))^T$$

Eqn. 3.7 is the problem of characteristic numbers for the matrix operator-valued function $\mathcal{K}_V(\mathbf{z})$ of the problem V of normal waves. We shall consider $\mathcal{K}_V(\mathbf{z})$ on the pair of weighted Hölder-type Banach spaces $\mathbf{X},\mathbf{Y}$ of the two-component vector-functions ϕ. This problem is equivalent to the problem V if

$$\eta \notin E = \cup_{i,m=1}^{2} E_m^i, \quad E_m^i = \{\eta_{nm}^i = \epsilon_m - \mu_{nm}^i\}_{n=0}^{\infty}$$

where $\{\mu_{nm}^i\}_{n=0}^{\infty}$ are the eigenvalues of the first $(i=1)$ or the second $(i=2)$ boundary-value problem for Laplace's equation in the domains Ω^m (for shielded slot lines) or of the domain Ω^2 (in the case corresponding to the family NB_p). E is the set of poles of the finite-meromorphic operator-valued function $\mathcal{K}_V$ considered with respect to η which coincides with the set of poles of Green's functions.

Let us note that in the case corresponding to the family NB_p

$$\overset{0}{G_1^2} = G_1^2(\mathbf{r}^0,\mathbf{q}^0;g) = \frac{j}{4}H_0^{(1)}(\sqrt{g}|t-s|)$$

and, consequently, $\mathcal{K}_V$ is considered as the operator-valued function of g admits analytical continuation to the infinite-sheet Riemann surface $\mathcal{D}^{(0)} = \Lambda$ (of the function $\ln g$) with one logarithmic branch point $g=0$ [or to the Riemann surface $\mathcal{D}_\eta^{(0)}$ of the function $\ln \lambda(\epsilon_m - \eta)$, $\lambda \neq 0$, with two logarithmic branch points $\eta = \eta_m = \epsilon_m$ if to consider $\mathcal{K}_V$ as the operator-valued function of η] except for the real poles forming the set E. The problem of characteristic numbers given by eqn. 3.7 may be considered either on Λ or on $\mathcal{D}_\eta^{(0)}$.

Let us take the canonical structure for the families BD_u^r (BD_b^r), the rectangular shielded slot line (described in Section 1.1), whose cross-section is formed by two rectangular domains $\Omega^m = \Pi_{ab_m}$ ($m=1,2$) with the slot

$L = \{\mathbf{r} : x_2 = 0, \frac{a}{2} - w = d_1 < x_1 < d_2 = \frac{a}{2} + w\}$. Here one may represent the kernel of the operator-valued function in the following explicit form:

$$\mathcal{K}_V(t, s; \mathbf{z}) = \sum_{m=1}^{2} \begin{pmatrix} (\frac{\epsilon_m}{\eta} - 1)\overset{0}{G}_m & \overset{0}{G}_m \\ \overset{0}{G}_m & \epsilon_m R_m - \overset{0}{G}_m \end{pmatrix}$$

where

$$\overset{0}{G}_m = \overset{0}{l}(t, s) - \frac{2}{a} \sum_{n=0}^{\infty} (\alpha_n^m - \frac{a}{\pi n}) \psi_n(t) \psi_n(s)$$

$$R_m = R_m(t, s; \mathbf{z}) = -\frac{2a}{\pi^2} \sum_{n=1}^{\infty} \frac{\alpha_n^m}{n^2} \psi_n(t) \psi_n(s)$$

$$\overset{0}{l}(t, s) = -\frac{2}{\pi} \sum_{n=1}^{\infty} \frac{\psi_n(t) \psi_n(s)}{n}$$

$$= \frac{1}{\pi} \left[\ln \left| 2 \sin \frac{\pi|t - s|}{2a} \right| + \ln \left| 2 \sin \frac{\pi|t + s|}{2a} \right| \right]$$

$$\equiv \frac{1}{\pi} \ln \frac{1}{|t - s|} + l_1(t, s)$$

$$\alpha_n^m = \frac{\coth b_m \sqrt{\eta - (\epsilon_m - r_n)}}{\sqrt{\eta - (\epsilon_m - r_n)}}$$

$$r_n = n^2 r_1 = \frac{\pi^2 n^2}{a^2}$$

the series for $\overset{0}{G}_m$ and R_m converge absolutely and uniformly with respect to $s, t \in [0, a]$ and $l_1(t, s)$ is once continuously differentiable with respect to its variables.

As in problem H we may show that $K_V(\mathbf{z})$ admits separation of a singular (invertible) part and pole pencils at every simple pole η_{nm}^i:

$$K_V(\mathbf{z}) = G(\mathbf{z})\Upsilon(w) + N(\mathbf{z}, w)$$

$$\equiv G(\mathbf{z})l(w) + \frac{A^i_{-nm}}{\eta - \eta_{nm}^i} + U_{nm}^i(\mathbf{z}, w)$$

$$\equiv P_{nm}^i(\mathbf{z}, w) + U_{nm}^i(\mathbf{z}, w) \tag{3.8}$$

where

$$G(\mathbf{z}) = \begin{pmatrix} \dfrac{\epsilon_1 + \epsilon_2}{\eta} - 2 & 2 \\ 2 & -2 \end{pmatrix} \qquad \Upsilon(w)\phi = \frac{1}{\pi} \int\limits_L \ln \frac{1}{|t-s|} \phi(s)\,\mathrm{d}s \qquad (3.9)$$

$T(\mathbf{z}, w) = G(\mathbf{z})\Upsilon(w)$ is invertible and $N(\mathbf{z}, w)$ is completely continuous on the pair of Banach spaces $\mathbf{X}, \mathbf{Y}$; A^i_{-nm} is a two-dimensional operator of the pole pencil P^i_{nm} corresponding to the simple pole η^i_{nm}, $n = 0, 1, 2 \ldots$, $i, m = 1, 2$.

The above representations allow us to prove that $K_V(\mathbf{z})$ is a Fredholm finite-meromorphic operator-valued function (as a sum of the invertible operator defined by eqns. 3.8 and 3.9 and completely continuous operator defined by eqn. 3.8). Under the assumption of the absence of the point spectrum in a deleted vicinity of the origin in the complex η-plane (which is proved by the "continous perturbation" method described in Section 3.1) we can prove that the problem V for the rectangular shielded slot line has not more than a finite number of the elements of the point spectrum in every finite circle $\{\eta | |\eta| < r\}$.

To prove the existence of the point spectrum (and the existence of real eigenvalues) we can use similarly to the problem H the generalised theorem about the pole pencils and show the required result at least for sufficiently small values of the width of the slot. However, the techniques of this proof for the integral operator-valued function $K_V(\mathbf{z})$ is more complicated and bulky.

Similar considerations carried out for slot lines whose cross-sections are formed by arbitrary domains from the families NB_p, BD^r_u and BD^r_b (with the use of the assumptions concerning the properties of the analytical continuations of Green's functions presented in Section 1.7) enables one to reduce the problem V to determination of the characteristic numbers of a Fredholm finite-meromorphic operator-valued function admitting as in eqn. 3.8 the separation of pole pencils and thus to prove the discreteness of the normal waves spectrum.

We will not proceed here with this analysis since the discreteness and existence of the normal waves spectrum will be proved in the following Section for arbitrary waveguides (corresponding to the family BD_c) on the basis of a different technique.

3.3 Existence of the discrete spectrum and completeness of the system of normal waves

In this Section we will consider the problem of normal waves in shielded waveguides whose transversal cross-sections are formed by the domains of the family BD_c. In order to simplify the analysis we will use a different notation for the domains of this family. Assume that the bounded domains $Q, S_1, S_2 \subset \mathbb{R}^2$ forming waveguide's cross-section satisfy the following properties: $\overline{S}_1 \subset Q$, $S_2 = Q \setminus \overline{S}_1$ and the boundaries $\partial S_{1,2} \in C^\infty$. Let $P_i \in \partial S_1$ be the arbitrary $2N$ points dividing ∂S_1 into the parts Γ and Γ' such that $\Gamma = \partial S_1 \setminus \overline{\Gamma}'$, $\Gamma' = \partial S_1 \setminus \overline{\Gamma}$, $\Gamma \cup \Gamma' \cup_i P_i = \partial S_1$ (if $N = 0$ then $\Gamma = \partial S_1$, $\Gamma' = \emptyset$). We will also assume that

$$S' = S_1 \cup S_2, \quad S = S' \cup \Gamma, \quad \Gamma_0 = \partial Q \cup \Gamma' \tag{3.10}$$

The domain S described in eqn. 3.10 satisfies the cone property (Adams, 1975) which allows us to apply the embedding and trace theorems in the Sobolev spaces.

We will assume also that waveguide's filling contains two isotropic media with the dielectric permeabilities $\epsilon(\mathbf{r}) = \epsilon_i$, $\mathbf{r} \in S_i$ $(i = 1, 2)$ and the gap lines coincide with Γ.

The problem of normal waves has been reduced in Section 3.1 to the boundary eigenvalue problem V for the system of Helmholtz equations. Following Smirnov (1987) and Smirnov and Shestopalov (1990) we will consider this problem with respect to the (complex) spectral parameter γ coinciding with the longitudinal wavenumber of a normal wave and rewrite its statement in the form more convenient for further analysis

$$\Psi = \Psi(\mathbf{r}) = H_{x_3}(\mathbf{r}), \quad \Pi = \Pi(\mathbf{r}) = E_{x_3}(\mathbf{r}) \tag{3.11}$$

as the sought for (potential) functions:

$$\Delta\Pi + \tilde{k}^2\Pi = 0$$
$$\Delta\Psi + \tilde{k}^2\Psi = 0 \tag{3.12}$$
$$\mathbf{r} \in S'$$

$$\Pi, \Psi \in C^2(S') \cap C^1(\overline{S_1 \setminus \Gamma_\delta}) \cap C^1(\overline{S_2 \setminus \Gamma_\delta}) \tag{3.13}$$

where Γ_δ denote δ-vicinities of the points P_i ("edges")

$$\tilde{k}^2 = \tilde{k}_i^2 = k^2 \epsilon_i - \gamma^2 \quad (i = 1, 2) \tag{3.14}$$

with the boundary-value and transmission conditions

$$\Pi|_{\Gamma_0} = 0 \tag{3.15}$$

$$\left.\frac{\partial \Psi}{\partial \mathbf{n}}\right|_{\Gamma_0} = 0 \tag{3.16}$$

$$[\Pi]|_\Gamma = [\Psi]|_\Gamma = 0 \tag{3.17}$$

$$\begin{aligned}
\gamma \left[\frac{1}{\tilde{k}^2}\frac{\partial \Psi}{\partial \tau}\right]\Bigg|_\Gamma + k \left[\frac{\epsilon}{\tilde{k}^2}\frac{\partial \Pi}{\partial \mathbf{n}}\right]\Bigg|_\Gamma = 0 \\
\gamma \left[\frac{1}{\tilde{k}^2}\frac{\partial \Pi}{\partial \tau}\right]\Bigg|_\Gamma - k \left[\frac{1}{\tilde{k}^2}\frac{\partial \Psi}{\partial \mathbf{n}}\right]\Bigg|_\Gamma = 0
\end{aligned} \tag{3.18}$$

where $[\phi]$ denotes the break of the limiting values of the function ϕ on a corresponding curve, eqn. 3.16 holds on each side of Γ', $\mathbf{n}$ and τ denote the (exterior) normal and tangential unit vectors; the solution satisfies the energy ("edge") condition

$$\int\limits_S (|\nabla\Pi|^2 + |\nabla\Psi|^2 + |\Pi|^2 + |\Psi|^2)\, \mathrm{d}s < \infty \tag{3.19}$$

The problem given by eqns. 3.11–3.19 considered in terms of the potential functions Π and Ψ is equivalent to the problem V given by eqns. 3.2–3.5 (see Section 3.1) and we will not distinguish between them in what follows.

We will consider the problem V in the generalised statement and give its variational formulation. With this purpose let us introduce following Adams (1975) the Sobolev spaces

$$H_0^1(S) = \{f : f \in H^1(S),\ f|_{\Gamma_0} = 0\}$$
$$\hat{H}^1(S) = \{f : f \in H^1(S),\ \int_S f\, \mathrm{d}s = 0\}$$

with the inner product and norm

$$(f, g) = \int\limits_S \nabla f \nabla \bar{g}\, \mathrm{d}s, \quad \|f\|^2 = (f, f)$$

and apply Green's formula in the domains S_i $(i = 1, 2)$ multiplying eqns. 3.12 by arbitrary functions

$$\bar{u} \in H_0^1(S), \quad \bar{v} \in \hat{H}^1(S)$$
$$\bar{u}, \bar{v} \in C^1(\overline{S}_1) \cap C^1(\overline{S}_2)$$

and taking into account eqns. 3.15 and 3.16:

$$\int_{S_i} \nabla \Pi \nabla \bar{u} \, ds - \tilde{k}_i^2 \int_{S_i} \Pi \bar{u} \, ds = (-1)^i \int_\Gamma \left. \frac{\partial \Pi_i}{\partial \mathbf{n}} \right|_\Gamma \bar{u} \, d\tau \quad (i = 1, 2) \qquad (3.20)$$

$$\int_{S_i} \nabla \Psi \nabla \bar{v} \, ds - \tilde{k}_i^2 \int_{S_i} \Psi \bar{v} \, ds = (-1)^i \int_\Gamma \left. \frac{\partial \Psi_i}{\partial \mathbf{n}} \right|_\Gamma \bar{v} \, d\tau \quad (i = 1, 2) \qquad (3.21)$$

where Π_i and Ψ_i are the restrictions of Π and Ψ on S_i $(i = 1, 2)$. Substituting the normal derivatives from eqns. 3.20 and 3.21 to eqn. 3.18 we obtain the variational expression

$$\int_{S_i} \frac{1}{\tilde{k}^2} (\epsilon \nabla \Pi \nabla \bar{u} + \nabla \Psi \nabla \bar{v}) \, ds - \int_{S_i} (\epsilon \Pi \bar{u} + \Psi \bar{v}) \, ds$$
$$- \frac{\gamma}{k} \left[\frac{1}{\tilde{k}} \right] \int_\Gamma \left(\frac{\partial \Pi}{\partial \tau} \bar{v} - \frac{\partial \Psi}{\partial \tau} \bar{u} \right) \, d\tau = 0 \qquad (3.22)$$
$$\forall u \in H_0^1(S) \qquad \forall v \in \hat{H}^1(S)$$

which is derived for the smooth functions $\Pi, \Psi (u, v)$. We will prove the continuity of the quadratic forms defined by the integrals in eqn. 3.22. Hence eqn. 3.22 can be extended to arbitrary functions $\Pi \in H_0^1(S)$, $\Psi \in \hat{H}^1(S)$ $[u \in H_0^1(S), v \in \hat{H}^1(S)]$ For $v = 1$, $u = 0$ we get in a similar way

$$\int_S \Psi \, ds = - \int_\Gamma \left[\frac{1}{\tilde{k}} \frac{\partial \Psi}{\partial \mathbf{n}} \right] \Big|_\Gamma d\tau = -\frac{\gamma}{k} \left[\frac{1}{\tilde{k}} \right] \int_\Gamma \left. \frac{\partial \Pi}{\partial \tau} \right|_\Gamma d\tau = 0 \qquad (3.23)$$

The conditions given by eqns. 3.15, 3.17 and 3.19 yield $\Pi \in H_0^1(S)$, $\Psi \in \hat{H}_0^1(S)$. Thus we have proved that if the conditions of the problem V are fulfilled and

$$\Pi, \Psi \in C^2(S') \cap C^1(\overline{S_1 \setminus \Gamma_\delta}) \cap C^1(\overline{S_1 \setminus \Gamma_\delta})$$

then $\Pi \in H_0^1(S)$, $\Psi \in \hat{H}^1(S)$ and the variational expression given by eqn. 3.22 holds for arbitrary functions $u \in H_0^1(S)$, $v \in \hat{H}^1(S)$.

The inverse assertion is true. In fact, it follows from eqn.3.22 that eqns. 3.11–3.18 are fulfilled in terms of distributions (Lions and Magenes, 1968). Then using the *a priori* estimates technique (Ladyzhenskaya and Uraltzeva, 1964) one can show that

$$\Pi, \Psi \in C^2(S_1 \cup S_2) \cap C^1(\overline{S_1 \setminus \Gamma_\delta}) \cap C^1(\overline{S_1 \setminus \Gamma_\delta})$$

and eqns. 3.11–3.18 are satisfied in the "classical" sense.

Let us give the following:

Definition 3.1: The pair of functions

$$\Pi \in H_0^1(S), \quad \Psi \in \hat{H}^1(S), \quad (\|\Pi\| + \|\Psi\| \neq 0)$$

is called the eigenvector of the problem V corresponding to the eigenvalue γ_0 if the variational expression given by eqn. 3.22 holds for $\gamma = \gamma_0$ and arbitrary functions $u \in H_0^1(S)$, $v \in \hat{H}^1(S)$.

Multiplying eqn. 3.22 by $\tilde{k}_1^2 \tilde{k}_2^2$ we will rewrite it in the form

$$
\gamma^4 \int_S (\epsilon \Pi \bar{u} + \Psi \bar{v}) \, ds + \gamma^2 \left[\int_S (\epsilon \nabla \Pi \nabla \bar{u} + \nabla \Psi \nabla \bar{v}) \, ds \right.
$$
$$
\left. -(\epsilon_1 + \epsilon_2) k^2 \int_S (\epsilon \Pi \bar{u} + \Psi \bar{v}) \, ds \right] + (\epsilon_1 - \epsilon_2) k \gamma \int_S \left(\frac{\partial \Pi}{\partial \tau} \bar{v} - \frac{\partial \Psi}{\partial \tau} \bar{u} \right) \, d\tau \tag{3.24}
$$
$$
+ \epsilon_1 \epsilon_2 \left[k^4 \int_S (\epsilon \Pi \bar{u} + \Psi \bar{v}) \, ds - k^2 \int_S \left(\epsilon \nabla \Pi \nabla \bar{u} + \frac{1}{\epsilon} \nabla \Psi \nabla \bar{v} \right) \, ds \right] = 0
$$
$$
\forall u \in H_0^1(S), \quad \forall v \in \hat{H}^1(S)
$$

Let $H = H_0^1(S) \times \hat{H}^1(S)$ be the Cartesian product of the Hilbert spaces with the inner product and the norm

$$(\mathbf{f}, \mathbf{g}) = (f_1, g_1) + (f_2, g_2), \quad \|\mathbf{f}\|^2 = \|f_1\|^2 + \|f_2\|^2$$

where

$$\mathbf{f}, \mathbf{g} \in H, \quad \mathbf{f} = (f_1, f_2)^T, \quad \mathbf{g} = (g_1, g_2)^T$$
$$f_1, g_1 \in H_0^1(S), \quad f_2, g_2 \in \hat{H}^1(S)$$

Then the integrals in eqn. 3.24 can be considered as the quadratic forms on C, defined in H with respect to vector-functions

$$\mathbf{f} = (\Pi, \Psi)^T, \quad \mathbf{g} = (u, v)^T$$

These forms (if they are bounded) define, in accordance with the results of Lions and Magenes (1968), linear bounded operators $T : H \to H$:

$$t(\mathbf{f}, \mathbf{g}) = (T\mathbf{f}, \mathbf{g}) \quad \forall, \quad \mathbf{g} \in H$$

Let us consider the following quadratic forms and corresponding operators

$$a_1(\mathbf{f}, \mathbf{g}) \equiv \int_S (\epsilon \nabla f_1 \nabla \bar{g}_1 + \nabla f_2 \nabla \bar{g}_2) \, ds = (A_1 \mathbf{f}, \mathbf{g}), \quad \forall \mathbf{g} \in H$$

$$a_2(\mathbf{f}, \mathbf{g}) \equiv \int_S \left(\nabla f_1 \nabla \bar{g}_1 + \frac{1}{\epsilon} \nabla f_2 \nabla \bar{g}_2 \right) \, ds = (A_2 \mathbf{f}, \mathbf{g}), \quad \forall \mathbf{g} \in H$$

$$\tag{3.25}$$

$$c(\mathbf{f}, \mathbf{g}) \equiv \int_S (\epsilon_1 f_1 \bar{g}_1 + f_2 \bar{g}_2) \, ds = (K \mathbf{f}, \mathbf{g}), \quad \forall \mathbf{g} \in H$$

$$s(\mathbf{f}, \mathbf{g}) \equiv \int_\Gamma \left(\frac{\partial f_1}{\partial \tau} \bar{g}_1 + \frac{\partial f_2}{\partial \tau} \bar{g}_2 \right) \, dl = (\Omega \mathbf{f}, \mathbf{g}), \quad \forall \mathbf{g} \in H$$

It is easy to see that the forms $a_1(\mathbf{f}, \mathbf{g})$ and $a_2(\mathbf{f}, \mathbf{g})$ are bounded. The same property for the form $c(\mathbf{f}, \mathbf{g})$ follows from Poincaré's inequality (Adams, 1975).

Let us prove that the form $s(\mathbf{f}, \mathbf{g})$ is bounded too. Assume that functions $f_1, f_2, g_1, g_2 \in C^1(\overline{S}_1) \cap C^1(\overline{S}_2)$. Then

$$\int_\Gamma \left(\frac{\partial f_1}{\partial \tau} \bar{g}_2 - \frac{\partial f_2}{\partial \tau} \bar{g}_1 \right) \, dl = \frac{1}{2} \int_S \xi \left(\frac{\partial f_1}{\partial x_1} \frac{\partial \bar{g}_2}{\partial x_1} - \frac{\partial f_1}{\partial x_1} \frac{\partial \bar{g}_2}{\partial x_1} \right.$$

$$\left. + \frac{\partial f_2}{\partial x_1} \frac{\partial \bar{g}_1}{\partial x_1} - \frac{\partial f_1}{\partial x_1} \frac{\partial \bar{g}_1}{\partial x_1} \right) \, ds$$

where

$$\xi = \begin{cases} 1 & \text{in } S_1 \\ -1 & \text{in } S_2 \end{cases}$$

and using the Schwartz inequality we finally obtain

$$|s(\mathbf{f}, \mathbf{g})| \leq \frac{1}{2} \|\mathbf{f}\| \cdot \|\mathbf{g}\| \tag{3.26}$$

The required property of the quadratic form may be easily obtained if to extend the estimate given by eqn. 3.26 for arbitrary functions $\mathbf{f}, \mathbf{g} \in H$ using the continuity.

Let us note that for the traces of $g_1, g_2, \frac{\partial f_1}{\partial \tau}$ and $\frac{\partial f_2}{\partial \tau}$ in the integrals occurring in the definition of s we assume that

$$g_i|_{\partial S_1} \in H^{1/2}(\partial S_1), \quad \frac{\partial f_i}{\partial \tau}\bigg|_{\partial S_1} \in H^{-1/2}(\partial S_1) \quad (i = 1, 2)$$

and integration by parts

$$\int_\Gamma \frac{\partial f_1}{\partial \tau} \bar{g}_2 \, \mathrm{d}l = -\int_\Gamma \frac{\partial \bar{g}_2}{\partial \tau} f_1 \, \mathrm{d}l = \int_\Gamma \frac{\partial f_2}{\partial \tau} \bar{g}_1 \, \mathrm{d}l = -\int_\Gamma \frac{\partial \bar{g}_1}{\partial \tau} f_2 \, \mathrm{d}l \qquad (3.27)$$

is possible since

$$f_1|_{\Gamma'} = 0, \quad g_1|_{\Gamma'} = 0$$

and $\operatorname{supp} f_1 \subset \overline{\Gamma}$, $\operatorname{supp} g_1 \subset \overline{\Gamma}$.

Now the variational problem given by eqn. 3.24 may be written in the operator form

$$(L\mathbf{f}, \mathbf{g}) = 0 \quad \forall \mathbf{g} \in H$$

which is equivalent to the following equation for the operator-valued pencil

$$L(\gamma)\mathbf{f} = 0, \qquad L = L(\gamma) : H \to H$$
$$L(\gamma) \equiv \gamma^4 K + \gamma^2 [A_1 - (\epsilon_1 + \epsilon_2)k^2 K] \qquad (3.28)$$
$$+ (\epsilon_1 - \epsilon_2)k\gamma\Omega + \epsilon_1\epsilon_2(k^4 K - k^2 A_2)$$

where all the operators are bounded.

According to definition 3.1 the eigenvalues and eigenvectors of $L(\gamma)$ coincide with the the eigenvalues and eigenfunctions of the problem V. Thus the latter is reduced to the eigenvalue problem (or to the problem on characteristic numbers) for the operator pencil $L(\gamma)$.

Let us formulate some important properties of pencil's operators which have been proved by Smirnov (1990, 1991).

Lemma 3.1 The operators A_i $(i = 1, 2)$ are uniformly positive:

$$I \leq A_1 \leq \epsilon_{max} I, \quad \epsilon^{-1} \leq A_2 \leq I \qquad (3.29)$$

where I is the unit operator in H.

Lemma 3.2: The operator Ω is self-adjoint, $\Omega = \Omega^*$, and the following inequalities hold

$$-\frac{1}{2}I \leq \Omega \leq \frac{1}{2}I \qquad (3.30)$$

Lemma 3.3: The operator K is positive, $K > 0$, and compact. The following estimate holds for its eigenvalues

$$\lambda_n(K) = O(n^{-1}), \quad n \to \infty \qquad (3.31)$$

Lemma 3.4: The operator pencil $L(\gamma)$ is self-adjoint:

$$L^*(\gamma) = L(\bar{\gamma}) \qquad (3.32)$$

Lemma 3.5: Let P be the orthogonal projection such that

$$P(f_1, f_2)^T = (-f_1, f_2)^T$$

Then $A_1 = PA_1P$, $A_2 = PA_2P$, $K = PKP$, $\Omega = -P\Omega P$ and the following representation holds

$$L(-\gamma) = PL(\gamma)P \qquad (3.33)$$

Let us note that the proof of the last assertion directly follows from the explicit form of the variational expression given by eqn. 3.24.

We will denote by $\rho(L)$ the resolvent set of $L(\gamma)$ (consisting of all complex values of γ where there exists a bounded inverse operator $L^{-1}(\gamma)$) and by $\sigma(L) = \mathbb{C} \setminus \rho(L)$ the spectrum of $L(\gamma)$.

In some cases it will be convenient for us to consider the regularised pencil

$$\begin{aligned}
\tilde{L}(\gamma) &= A_1^{-1/2}L(\gamma)A_1^{-1/2} = \gamma^4\tilde{K} + \gamma^2(I - (\epsilon_1 + \epsilon_2)k^2\tilde{K}) \\
&\quad + (\epsilon_1 - \epsilon_2)k\gamma\tilde{\Omega} + \epsilon_1\epsilon_2(k^4\tilde{K} - k^2\tilde{A}_2)
\end{aligned} \qquad (3.34)$$

where

$$\tilde{K} = A_1^{-1/2}KA_1^{-1/2}$$
$$\tilde{\Omega} = A_1^{-1/2}\Omega A_1^{-1/2}$$
$$\tilde{A}_2 = A_1^{-1/2}A_2A_1^{-1/2}$$

It is easy to see that $\sigma(L) = \sigma(\tilde{L})$ and the following relation holds for eigenvectors

$$\varphi_0(L) = A_1^{-1/2}\varphi_0(\tilde{L}) \tag{3.35}$$

The operators $\tilde{K}, \tilde{\Omega}$ and $\tilde{A}_2$ keep all properties of the operators K, Ω and A given in lemmas 1–3 with the estimates

$$-\frac{1}{2}I \le \tilde{\Omega} \le \frac{1}{2}I, \quad \epsilon_{max}^{-2} \le \tilde{A}_2 \le I \tag{3.36}$$

Properties of the spectrum of the pencil $L(\gamma)$ are given by the following theorems.

Theorem 3.1:

$$\sigma(L) \subset \Pi_l = \{\gamma : |\operatorname{Re}\gamma| < l\}$$

i.e. for a certain $l > 0$ the spectrum of the pencil $L(\gamma)$ lies in the strip Π_l.

In order to prove this theorem, assume that $l > k\sqrt{\epsilon_1 + \epsilon_2}$ and consider the operator-valued function

$$F(\gamma) = [\gamma^2 - k^2(\epsilon_1 + \epsilon_2)]^{-1}\tilde{L}(\gamma) = \gamma^2\tilde{K} + I + \gamma^{-1}T(\gamma)$$

in the domain $D_0 = \{\gamma : |\gamma| > l\}$, where

$$\begin{aligned} T(\gamma) = {}& \gamma[\gamma^2 - (\epsilon_1 + \epsilon_2)k^2]^{-1} \\ & \times [(\epsilon_1 + \epsilon_2)k^2 I + (\epsilon_1 - \epsilon_2)k\gamma\tilde{\Omega} \\ & + \epsilon_1\epsilon_2(k^4\tilde{K} - k^2\tilde{A}_2)] \end{aligned} \tag{3.37}$$

is a holomorphic and bounded operator-valued function in the domain D_0: $\|T(\gamma)\| < T_0$ for $\gamma \in D_0$. One can see that $\sigma(\tilde{L}) \cap D_0 = \sigma(F) \cap D_0$. If $|\operatorname{Re}\gamma| > l$, there exists a bounded operator

$$R(\gamma) = (\gamma^2\tilde{K} + I)^{-1}$$

the estimates of Cohberg and Krein (1965) hold for its norm

$$\|R(\gamma)\| \le \frac{|\gamma^{-2}|}{|\operatorname{Im}\gamma^{-2}|} \le \frac{1}{2}\left(1 + \frac{|\gamma|}{l}\right)$$

when $|\gamma''| > |\gamma|$ and $\|R(\gamma)\| = 1$ for $\|\gamma''\| \leq |\gamma'|$ where $\gamma = \gamma' + j\gamma''$. Choosing the value $l > T_0$ we obtain

$$\|\gamma^{-1}T(\gamma)R(\gamma)\| \leq \frac{1}{2}\left(1 + \frac{|\gamma|}{l}\right)\frac{T_0}{|\gamma|} < 1$$

hence there exists a bounded operator

$$F^{-1}(\gamma) = R(\gamma)(I + \gamma^{-1}T(\gamma)R(\gamma))^{-1}$$

what yields the existence of bounded operators $\tilde{L}^{-1}(\gamma)$ and $L^{-1}(\gamma)$ for $\gamma \in \Pi_l = \{\gamma : |\operatorname{Re}\gamma| < l\}$.

Theorem 3.2: The spectrum of the pencil $L(\gamma)$ is symmetric with respect to the real and imaginary axes: $\sigma(L) = \overline{\sigma(L)} = -\sigma(L)$. Each nonreal point $\gamma_0 \in \sigma(L)$ is an eigenvalue of $L(\gamma)$, i.e. γ_0 is an isolated point having finite algebraic multiplicity. There exists the countable set of eigenvalues with the only one accumulation point at infinity.

The first assertion of theorem 3.2 follows from lemmas 3.4 and 3.5 (eqns. 3.32 and 3.33).

Assume now that $\gamma = \gamma' + j\gamma''$, $\gamma'' \neq 0$. Then the following relations hold for the imaginary part of the operator expression (considered with respect to γ)

$$\operatorname{Im}\left\{\frac{1}{\gamma''}\left[\gamma A_1 + (\epsilon_1 - \epsilon_2)k\Omega - \frac{\epsilon_1\epsilon_2 k^2}{\gamma}A_2\right]\right\}$$
$$= A_1 + \frac{\epsilon_1\epsilon_2 k^2}{\gamma}A_2 \geq I$$

hence according to Cohberg and Krein (1965) the operator

$$L_0(\gamma) = \gamma^2 A_1 + (\epsilon_1 - \epsilon_2)k\gamma\Omega - \epsilon_1\epsilon_2 k^2 A_2$$

has the bounded inverse and, consequently, $L(\gamma)$ is a Fredholm operator what proves the second assertion of theorem 3.2 if to take into account general properties of the spectrum of Fredholm operator-valued functions.

The third assertion follows from theorem 3.3 which will be proved next.

The existence of real eigenvalues γ may be proved by the method of integral operator-valued functions (see Section 3.1) for the waveguies with rectangular cross-sections. But theorem 3.2 states that, unlike in the case of shielded resonators (considered in Chapter 1) they never exhaust all eigenvalues of the problem of normal waves and, what is more, there always exist an infinite number of complex eigenvalues and only a finite number of real eigenvalues. The former have been predicted and determined by various heuristic, asymptotic and numerical methods (Weinstein, 1957) but without a proper justification.

Let us note that the points $\gamma = \pm\gamma_i = \pm k\sqrt{\epsilon_i} \in \sigma(L)$ where the coefficient of eqns. 3.12 vanishes play a special role and may be called the degeneration points of the pencil $L(\gamma)$, where the dimension of the kernel $L(\pm\gamma_i) = \infty$.

Now let us prove the completeness of the system of eigenvectors and associated vectors of the pencil $L(\gamma)$.

In order to solve excitation problems it is important to know the properties of the whole family of normal waves. The analysis carried out, e.g. by Weinstein (1957) shows that the system of normal waves is not complete in some waveguides which belong to the general family considered here and "associated" waves may appear which are defined as solutions of the system

$$\operatorname{curl} \mathbf{E}^{(p)} = -jk\mathbf{H}^{(p)} + j[\mathbf{E}^{(p-1)} \times \hat{i}_3]$$

$$\operatorname{curl} \mathbf{H}^{(p)} = jk\mathbf{E}^{(p)} + j[\mathbf{H}^{(p-1)} \times \hat{i}_3]$$

$$\mathbf{E}_{\perp}^{(p)} = \frac{1}{k}[\gamma\nabla\Pi_p - k\nabla \times (\Psi_p \hat{i}_3) + \gamma\mathbf{E}_{\perp}^{(p-1)} - k\mathbf{H}_{\perp}^{(p-1)} \times \hat{i}_3] \qquad (3.38)$$

$$\mathbf{H}_{\perp}^{(p)} = \frac{1}{k}[\gamma\nabla\Psi_p + \epsilon k\nabla \times (\Pi_p \hat{i}_3) + \gamma\mathbf{H}_{\perp}^{(p-1)} + \epsilon k\mathbf{E}_{\perp}^{(p-1)} \times \hat{i}_3]$$

where $p \geq 1$ and

$$\Pi_0 \equiv \Pi, \quad \Psi_0 \equiv \Psi$$

The above analysis enables us to reduce the study of the completeness of normal waves of shielded waveguides (which include "associated" waves) to the problem of completeness of the system of eigenvectors and associated vectors of the pencil $L(\gamma)$ which are defined as follows

$$L(\gamma)\varphi_0^k = 0, \quad k \geq 1$$

$$L(\gamma)\varphi_p^{(k)} + \frac{1}{1!}\frac{\partial L(\gamma)}{\partial\gamma}\varphi_{p-1}^{(k)} + \ldots + \frac{1}{k!}\frac{\partial^k L(\gamma)}{\partial\gamma^k}\varphi_0^{(k)} = 0, \quad p \geq 1$$

where the multiplicity of the vector $\varphi_0^{(k)}$ is not less then the multiplicity of the vector $\varphi_0^{(k+1)}$. If the eigenvalue γ has the finite algebraic multiplicity then the chain $\varphi_0^{(k)}, \varphi_1^{(k)}, \ldots, \varphi_{p_k}^{(k)}$ is finite. Following Keldysh (1971) we will call this system complete with the power n if any set of the vectors $f_0, f_1, \ldots, f_{n-1}$ can be represented as a limit with respect to the norm of the linear combination of the elements of the system

$$f_{\nu,N} = \sum_{k=1}^{N} \sum_{(p)} a_{p,N}^{(k)} \varphi_p^{(k,\nu)}, \quad \nu = 0, 1, \ldots, n-1$$

where the coefficients $a_{p,N}^{(k)}$ do not depend on ν,

$$\varphi_p^{(k,\nu)} = \left[\frac{\mathrm{d}\nu}{\mathrm{d}t^\nu} \exp(\gamma_k t) \left(\varphi_p^{(k)} + \frac{t}{1!} \varphi_{p-1}^{(k)} + \ldots + \frac{t^p}{p!} \varphi_0^{(k)} \right) \right]\Bigg|_{t=0}$$

and γ_k are the eigenvalues of the pencil $L(\gamma)$.

For $n = 2$ this system is called double complete.

Theorem 3.3: The system of eigenvectors and associated vectors of $L(\gamma)$ corresponding to the eigenvalues γ_k such that for any $\eta > 0$ they lie in the domain $D_0 = \{\gamma : |\gamma| > \eta\}$ is double complete with the finite defect in $H \times H$.

Let us represent the pencil $L(\gamma)$ as the perturbation with respect to the holomorphic operator-valued function of the Keldysh pencil. We will consider the pencil $F(\gamma)$ in the form 3.37 choosing $\eta > k\sqrt{\epsilon_1 + \epsilon_2}$. Taking into account the spectral properties of the pencils $F(\gamma)$ and $\tilde{L}(\gamma)$ indicated in the proof of theorem 3.1 one can see that it is sufficient to prove the assertion of theorem 3.3 for the pencil $F(\gamma)$.

For $T_1(\gamma) = \gamma^{-1} T(\gamma)$ we have $\lim_{\gamma \to \infty} \|T_1(\gamma)\| = 0$. Using lemma 3.3 we conclude that $\tilde{K} > 0$ is the Hilbert–Schmidt operator and the assertion of theorem 3 follows then from the results of Radzievsky (1973).

Let us formulate the sufficient conditions for the completeness without defect under some inevitable restrictions on the parameters of the problem. First we shall rewrite the pencil $\tilde{L}(\gamma)$ in the form

$$\begin{aligned} \tilde{L}(\gamma) &= (p^2 - \gamma^2)((p^2 - \gamma^2)\tilde{K} - I) + \delta B(\gamma) \\ &= F_1(\gamma) + \delta B(\gamma) = (p^2 - \gamma^2)F_0(\gamma) + \delta B(\gamma) \\ B(\gamma) &= 2k\gamma\tilde{\Omega} - \delta k^4 \tilde{K} + k^2 \tilde{A}_1' \end{aligned} \tag{3.39}$$

where

$$p = k\sqrt{(\epsilon_1 + \epsilon_2)/2}, \quad \delta = (\epsilon_1 - \epsilon_2)/2$$

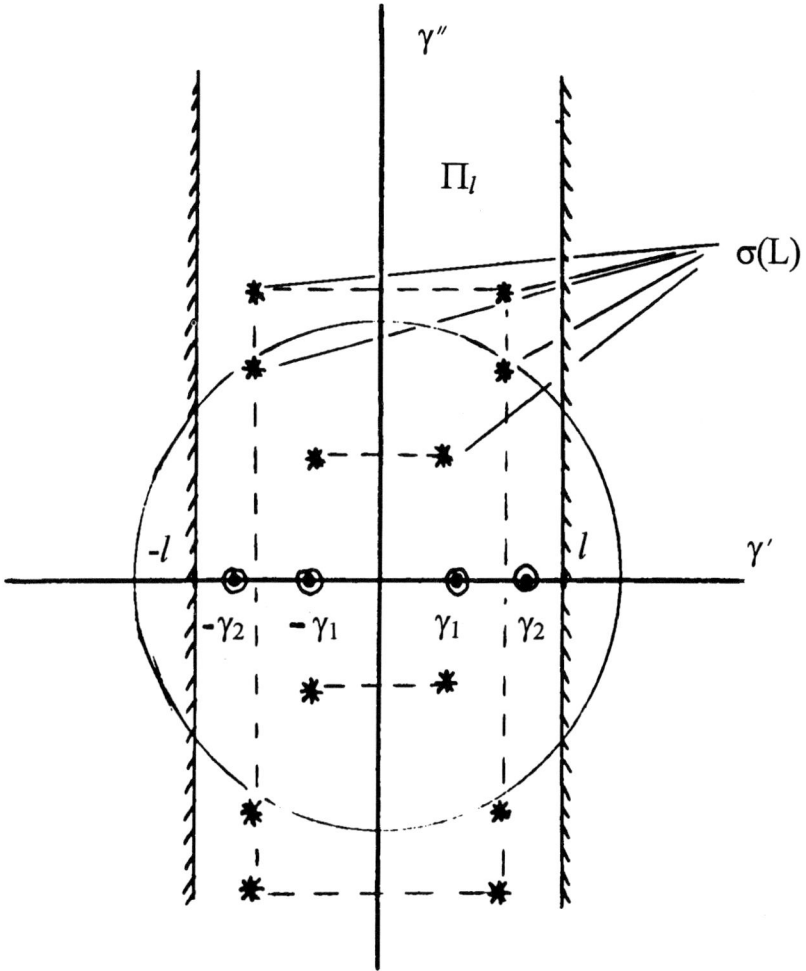

Figure 3.1   *Distribution of the spectrum of normal waves on the*
complex plane of the spectral parameter (longitudinal
wavenumber)

$$\tilde{A}'_1 = A_1^{-1/2} A'_1 A_1^{-1/2}$$

and the operator A'_1 is produced by the form

$$\int_S \xi(\epsilon \nabla f_1 \nabla \bar{g}_1 + \nabla f_2 \nabla \bar{g}_2) \, \mathrm{d}s = (A'_1 \mathbf{f}, \mathbf{g}) \qquad \forall, \quad \mathbf{g} \in H$$

where

$$\xi = \begin{cases} 1 & \text{in } S_1 \\ -1 & \text{in } S_2 \end{cases} \quad -I < \tilde{A}_1' < I$$

In order to factorise the pencil $\tilde{L}(\gamma)$ we shall select the regions on the complex γ-plane containing the fixed number of solutions to the equation

$$(\tilde{L}\mathbf{f}, \mathbf{f}) = 0 \tag{3.40}$$

where the form $(\tilde{L}\mathbf{f}, \mathbf{f})$ is considered for a fixed element $\mathbf{f}$ as a function of γ.

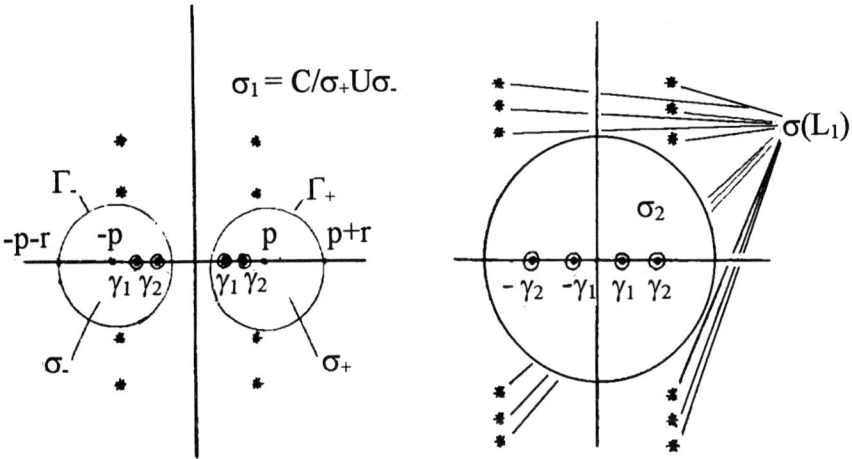

Figure 3.2 *Distribution of the spectrum of normal waves on the complex plane of the spectral parameter*

Let us separate two cases when $p^2 \geq 0.25 N_K^{-1}$ and $0 < p^2 < 0.25 N_K^{-1}$ where $N_K = \|\tilde{K}\|$. In the first case we will introduce two circles $\Gamma_\pi = \{\gamma : |\gamma \mp p| = r\}$ where the radius r is chosen such that

$$r = p - \sqrt{p^2 - \frac{1}{2}N_K^{-1}} \quad \text{for} \quad p^2 > \frac{2}{3}N_K^{-1}$$

and

$$r = p \quad \text{for} \quad \frac{1}{4}N_k^{-1} \leq p^2 \leq \frac{2}{3}N_k^{-1}$$

The explicit form of the pencils $B(\gamma)$ and $F(\gamma)$ yields estimates (uniform with respect to γ and φ)

$$\frac{1}{|p^2 - \gamma^2|}|(B(\gamma)\varphi, \varphi)| \le 2N_K(k + k^2 + |\delta|N_K k^4)$$

$$|(F_0(\gamma)\varphi, \varphi)| \ge \frac{1}{3\sqrt{3}} \qquad \gamma \in \Gamma_1 = \Gamma_+ \cup \Gamma_- \quad \|\varphi\| = 1$$

Hence if the following inequality holds

$$6\sqrt{3}N_K|\delta|(k + k^2 + |\delta|N_K k^4) < 1 \tag{3.41}$$

then the operator $\tilde{L}(\gamma)$ has the bounded inverse for $\gamma \in \Gamma_1$ and

$$|(F_1(\gamma)\varphi, \varphi)| > |\delta(B(\gamma)\varphi, \varphi)|, \quad \gamma \in \Gamma \tag{3.42}$$

Using Rouchet's theorem we can verify that there is only one solution of eqn. 3.40 in every region

$$\sigma_\pi = \{\gamma : |p \mp \gamma| < r\}$$

as well as of the equation $(F_1(\gamma)\varphi, \varphi) = 0$ and

$$\inf_{\|\varphi\|=1} \inf_{\gamma \in \Gamma} |(\tilde{L}(\gamma)\varphi, \varphi)| > 0 \tag{3.43}$$

Thus the spectrum of the pencil $\tilde{L}(\gamma)$ may lie in three regions σ_+, σ_- and $\sigma_1 = \mathbb{C} \setminus (\bar{\sigma}_+ \cup \bar{\sigma}_-)$. Taking into account that the coefficients of the form $(\tilde{L}\varphi, \varphi)$ are real we conclude that

$$\sigma(L) \cap \sigma_+ \subset (p - r, p + r), \quad \sigma(L) \cap \sigma_- \subset (-p - r, -p + r)$$

The pencil $\tilde{L}(\gamma)$ admits the factorisation with respect to the contour Γ_1 (Markus and Matzaev, 1975):

$$\tilde{L}(\gamma) = L_1(\gamma)L_2(\gamma) \tag{3.44}$$

where

$$L_1(\gamma) = \gamma^2 \tilde{K} - \gamma \tilde{K} B_1 + I - \tilde{K}(k^2(\epsilon_1 + \epsilon_2)I + B_2 - B_1^2)$$
$$L_2(\gamma) = \gamma^2 I + \gamma B_1 + b_2$$

B_i $(i = 1, 2)$ are bounded operators and

$$\sigma_2 = \sigma_+ \cup \sigma_-, \quad \sigma(L_1) \subset \sigma_1, \quad \sigma(L_2) \subset \sigma_2$$

In the second case we will choose the circles

$$\Gamma_2 = \{\gamma : |\gamma| = r\}, \quad \sigma_1 = \{\gamma : |\gamma| > r\}, \quad \sigma_2 = \mathbb{C} \setminus \bar{\sigma}_1$$

where $r = (2N_K)^{-1/2}$. Then the following estimates hold (uniform with respect to γ and φ)

$$\frac{1}{|p^2 - \gamma^2|} |(B(\gamma)\varphi, \varphi)| \leq 4N_K ((2N_K)^{-1/2}k + k^2 + |\delta|N_K k^4)$$

$$|(F_0(\gamma)\varphi, \varphi)| \geq \frac{1}{4}$$

(3.45)

Now if we assume that

$$16 N_K |\delta| ((2N_K)^{-1/2}k + k^2 + |\delta|N_K k^2) < 1 \tag{3.46}$$

the estimates given by eqns. 3.42 and 3.43 are fulfilled for $\gamma \in \Gamma_2$, the region σ_2 contains two solutions of eqn. 3.40 and

$$\sigma(L) \cap \sigma_2 \subset (-r, r)$$

Hence the pencil $\tilde{L}(\gamma)$ admits the factorisation given by eqn. 3.44 with respect to Γ_2 and

$$\sigma(L_1) \subset \sigma_1, \quad \sigma(L_2) \subset \sigma_2$$

The system of eigenvectors and associated vectors of the pencil $L_1(\gamma)$ is double complete by the Keldysh theorem (Keldysh, 1971) and, consequently, the system of eigenvectors and associated vectors of the pencil $L(\gamma)$ corresponding to the eigenvalues $\gamma' \in \sigma_1$ is complete too. Thus we have proved the following theorem:

Theorem 3.4: Assume that the following inequality holds

$$16 N_K |\delta| (k \max(1, (2N_K)^{-1/2}) + k^2 + |\delta|N_K k^4) < 1 \tag{3.47}$$

Then the system of eigenvectors and associated vectors of the pencil $L(\gamma)$ corresponding to the eigenvalues $\gamma_n \neq \pm k\sqrt{\epsilon_i}$ $(i = 1, 2)$ is doubly complete.

The points $\pm k\sqrt{\epsilon_i}$ should be naturally excluded from consideration because for $\gamma = \pm k\sqrt{\epsilon_i}$ the boundary eigenvalue problem V is not equivalent to the eigenvalue problem for the operator pencil. Anyway, the points $\pm k\sqrt{\epsilon_i}$ ($i = 1, 2$) are always contained in the region σ_2. We have chosen the estimate given by eqn. 3.47 which is stronger than that given by eqns. 3.41 or 3.46.

In fact, we have proved more than it is asserted in theorem 3.4. Namely, to construct the complete system of eigenvectors and associated vectors of the pencil $L(\gamma)$ it is not necessary to take the eigenvectors corresponding to the eigenvalues $\gamma_n \in \sigma_2$. Such eigenvalues and eigenvectors correspond to the so-called dominant modes.

In empty regular waveguides it is sufficient to take the whole system of normal waves which forms the complete set (Samarsky and Tikhonov, 1947).

Of course, the present results generalise the completeness of normal waves in empty waveguides established by Samarsky and Tikhonov (1947).

It turns out that the double completeness of the longitudinal components in the space $H(S)$ implies the completeness (in the usual sense) of the cross-sectional components of the system of eigenvectors and associated vectors in the space $L_2^4(S)$ (coinciding with the product of four copies of $L_2(S)$). This assertion follows from eqn. 3.38 and the relations

$$\mathbf{E}_\perp = \frac{1}{k^2}[\gamma\nabla\Pi - k\nabla \times (\Psi\hat{i}_3)]$$

$$\mathbf{H}_\perp = \frac{1}{k^2}[\epsilon\gamma\nabla \times (\Pi\hat{i}_3) - \gamma\nabla\Psi]$$

$$\Pi = jE_{x_3}(\mathbf{r}), \quad \Psi = jH_{x_3}(\mathbf{r})$$

giving the connection between cross-sectional and longitudinal components.

Part 2 Eigenoscillations and excitation of open waveguide resonators

This part deals with the features of analytical continuations with respect to the complex frequency parameter ω of the resolvents of the waves diffraction problems on inhomogeneities in open waveguide resonators. The domains of the fundamental frequencies localisation have been described and their existence has been shown. The local expansion theorems for the diffraction characteristics in the vicinities of the resolvents' isolated singularities (connected with fundamental frequencies) and branch points (connected with threshold frequencies of regular waveguides) have been proved.

Numerical algorithms for the determination of fundamental frequencies (as functions of the waveguides' parameters) and eigenfields have been constructed and justified. A series of systematical calculations have been carried out for basic structures. The connections between the spectral and resonance characteristics for two groups of open waveguide resonators are established: cylindrical resonators formed by the segments of regular waveguides whose cross-sections are greater than radiation channels and resonators containing irregular communication domains. It is shown that total resonant reflection and transmission phenomena result from excitation of forced oscillation close to resonators' eigenoscillations. The Morse critical points and the interaction of oscillations are considered.

Chapter 4

Qualitative spectra characteristics of open waveguide resonators and the local expansion theorems

Various inhomogeneities play an essential role as the elements of open waveguide resonators. Some of inhomogeneities are used intentionally for co-ordinating and filtering modes and frequencies, branching a part of energy etc. while some of them, like waveguide breaks and junctions, are caused by construction requirements. Mathematical and computer models here must provide the means to obtain data in real time which would enable one to bridge the gap between the technical task for a device and its realisation within the minimal period of time.

Such models and electrodynamic theories must satisfy the following demands: rigorous statements of the problems which enable one to take into account the influence of various idealisations used within the frame of a given mathematical model; universality and efficiency of numerical algorithms for obtaining both qualitative and quantitative characteristics with given accuracy.

In this way the theory of waves diffraction based on the semi-inversion method developed by Shestopalov (1971, 1983) and Shestopalov *et al.* (1984) is practically completed. Its main achievement is mastering the resonance frequency range which is the most difficult from the mathematical viewpoint and the most perspective for practical applications. This theory is an effective tool which allows one to obtain reliable data describing diffraction properties of various scatterers in an arbitrary wavelength range. But within the diffraction theory (stationary scattering) one studies only "simple" time-harmonic solutions to Maxwell's equations. The resolvent of a non-stationary scattering problem is connected with the resolvent of the corresponding diffraction problem by means of the Laplace transform (Muravej, 1978) and the solution to the latter is completely determined by the solution to the former. But this simple connection may be effectively applied only with the help of the contour integration when all the singularities of the analytical continuations of both resolvents are known.

As a result of the above considerations, one naturally comes to the requirement to develop the theory of quasi-stationary free oscillations in open waveguide resonators with inhomogeneities. These free oscillations are tightly connected with the residues in the poles of Green's functions and are described by non-trivial solutions of Maxwell's equations which may occur at complex frequencies. The spatial structure of eigenfields is constant with respect to time t and the amplitude factor may exponentially decrease ($\mathrm{Im}\,\omega < 0$) or increase ($\mathrm{Im}\,\omega > 0$) when $t \to \infty$. Within the frame of this theory one can consider any domain violating the regularity of a (regular) waveguide as an open waveguide resonator which differs from an open resonator by the fact that its diffraction losses are caused by the radiation to semi-infinite waveguide channels (and not to the free space). Therefore quasi-stationary free oscillations (eigenfrequencies) are called eigenoscillations (fundamental frequencies) of open waveguide resonators.

Unlike the closed (volume) resonators, the eigenoscillations spectra and the types of eigenoscillations in open resonators essentially depend on the radiation conditions determining the diffraction communication between the volume of a resonator itself and the surrounding space. The radiation conditions for open waveguide resonators will be formulated on the basis of a physically justified criterion of the absence of waves coming to a resonator from infinity for $\mathrm{Im}\,\omega = 0$.

The spectral theory (or the theory of quasi-stationary free oscillations) contributes to the diffraction theory itself by establishing, in particular, the boundaries of applicability for the limiting absorption principle. In fact, the existence and uniqueness theorems are proved in the diffraction theory under the conditions of a low-loss surrounding medium with $\epsilon_0 = \epsilon'_0 + j\epsilon''_0$ ($|\epsilon''_0| \ll 1$)

and a correct limiting transition to the real values of the frequency parameter is possible only if the corresponding limiting values are not the fundamental frequencies. Hence, if the real fundamental frequencies exist one has to put additional requirements separating unique (true) solutions of the diffraction problems. The second problem is in delivering from "naive" physical interpretations of resonant and anomalous phenomena, like total transparence, total reflection, total mode transformation etc. In accordance with the results of spectral theory, they have a resonant nature and are connected with excitation of oscillations which are close to the eigenoscillations of inhomogeneities. The spectral theory allows us to answer various practical questions arising in the attempts to explain such "exotic" phenomena by means of the specially created analytical and numerical methods.

One cannot say that similar problems were not solved previously. But there was a lack of a universal approach and most of the existing results might be considered as the links of a chain which should be put together by the unified spectral theory developed in this book.

The main results are formulated in what follows for the TE oscillations in two-dimensional open waveguide resonators loaded with two parallel-plane waveguides and may be generalised for the TE and TM oscillations in the resonators with an arbitrary number of radiation channels (see Figure 4.1). The results remain valid for circular, rectangular and other types of regular waveguides if there is no polarisation communication of oscillations.

4.1 Statements of diffraction problems for open waveguide resonators

It is known (Muravej, 1978) that scattering problems are connected with diffraction problems by means of the Laplace transform with respect to the frequency parameter $\kappa = \omega\sqrt{\epsilon_0\mu_0}$ where the originals are the solutions to the diffraction problems with the right-hand sides taken as combinations of the initial data of the scattering problems. In other words, the resolvent R_D of the former determine the resolvent R_S of the latter. But can one realise this connection using only the (regular) solutions to diffraction problems at real frequencies? Most probably not even taking into account high efficiency of

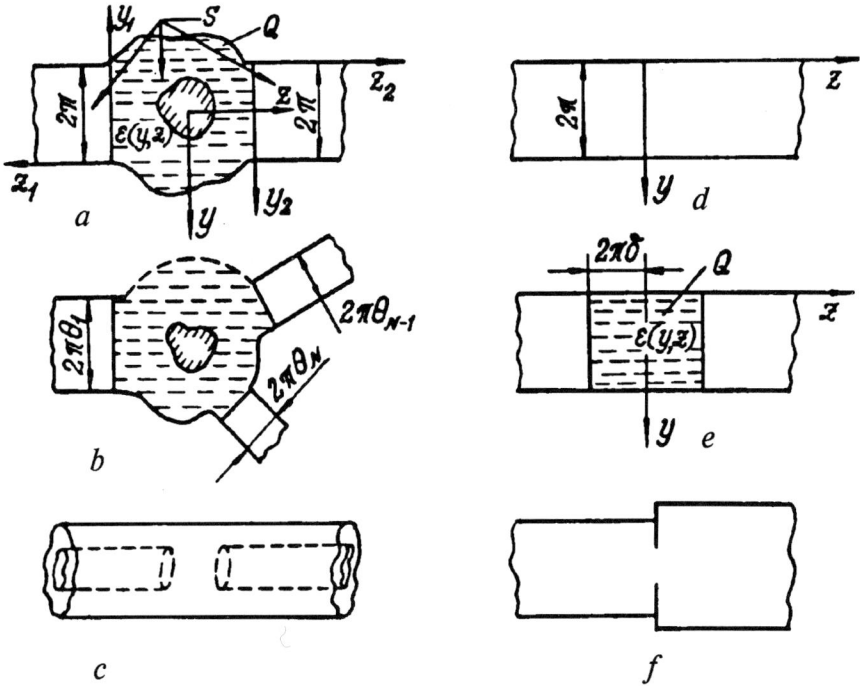

Figure 4.1 *Two-dimensional open waveguide resonators*

the modern computers and software.

Application of the complex analysis based on contour integration in the complex plane κ enables one to solve this problem by calculating the residues in the singularities of the analytical continuations of R_D with respect to κ. It is obvious then that the main attention should be paid not to the regular solutions but to these singularities (in particular, poles of Green's functions).

Assume first that the frequency is a real value. Let us carry out a complete statement of the excitation problem of the parallel-plane waveguide presented in Figure 4.1 denoting by Q $(z_1 < 0, z_2 < 0)$ the communication domain situated inside the waveguide. We put $u = u(\mathbf{r}) = E_{x_1}$ (TE waves) or $u = u(\mathbf{r}) = H_{x_1}$ (TM waves) where $\mathbf{r} = (x_2, x_3)$, E_{x_1} and H_{x_1} are the components of the vectors $\mathbf{E}$ and $\mathbf{H}$ parallel to waveguide's axis. Then inhomogeneous Maxwell's equations are reduced to the inhomogeneous Helmholtz equation

$$\Delta u + \kappa^2 \epsilon u = f, \quad \mathbf{r} \in S \tag{4.1}$$

where $\epsilon = \epsilon(\mathbf{r})$ is a complex-valued piece-wise continuous function (a piece-wise constant function in the TM case) characterising the permittivity of the medium in Q such that $\epsilon(\mathbf{r}) \equiv 1$ $(\mathbf{r} \in S \setminus Q)$, $\mathrm{Re}\,\epsilon \geq 1$, $\mathrm{Im}\,\epsilon \geq 0$. $f = f(\mathbf{r})$ is the source function with a compact support $\mathrm{supp}\,f \in Q$ $(f = 0$ if $\mathbf{r} \in S \setminus \mathrm{supp}\,f)$. The solutions of eqn. 4.1 satisfy the following generalised boundary conditions:

$E_{tg} = 0$ on the perfectly conducting boundary ∂S of the domain S;

E_{tg} and H_{tg} are continuous on the break lines of ϵ;

for every compactum $B \subset S \setminus \mathrm{supp}\,f$

$$\iint_B \left(|u|^2 + \left| \frac{\partial u}{\partial x_2} \right|^2 + \left| \frac{\partial u}{\partial x_3} \right|^2 \right) \mathrm{d}\mathbf{r} < \infty \qquad (4.2)$$

$$u(\mathbf{r}) = \sum_{n=-\infty}^{\infty} a_{nl} \exp(j\omega_n x_3)\phi_n^l(x_2), \quad x_3 > 0 \ (l = 1, 2)$$

where $\phi_n^2(t) = \cos\frac{1}{2}nt$ $(n = 0, 1, 2\ldots)$ in the TM case, $\phi_n^1(t) = \sin\frac{1}{2}nt$ $(n = 1, 2\ldots)$ in the TE case, $\omega_n = \sqrt{\kappa^2 - \frac{n^2}{4}}$ $(\mathrm{Im}\,\omega_n \geq 0$, $\mathrm{Re}\,\omega_n \geq 0$ for $\mathrm{Re}\,\kappa > 0$ and $\mathrm{Re}\,\omega_n \leq 0$ for $\mathrm{Re}\,\kappa < 0$).

The last expressions are a mathematical formulation of the radiation conditions at infinity which provide the absence of waves coming from the semi-infinite parts of the regular waveguides.

When constructing an analytical continuation of the problem given by eqns. 4.1 and 4.2 to the domain H of the complex values of κ, one has to determine the natural boundaries of H and to re-formulate the radiation conditions. With this purpose let us treat the open waveguide resonator as a perturbed parallel-plane resonator [with $\epsilon(\mathbf{r}) \equiv 1$]. Then it is natural to assume that H coincides with the domain of analytical continuation of Green's function of the unperturbed problem (from the real axis of the frequency parameter)

$$G(\mathbf{r}, \mathbf{r}_0) = \frac{1}{2\pi} \sum_{n=-\infty}^{\infty} \exp(j\omega_n |x_3 - x_{30}|)\phi_n(x_2)\phi_n(x_{20})\omega_n^{-1}, \quad \mathrm{Im}\,\kappa = 0 \quad (4.3)$$

satisfying eqn. 4.1 with $f(\mathbf{r}) = -\delta(\mathbf{r} - \mathbf{r}_0)$, $\mathbf{r}_0 = (x_{20}, x_{30})$ [Morse and Feschbach (1958) called it canonical Green's function of the domain formed by two parallel planes situated at the distance 2π from each other]. Due to the explicit form of $\omega_n(\kappa)$ it is easy to show that H is the infinite-sheet Riemann surface with the branch points κ_n determined by the condition $\omega_n(\kappa) = 0$ ($n = 0, 1, 2 \ldots$ in the TM case and $n = 1, 2 \ldots$ in the TE case).

Theorem 4.1: Assume that the radiation conditions

$$\operatorname{Im}\omega_n \geq 0, \quad \operatorname{Re}\omega_n \geq 0 \quad \text{for} \quad \operatorname{Re}\kappa > 0$$
$$\operatorname{Re}\omega_n \leq 0 \quad \text{for} \quad \operatorname{Re}\kappa < 0 \tag{4.4}$$

hold and they are co-ordinated with the demand of the absence of waves coming from infinity. Then Green's function $G(\mathbf{r}, \mathbf{r}_0)$ admits a natural analytical continuation from the real axis κ to the Riemann surface whose sheets are the complex planes κ cut along the curves

$$d_m(\kappa) = (\operatorname{Re}\kappa)^2 - (\operatorname{Im}\kappa)^2 - \frac{m^2}{4}, \quad \operatorname{Im}\kappa \leq 0 \tag{4.5}$$

The proof of this theorem obtained by Sirenko *et al.* (1985a) is based on the uniform convergence of the series in eqn. 4.3 for every value $\kappa \in H$, $\kappa \neq \kappa_n$ because it turns out that always a finite number of the functions $\omega_n(\kappa)$ have a negative imaginary part. The first sheet of H is completely determined by eqn. 4.4 and the cuts given by eqn. 4.5. The successive sheets differ from the first sheet by a finite number of opposite signs of $\omega_n(\kappa)$.

Theorem 4.1 defines the domain of variation of the spectral parameter κ for the diffraction problems at complex frequencies and establishes the radiation conditions for their solutions $u(\mathbf{r})$ which means that $\omega_n(\kappa)$ in eqn. 4.2 are uniquely defined. In fact, this representation holds only in the regular points of H where the diffraction problem is uniquely solvable. For some values of $\kappa \in H$ the series in eqn. 4.2 may diverge. Such singular values of κ (the spectral points) which complement the set of regular points to the whole set H as well as the character of the solutions in these points are the main subjects of our consideration.

It is easy to see that there is no physically justified criterion for the correct consideration of the problems at complex frequencies other than the radiation conditions given by eqn. 4.4. This predetermines the continuation to the Riemann surface H. Limiting oneself to the consideration of the first sheet of H, it would not be possible to choose the cuts in a proper way and one would have to introduce the notion of the continuous spectrum (connected with the necessity to integrate along the cuts) which has no direct physical meaning. Eqn. 4.4 determines only the (third and the fourth) quarters of the complex plane κ where the cuts must go. They may have an arbitrary configuration on the whole surface H if to consider the complete statement of the problem given by eqns. 4.1 and 4.2 (for $\kappa \in H$). The choice of cuts depends on a conditional relation of an arbitrary point κ with a certain sheet of H and their determination by eqn. 4.5 is caused by the convenience for further analytical investigations of the problem.

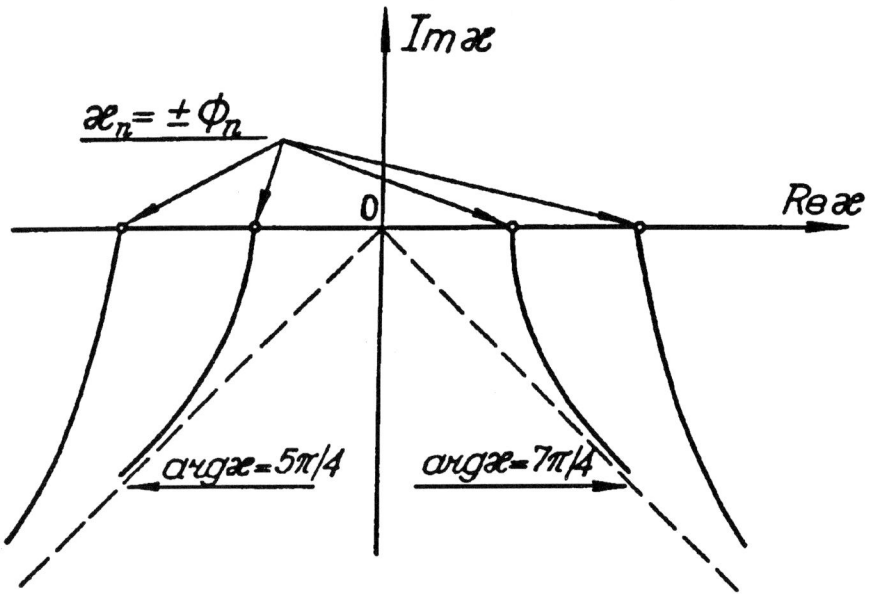

Figure 4.2 *The first sheet of the Riemann surface for a two-dimensional open waveguide resonator*

Let us consider in detail the first sheet of H (see Figure 4.2). For $0 < \arg \kappa < \pi - \operatorname{Im} \omega_n > 0$ and $\operatorname{Re} \omega_n \geq 0$, for $0 < \arg \kappa \leq \frac{\pi}{2}$ $\operatorname{Re} \omega_n \leq 0$ and for $\frac{\pi}{2} < \arg \kappa < \pi$ $\operatorname{Re} \omega_n \geq 0$ for all n. For the points κ from the fourth quarter $\frac{3\pi}{2} \leq \arg \kappa \leq 2\pi$ the values of a finite number of the functions ω_n (with the indices n satisfying the condition $d_m(\kappa) = (\operatorname{Re} \kappa)^2 - (\operatorname{Im} \kappa)^2 - \frac{m^2}{4} > 0$) are determined by the inequalities $\operatorname{Im} \omega_n < 0$ and $\operatorname{Re} \omega_n > 0$ while for the rest of the functions ω_n $\text{Im}\,\omega_n \geq 0$. The same conditions are stated in the domain $\pi < \arg \kappa \leq \frac{3\pi}{2}$ if the sign of $\text{Re}\,\omega_n$ is changed to the opposite one.

Such a structure of the first sheet of the Riemann surface H enables us to conclude that the series in the radiation conditions given by eqn. 4.2 are formed by the terms corresponding to the following types of waves:

(i) The waves which do not come to an inhomogeneity from infinity, grow in time and attenuate in space ($\text{Re}\,\omega_n\,\text{Re}\,\kappa \geq 0$, $\text{Im}\,\omega_n > 0$, $\text{Im}\,\kappa > 0$).

(ii) The waves which do not go from an inhomogeneity to infinity, grow in time and space ($\text{Re}\,\omega_n\,\text{Re}\,\kappa \leq 0$, $\text{Im}\,\omega_n < 0$, $\text{Im}\,\kappa > 0$).

(iii) The waves which do not come to an inhomogeneity from infinity, grow in space and attenuate in time ($\text{Re}\,\omega_n\,\text{Re}\,\kappa \geq 0$, $\text{Im}\,\omega_n < 0$, $\text{Im}\,\kappa \leq 0$).

(iv) The waves which do not go from an inhomogeneity to infinity, attenuate in space and time ($\text{Re}\,\omega_n\,\text{Re}\,\kappa \leq 0$, $\text{Im}\,\omega_n > 0$, $\text{Im}\,\kappa < 0$).

4.2 Uniqueness theorems

It is has been pointed out above that a part of the spectrum of the problem given by eqns. 4.1 and 4.2 consists of the points κ where its uniqueness is violated, i.e. where the homogeneous problem

$$\Delta v + \kappa^2 \epsilon v = 0, \quad \mathbf{r} \in S \tag{4.6}$$

has non-trivial solutions satisfying the generalised boundary conditions and the radiation conditions at infinity

$$v(\mathbf{r}) = \sum_{n=-\infty}^{\infty} b_{nl} \exp(j\omega_n x_3)\phi_n^l(x_2), \quad x_3 > 0 \quad (l = 1, 2) \tag{4.7}$$

These points practically exhaust the spectrum (see Chapter 5). Let us note that if the problem given by eqns. 4.6 and 4.7 has only a trivial solution for a certain (regular) value $\kappa \in H$, the corresponding inhomogeneous problem (given by eqns. 4.1 and 4.2) is uniquely solvable. In this Section we will describe some domains in H which consist only of the regular points establishing thus the uniqueness of the inhomogeneous problem and carrying out a rough description of the spectrum location.

Every non-trivial solution of the problem given by eqns. 4.6 and 4.7 yields the relation

$$\int_{\partial V} [\mathbf{E}, \mathbf{H}] \mathbf{n} \, dS = j \int_V (\omega \mu_0 |\mathbf{H}|^2 - \omega^* \epsilon^* \epsilon_0^* |\mathbf{E}|^2) \, dV \qquad (4.8)$$

which is known as the complex power theorem (Weinstein, 1951). Here V is the volume of a (unit-width along the x_1-axis) domain Q, ∂V is the boundary of V, $\mathbf{n}$ is a unit vector normal to ∂V.

In the TE case

$$H_{x_1} = E_{x_2} = E_{x_3} = 0$$

$$H_{x_2} = \frac{1}{j\omega\mu_0} \frac{\partial E_{x_1}}{\partial x_3}, \quad H_{x_3} = -\frac{1}{j\omega\mu_0} \frac{\partial E_{x_1}}{\partial x_2}$$

and

$$[\mathbf{E}, \mathbf{H}^*] = E_{x_1}(H_{x_2}^* \hat{i}_3 - H_{x_3}^* \hat{i}_2)$$

vanishes on ∂V.

In the TM case

$$E_{x_1} = E_{x_2} = H_{x_3} = 0$$

$$E_{x_2} = -\frac{1}{j\omega\epsilon\epsilon_0} \frac{\partial H_{x_1}}{\partial x_3}, \quad E_{x_3} = \frac{1}{j\omega\epsilon\epsilon_0} \frac{\partial H_{x_1}}{\partial x_2}$$

$$[\mathbf{E}, \mathbf{H}^*] = H_{x_1}^*(E_{x_3}^* \hat{i}_2 - E_{x_3} \hat{i}_3)$$

$[\mathbf{E}, \mathbf{H}^*]\mathbf{n}$ vanishes on ∂V and the following relations hold for the waveguide resonators with equal waveguide channels

$$\frac{\pi}{\epsilon_0 \mu_0} \sum_{m=1}^{\infty} \sum_{l=1}^{2} |b_{ml}|^2 \operatorname{Re} \omega_m$$

$$= \int_V \{ \operatorname{Im} \epsilon [(\operatorname{Re} \omega)^2 - (\operatorname{Im} \omega)^2] + 2 \operatorname{Re} \omega \operatorname{Im} \omega \operatorname{Re} \epsilon \} |\mathbf{E}|^2 \, dV$$

$$\frac{\pi}{\mu_0} \sum_{m=1}^{\infty} \sum_{l=1}^{2} |b_{ml}|^2 \operatorname{Im} \omega_m$$

$$= \int_V \{ -|\omega|^2 \mu_0 |\mathbf{H}|^2 + \epsilon_0 \operatorname{Re} \omega [(\operatorname{Re} \omega)^2 - (\operatorname{Im} \omega)^2]$$

$$- 2\epsilon_0 \operatorname{Re} \omega \operatorname{Im} \omega \operatorname{Im} \epsilon \} |\mathbf{E}|^2 \, dV$$

Similar energy relations hold in the TM case:

$$\frac{\pi}{\epsilon_0} \sum_{m=1}^{\infty} (1 + \delta_m^0) \sum_{l=1}^{2} |b_{ml}|^2 \operatorname{Re} \omega_m$$

$$= - \int_V \{\omega^2 \operatorname{Im} \epsilon(\epsilon_0 |\mathbf{E}|^2) + 2 \operatorname{Re} \omega \operatorname{Im} \omega(\mu_0 |\mathbf{H}|^2)\} \, \mathrm{d}\,V$$

$$\frac{\pi}{\epsilon_0} \sum_{m=0}^{\infty} (1 + \delta_m^0) \sum_{l=1}^{2} |b_{ml}|^2 \operatorname{Im} \omega_m$$

$$= \int_V \{\mu_0[(\operatorname{Re}\omega)^2 - (\operatorname{Im}\omega)^2]|\mathbf{H}|^2 - \omega^2 \operatorname{Re} \epsilon(\epsilon_0 |\mathbf{E}|^2)\} \, \mathrm{d}\,V$$

An analysis of the energy relations (separation of their real and imaginary parts together with comparison of signs of the left-hand and right-hand sides in the relations obtained) enables us to prove the following results which play the role of uniqueness theorems for diffraction problems in waveguides with inhomogeneities and are valid both in the TE and TM cases.

Theorem 4.2: Assume that $\omega \neq 0$, $V \neq \emptyset$ and $\operatorname{Im} \epsilon = 0$. Then the following situations are possible:

(i) For $\operatorname{Re}\omega \neq 0$, $\operatorname{Im}\omega \neq 0$ there are no non-trivial solutions of the problem given by eqns. 4.6 and 4.7 such that

$$\operatorname{Re}\omega \operatorname{Im}\omega \sum_{m=0}^{\infty} \sum_{l=1}^{2} |b_{ml}|^2 \operatorname{Re}\omega_m > 0$$

In particular, there are no "outgoing" solutions ($\operatorname{Re}\omega \operatorname{Re}\omega_n < 0$ for all n) for $\operatorname{Im}\omega > 0$ and "incoming" solutions ($\operatorname{Re}\omega \operatorname{Re}\omega_n > 0$ for all n) for $\operatorname{Im}\omega < 0$.

(ii) For $\operatorname{Re}\omega \neq 0$, $\operatorname{Im}\omega = 0$ there are no non-trivial solutions such that $\sum_{m=0}^{\infty} \sum_{j=1}^{2} |b_{mj}|^2 \operatorname{Re}\omega_m \neq 0$. It means, in particular, that "non-outgoing" solutions ($\operatorname{Re}\omega \operatorname{Re}\omega_n \leq 0$ for all n) and "non-incoming" solutions ($\operatorname{Re}\omega \operatorname{Re}\omega_n \geq 0$ for all n) may have the non-zero amplitudes b_{mj} only if $\operatorname{Re}\omega_m = 0$.

(iii) For $\operatorname{Re}\omega = 0$, $\operatorname{Im}\omega \neq 0$ there are no non-trivial solutions such that $\operatorname{Im}\omega_n > 0$ for all n.

(iv) For $|\operatorname{Im}\omega| \geq |\operatorname{Re}\omega|$ there are no non-trivial solutions such that $\sum_{m=0}^{\infty}\sum_{j=1}^{2}|b_{mj}|^{2}\operatorname{Im}\omega_{m} \geq 0$. In particular, there are no solutions for which $\operatorname{Im}\omega_{n} \geq 0$ for all n.

It follows from this theorem that the problem given by eqns. 4.6 and 4.7 has only trivial solutions in the domains $0 < \arg\kappa < \pi$ and $\frac{5\pi}{4} < \arg\kappa < \frac{7\pi}{4}$ on the first sheet of H. The real points κ where there may exist non-trivial solutions and which do not coincide with the threshold points $\kappa : \omega_{n}(\kappa) = 0$ and the branch points of H may belong in the TE case to the interval $(-0.5, 0.5)$.

Theorem 4.3: Assume that $\omega \neq 0$, $V \neq \emptyset$ and $\operatorname{Im}\epsilon \neq 0$. Then the following situations are possible:

(i) In the TE case there are no non-trivial solutions of the problem given by eqns. 4.6 and 4.7 such that $\{\operatorname{Im}\epsilon[(\operatorname{Re}\omega)^{2} - (\operatorname{Im}\omega)^{2} + 2\operatorname{Re}\omega\operatorname{Im}\omega\operatorname{Re}\epsilon]\}\sum_{m=0}^{\infty}\sum_{l=1}^{2}|b_{ml}|^{2}\operatorname{Re}\omega_{m} > 0$ everywhere in the domain Q (here it is assumed that ϵ is a function of the points of the domain Q). In particular, if $\operatorname{Re}\omega \neq 0$, $\operatorname{Im}\omega \neq 0$ and $|\operatorname{Re}\omega| \geq |\operatorname{Im}\omega|$, there are no "incoming" solutions for $\operatorname{Re}\omega < 0$, $\operatorname{Im}\omega < 0$ and "outgoing" solutions for $\operatorname{Re}\omega > 0$, $\operatorname{Im}\omega > 0$. If $|\operatorname{Re}\omega| \leq |\operatorname{Im}\omega|$, there are no "incoming" solutions for $\operatorname{Re}\omega > 0$, $\operatorname{Im}\omega > 0$ and "outgoing" solutions for $\operatorname{Re}\omega < 0$, $\operatorname{Im}\omega > 0$. There are no non-trivial solutions on the imaginary axis $\operatorname{Re}\omega = 0$ and on the real axis $\operatorname{Im}\omega = 0$ such that $\operatorname{Re}\omega_{n} \geq 0$ for all n.

(ii) In the TM case there are no non-trivial "outgoing" solutions of the problem given by eqns. 4.6 and 4.7 for $\operatorname{Re}\omega > 0$, $\operatorname{Im}\omega > 0$. There are no "incoming" solutions for $\operatorname{Im}\omega < 0$, $\operatorname{Re}\omega < 0$. There are no non-trivial solutions on the imaginary axis $\operatorname{Re}\omega = 0$ and on the real axis $\operatorname{Im}\omega = 0$ such that $\operatorname{Re}\omega_{n} \geq 0$ for all n.

For waveguides with plane inhomogeneities (an example is presented in Figure 4.1) when $V = \emptyset$ one can prove a similar assertion.

Theorem 4.4: Assume that $V = \emptyset$ (the waveguide considered has only plane inhomogeneities) Then there are no non-trivial solutions of the problem given by eqns. 4.6 and 4.7 such that

$$\sum_{m=0}^{\infty}\sum_{l=1}^{2}|b_{ml}|^2\,\mathrm{Re}\,\omega_m \neq 0 \quad \text{and (or)} \quad \sum_{m=0}^{\infty}\sum_{l=1}^{2}|b_{ml}|^2\,\mathrm{Im}\,\omega_m \neq 0$$

One easily obtains a positive result concerning the existence of spectral points as a corollary from theorem 4.4.

Theorem 4.5: Assume that the waveguide considered has only plane inhomogeneities. Then the problem given by eqns. 4.6 and 4.7 may have non-trivial solutions on the first sheet of H only in the threshold points $\kappa : \omega_n(\kappa) = 0$

In order to prove the last theorem it is sufficient to analyse the "suspicious" points only in the domains $0 < \arg\kappa < \frac{5\pi}{4}$ and $\frac{7\pi}{4} < \arg\kappa < 2\pi$. Let us consider here only the second domain where $\mathrm{Im}\,\omega_n < 0$, $\mathrm{Re}\,\omega_n > 0$, $|\mathrm{Re}\,\omega_n| > |\mathrm{Im}\,\omega_n|$, $d_n = (\mathrm{Re}\,\kappa)^2 - (\mathrm{Im}\,\kappa)^2 - \frac{n^2}{4} > 0$ for $n = n_1, n_2, \ldots n_k$ and $\mathrm{Im}\,\omega_n > 0$, $\mathrm{Re}\,\omega_n < 0$, $|\mathrm{Im}\,\omega_n| > |\mathrm{Re}\,\omega_n|$, $d_n < 0$ for all other values of n. Due to theorem 4.4, the condition necessary for the existence of a non-trivial solution in the point κ has the form of two equalities

$$\sum_{n=n_l}\sum_{l=1}^{2}|b_{nl}|^2\,\mathrm{Re}\,\omega_n = \sum_{n\neq n_l}\sum_{l=1}^{2}|b_{nl}|^2|\,\mathrm{Re}\,\omega_n| \quad (l = 1, 2, \ldots, k) \tag{4.9}$$

and

$$\sum_{n=n_l}\sum_{l=1}^{2}|b_{nl}|^2\,\mathrm{Im}\,\omega_n = \sum_{n\neq n_l}\sum_{l=1}^{2}|b_{nl}|^2|\,\mathrm{Im}\,\omega_n| \quad (l = 1, 2, \ldots, k) \tag{4.10}$$

which must hold simultaneously. But the latter is not possible since the left-hand side of eqn. 4.9 is greater than the left-hand side of eqn. 4.10 and the right-hand side of eqn. 4.9 is less than the right-hand side of eqn. 4.10.

4.3 Qualitative spectra characteristics

When studying the qualitative spectra characteristics one should consider mathematical objects (like the resolvents of the diffraction problems, operator-valued functions determining the equivalent operator equations or Green's functions) which admit direct investigation of the singularities of their analytical continuations with respect to the complex spectral parameter κ. In this way the approach based on the methods of the potential theory will be used here which enables one to consider both arbitrary inhomogeneities and communication domains and to obtain the Fredholm-type operator equations describing the spectral properties of the problems. In this Section we will consider in detail the case of the E-polarisation because the H-polarisation may be treated in a similar way.

Let us construct the solution $u = u(\mathbf{r}, \kappa)$ $(\mathbf{r} = (x_2, x_3))$ of the excitation problem given by eqns. 4.1 and 4.2 for the parallel-plane waveguide presented in Figure 4.1 assuming that $\epsilon = \epsilon(\mathbf{r})$ and $f = f(\mathbf{r})$ are continuously differentiable complex-valued functions. Following Titchmarsch (1961), we will introduce the sequence of functions $u_n = u_n(q, \kappa)$ such that

$$(\Delta + \kappa^2)u_0 = f$$
$$(\Delta + \kappa^2)u_1 = (1 - \epsilon)\kappa^2 u_0 + f$$

$$\cdots \tag{4.11}$$

$$(\Delta + \kappa^2)u_n = (1 - \epsilon)\kappa^2 u_{n-1} + f$$

and

$$u_0 = -\int_Q G_0 f \, d\mathbf{r}_0$$

$$u_1 = -\int_Q G_0[(1 - \epsilon)\kappa^2 u_0 + f] \, d\mathbf{r}_0$$

$$\cdots \tag{4.12}$$

$$u_n = -\int_Q G_0[(1 - \epsilon)\kappa^2 u_{n-1} + f] \, d\mathbf{r}_0$$

where $\mathbf{r}_0 = (x_{20}, x_{30})$ and $G_0 = G_0(\mathbf{r}, \mathbf{r}_0, \kappa)$ is Green's function given by eqn. 4.3. It turns out that this sequence is uniformly convergent. Hence it is natural to put $u = \lim_{n \to \infty} u_n$ and consider u as a solution to the excitation problem.

Lemma 4.1: There exists a domain $R \subset H$ such that the sequence defined by eqn. 4.12 uniformly (with respect to $\kappa \in R$ and $\mathbf{r} \in Q$) converges to the function

$$u(\mathbf{r}, \kappa) = -\int_Q G_0\{[1 - \epsilon(\mathbf{r}_0)]\kappa^2 u(\mathbf{r}_0, \kappa) + f(\mathbf{r}_0)\} \, d\mathbf{r}_0, \quad \mathbf{r} \in Q \tag{4.13}$$

Let us give the scheme of the proof of this lemma. Since $\int_Q |G_0|^2 d\mathbf{r}_0 \leq M(\kappa)$ for $\kappa \neq \kappa_n$ $(n = 1, 2, \ldots)$, it follows from the Schwartz inequality that $u_0(\mathbf{r}, \kappa)$ is bounded for all $\kappa \neq \kappa_0$ and $\mathbf{r} \in Q$ as well as all $u_n(\mathbf{r}, \kappa)$ $(n = 1, 2, \ldots)$. There exist a constant $\theta > 0$ such that

$$|u_n(\mathbf{r}_0, \kappa) - u_{n-1}(\mathbf{r}_0, \kappa)| \leq \theta$$

Then

$$|u_{n+1}(\mathbf{r}_0, \kappa) - u_n(\mathbf{r}_0, \kappa)| = \left| \int_Q G_0(1 - \epsilon)\kappa^2 (u_n - u_{n-1}) d\mathbf{r}_0 \right|$$
$$\leq \theta|\kappa|^2 \sqrt{\int_Q |G_0|^2 d\mathbf{r}_0 \int_Q |1 - \epsilon|^2 d\mathbf{r}_0}$$

for all $\mathbf{r} \in Q$ and, consequently,

$$|u_{n+1}(\mathbf{r}_0, \kappa) - u_n(\mathbf{r}_0, \kappa)| \leq 0.5\theta$$

for sufficiently small $|\kappa|$ $(\kappa \neq \kappa_n)$ what yields (due to the Cauchy criterion) the required uniform convergence of the sequence u_n. Taking the limit in eqn. 4.2 we get eqn. 4.13.

Now it follows from lemma 4.1 and the properties of $\epsilon(\mathbf{r})$ and $f(\mathbf{r})$ that $u(\mathbf{r}, \kappa)$ is a twice continuously differentiable solution of the problem given by eqns. 4.1 and 4.2. This function may be continued with the help of eqn. 4.13 to the complex domain R and considered as an analytical function of $\kappa \in R$ (an analytical solution of the problem given by eqns. 4.1 and 4.2).

Lemma 4.2: $u(\mathbf{r}, \kappa)$ may be continued from the domain $R \subset H$ to the whole surface H as a meromorphic function of κ where $u(\mathbf{r}, \kappa)$ keeps all its properties as a solution to the problem given by eqns. 4.1 and 4.2.

The proof of this lemma is based on the analysis of the operator-valued function

$$A(\kappa)[u] = \kappa^2 \int_Q g_0(\mathbf{r}, \mathbf{r}_0, \kappa) u(\mathbf{r}_0, \kappa) d\mathbf{r}_0, \quad \kappa \in H \tag{4.14}$$

where $g_0(\mathbf{r}, \mathbf{r}_0, \kappa) = G_0[1 - \epsilon(\mathbf{x}_0)]$. Let us show that $A(\kappa) : L_2(Q) \to L_2(Q)$, $\kappa \in H$ is a compact meromorphic operator-valued function. In fact, the kernel

$$g_0(\mathbf{r}, \mathbf{r}_0, \kappa) = B(\mathbf{r}, \mathbf{r}_0, \kappa) + C(\mathbf{r}, \mathbf{r}_0, \kappa)$$

where

$$B(\mathbf{r}, \mathbf{r}_0, \kappa) = \frac{D(\mathbf{r}, \mathbf{r}_0, \kappa)}{\ln \rho} P(\rho), \quad C(\mathbf{r}, \mathbf{r}_0, \kappa) = \frac{D(\mathbf{r}, \mathbf{r}_0, \kappa)}{\ln \rho}[1 - P(\rho)]$$

$$P(\rho) = \begin{cases} 0, & \rho \geq \theta \\ 1, & \rho \leq \frac{1}{2}\theta \end{cases} \quad \rho = \rho(\mathbf{r}, \mathbf{r}_0) = \sqrt{(x_2 - x_{20})^2 + (x_3 - x_{30})^2}$$

The operator $A_C(\kappa)[u] = \int_Q C(\mathbf{r}, \mathbf{r}_0, \kappa) u(\mathbf{r}_0, \kappa)\, d\mathbf{r}_0$ is a compact finite-mero-morphic operator-valued function in $L_2(Q)$ on the Riemann surface H. The L_2-norm of the operator $A_B(\kappa)[u] = \int_Q B(\mathbf{r}, \mathbf{r}_0, \kappa) u(\mathbf{r}_0, \kappa)\, d\mathbf{r}_0$ tends to zero for $\theta \to 0$ since

$$|A_B(\kappa)[u]|^2 = \left| \int_Q D(\mathbf{r}, \mathbf{r}_0, \kappa) \ln^{-1} \rho P(\rho) u(\mathbf{r}_0, \kappa)\, d\mathbf{r}_0 \right|^2$$

$$\leq \text{const} \int_{\rho \leq \theta} \ln^{-2} \rho\, d\mathbf{r}_0 \int_Q |u(\mathbf{r}_0, \kappa)|^2\, d\mathbf{r}_0$$

which yields

$$\|A_B(\kappa)[u]\| \leq \text{const}\, \theta \|u\|$$

Therefore, $A(\kappa) = A_C(\kappa) + A_B(\kappa)$ is is also a compact finite-meromorphic operator-valued function in $L_2(Q)$ because $A(\kappa)$ is the limiting operator (with respect to the L_2-norm) of the family of compact operators $A_C(\kappa)$.

It follows from lemma 4.2 and the results of Section 4.2 that the resolvent set of the problem given by eqns. 4.1, 4.2 (or of the corresponding operator defined in eqn. 4.13) is not empty, i. e. due to the analytical Fredholm theorem (Sanchez-Palencia, 1980) the resolvent $[I + A(\kappa)]^{-1}$ (I is a unit operator) is a holomorphic function defined in H except for not more than a countable set of (isolated) finite-order poles. Hence the solutions $u(\mathbf{r}, \kappa)$ of eqn. 4.13 (or, what is the same, of the problem given by eqns. 4.1, 4.2) are the functions (of κ) meromorphic in H.

The demands concerning smoothness of $\epsilon(\mathbf{r})$ may be lessened. In fact, eqn. 4.12 and 4.13 remain valid (as the equalities of continuous functions) if it is assumed that $\epsilon(\mathbf{r})$ is a bounded piece-wise continuous and piece-wise differentiable function in Q.

Now we can formulate the result concerning the spectrum discreteness.

Theorem 4.6: The spectrum of the problem given by eqns. 4.1 and 4.2 considered in a parallel-plane waveguide containing a dielectric inhomogene-ity with the piece-wise continuous permittivity consists of not more than a countable number of isolated points situated on the Riemann surface H and the resolvent of the problem has the finite-order poles in these points.

Let us rewrite eqn. 4.12 in the form

$$u_n(\mathbf{r}, \kappa) = -\int_Q G_n(\mathbf{r}, \mathbf{r}_0, \kappa) f(\mathbf{r}_0)\, d\mathbf{r}_0$$

where

$$G_n(\mathbf{r}, \mathbf{r}_0, \kappa) = G_0(\mathbf{r}, \mathbf{r}_0, \kappa) - \kappa^2 \int_Q G_0(\mathbf{r}, q_1, \kappa) G_{n-1}(\mathbf{r}_1, q, \kappa)[(1 - \epsilon(\mathbf{r}_1)]\, d\mathbf{r}_1$$

$$(4.15)$$

It may be shown similarly to the proof of lemma 4.1 that there exist a domain $R \subset H$ such that the sequence defined by eqn. 4.15 uniformly (with respect to $\kappa \in R$ and $\mathbf{r} \in Q$) converges to the function

$$G(\mathbf{r}, \mathbf{r}_0, \kappa) = G_0(\mathbf{r}, \mathbf{r}_0, \kappa) - \kappa^2 \int_Q G_0(\mathbf{r}, \mathbf{r}_1, \kappa) G(\mathbf{r}_1, \mathbf{r}, \kappa)[(1 - \epsilon(\mathbf{r}_1)]\, d\mathbf{r}_1 \quad (4.16)$$

and

$$u(\mathbf{r}, \kappa) = -\int_Q G(\mathbf{r}, \mathbf{r}_0, \kappa) f(\mathbf{r}_0)\, d\mathbf{r}_0 \qquad (4.17)$$

We will call $G(\mathbf{r}, \mathbf{r}_0, \kappa)$ the Green's function of the problem given by eqns. 4.1, 4.2.

Lemma 4.2 may be re-formulated for the function $G(\mathbf{r}, \mathbf{r}_0, \kappa)$ which may be continued to the whole surface H as a meromorphic function of κ for all $\mathbf{r}, \mathbf{r}_0 \in Q$.

Most of the properties of $G(\mathbf{r}, \mathbf{r}_0, \kappa)$ formulated below follow from the fact that this function is a limit (uniform with respect to all the parameters) solution to eqn. 4.15 which is a second-kind integral equation with a weak singularity of the kernel.

Lemma 4.3: The function $G(\mathbf{r}, \mathbf{r}_0, \kappa)$ satisfies the following conditions for all $\mathbf{r}, \mathbf{r}_0 \in Q$, $\mathbf{r} \neq \mathbf{r}_0$ and $\kappa \in R$.

(i) $G(\mathbf{r}, \mathbf{r}_0, \kappa) = G(\mathbf{r}_0, \mathbf{r}, \kappa) = G^*(\mathbf{r}, \mathbf{r}_0, -\kappa^*)$ and is an analytical function of κ.

(ii) $G(\mathbf{r}, \mathbf{r}_0, \kappa)$ is once continuously differentiable with respect to $\mathbf{r}, \mathbf{r}_0$ and

$$G = -\frac{1}{2\pi} \ln \rho + O(1), \quad \frac{\partial G}{\partial \rho} = -\frac{1}{2\pi\rho} + O(1), \quad \rho \to 0$$

(iii) $G(\mathbf{r}, \mathbf{r}_0, \kappa)$ satisfies the Helmholtz equation $(\Delta_\mathbf{r} + \kappa^2 \epsilon)G = 0$ and the conditions given by eqn. 4.2

In order to prove the symmetry of $G(\mathbf{r}, \mathbf{r}_0, \kappa)$ [the condition (i)] let us apply Green's identity in the domain Q:

$$
\begin{aligned}
(\kappa^2 - \kappa_1^2) & \int_Q G(\mathbf{r}_1, \mathbf{r}, \kappa) G(\mathbf{r}_1, \mathbf{r}_0, \kappa) \epsilon(\mathbf{r}_1) \, d\mathbf{r}_1 \\
&= \int_0^{2\pi} \left[G(\mathbf{r}_1, \mathbf{r}, \kappa) \frac{\partial}{\partial \mathbf{n}} G(\mathbf{r}_1, \mathbf{r}_0, \kappa) - G(\mathbf{r}_1, \mathbf{r}_0, \kappa) \frac{\partial}{\partial \mathbf{n}} G(\mathbf{r}_1, \mathbf{r}, \kappa) \right] d\mathbf{r}_1 \\
&\quad + G(\mathbf{r}_0, \mathbf{r}, \kappa) - G(\mathbf{r}, \mathbf{r}_0, \kappa_1)
\end{aligned}
\tag{4.18}
$$

and assume that $\kappa \to \kappa_1$. Then one gets $G(\mathbf{r}_0, \mathbf{r}, \kappa) = G(\mathbf{r}, \mathbf{r}_0, \kappa)$ since the first term in the right-hand side of eqn. 4.18 vanishes due to the radiation conditions.

The second equality in (i) is easily checked for canonical Green's function G_0

$$
\begin{aligned}
G_0(\mathbf{r}, \mathbf{r}_0, \kappa) &= \frac{j}{2\pi} \sum_{n=-\infty}^{\infty} \exp[j\omega_n(\kappa)|x_3 - x_{30}|] \left(\sin \frac{1}{2} n x_2 - \sin \frac{1}{2} n x_{20} \right) \omega_n^{-1}(\kappa) \\
&= -\left(\frac{j}{2\pi} \right)^* \sum_{n=-\infty}^{\infty} exp\{(j)^*[\omega_n(-\kappa^*)]^*|x_3 - x_{30}|\} \\
&\quad \times \left(\sin \frac{1}{2} n x_2 - \sin \frac{1}{2} n x_{20} \right) [\omega_n^*(-\kappa^*)]^{-1} \\
&= G_0^*(\mathbf{r}, \mathbf{r}_0, -\kappa^*)
\end{aligned}
\tag{4.19}
$$

and may be proved for $G(\mathbf{r}, \mathbf{r}_0, \kappa)$ if eqns. 4.16 and 4.19 directly apply.

Let us note that Green's function $G(\mathbf{r}, \mathbf{r}_0, \kappa)$ may be continued to H as a meromorphic function of κ such that all the results of lemma 4.3 remain valid.

Now we can formulate and prove the spectral theorem for the problem given by eqns. 4.1 and 4.2.

Theorem 4.7: The spectrum of the problem given by eqns. 4.1 and 4.2 considered in a parallel-plane waveguide containing perfectly conducting (metal) inhomogeneities bounded by piece-wise continuous contours Γ consists of not more than a countable number of isolated points situated on the Riemann surface H and the resolvent of the problem has the finite-order poles in these points.

In order to prove this theorem we will introduce Green's function $G(\mathbf{r}, \mathbf{r}_0, \kappa)$ of the problem given by eqns. 4.1 and 4.2 considered in a parallel-plane waveguide without perfectly conducting inhomogeneities and represent the solution to this problem in a waveguide with inhomogeneities as a sum of the double-layer potential on Γ and the volume potential connected with the source function

$$u(\mathbf{r}', \kappa) = \int_\Gamma \mu(\mathbf{r}_0) \frac{\partial}{\partial \mathbf{n}} G(\mathbf{r}', \mathbf{r}_0, \kappa) \, d\, l_{\mathbf{r}_0} - \int_Q G(\mathbf{r}', \mathbf{r}_0, \kappa) f(\mathbf{r}_0) \, d\, \mathbf{r}_0$$

Calculating the (external) limit for $\mathbf{r}' \to \mathbf{r} \in \Gamma$ and taking into account the boundary conditions for u and the properties of Green's function (see lemma 4.3) we obtain the boundary integral equation equivalent to the problem given by eqns. 4.1 and 4.2

$$-\mu(\mathbf{r}) = 2 \left[\int_\Gamma \mu(\mathbf{r}_0) \frac{\partial}{\partial \mathbf{n}} G(\mathbf{r}, \mathbf{r}_0, \kappa) \, d\, l_{\mathbf{r}_0} - \int_Q G(\mathbf{r}, \mathbf{r}_0, \kappa) f(\mathbf{r}_0) \, d\, \mathbf{r}_0 \right] \quad (4.20)$$

$$\mathbf{r} \in \Gamma \subset Q$$

where

$$B(\kappa)[\mu] = 2 \int_\Gamma \mu(\mathbf{r}_0) \frac{\partial}{\partial \mathbf{n}} G(\mathbf{r}, \mathbf{r}_0, \kappa) \, d\, l_{\mathbf{r}_0} \; : \; L_2(Q) \to L_2(Q)$$

is a compact finite-meromorphic operator-valued function of $\kappa \in H$. Due to the uniqueness theorem the resolvent set of the problem considered given by eqns. 4.1 and 4.2 (and of the corresponding operator defined in eqn. 4.20) is not empty, and hence the resolvent $[I + B(\kappa)]^{-1}$ is a holomorphic function defined in H except for not more than a countable set of (isolated) finite-order poles. The last assertion completes the proof of theorem 4.7.

The results of lemma 4.2 and theorem 4.6 may be applied for the determination of qualitative spectra characteristics for waveguides with non-ideal metal inhomogeneities conditionally replaced by dielectric inclusions with big $|\operatorname{Im} \epsilon|$.

Now we can assert that the whole spectrum of the problem given by eqns. 4.1 and 4.2 consists of the points $\kappa \in H$ where the homogeneous problem given by eqns. 4.6 and 4.7 has non-trivial solutions. In fact, the residues of Green's function calculated in its singularities (poles) coincide with the latter. On the other hand, if it is assumed that such a non-trivial solution exists for a certain value of $\kappa = \kappa_0$ which is not a pole of Green's function (i.e. κ_0 is the regular value), application of eqn. 4.17 leads to the contradiction.

Let us denote by $\tilde{G}_0$ Green's function of the problem given by eqns. 4.1 and 4.2 in a parallel-plane waveguide with arbitrarily perturbed rectilinear walls (see Figure 4.3) and assume that $\tilde{G}_0$ satisfies the same conditions as canonical Green's function G_0. Then we can formally define Green's function $\tilde{G}$ of the problem given by eqns. 4.1 and 4.2 in a perturbed parallel-plane waveguide with arbitrary dielectric inclusions in the communication domain Q (as if to repeat the procedure of constructing the convergent sequences described in lemma 4.1 where G_0 is replaced by $\tilde{G}_0$) and prove results similar to the ones stated in theorem 4.7.

Theorem 4.8: The spectrum of the problem given by eqns. 4.1 and 4.2 considered in an arbitrarily perturbed parallel-plane waveguide containing perfectly conducting (metal) and dielectric inhomogeneities consists of not more than a countable number of isolated points situated on the Riemann surface H and the resolvent of the problem has the finite-order poles in these points.

We will consider in what follows only the waveguide presented in Figure 4.3d assuming that the results may be applied for simpler configurations of other perturbed parallel-plane waveguides depicted in Figure 4.3.

Let $u = u(\mathbf{r}, \kappa)$ be a solution to the problem given by eqns. 4.1 and 4.2 considered in a parallel-plane waveguide mentioned in theorem 4.8

Lemma 4.4: $u(\mathbf{r}, \kappa)$ may have not more than a finite number of isolated finite-order poles (singularities of Green's function $\tilde{G}$) in any bounded domain $D \subset H$.

We will present the main stages of the proof of this lemma.

Let us divide the communication domain Q into two sub-domains Q_1 and Q_2 with a non-empty intersection bounded by the lines z_1 (situated inside in the regular waveguide's part) and z_2 (continuation of the walls of the regular waveguide's parts) and assume that $\text{supp} f \cap Q_2 = \emptyset$. Then one can obtain the following representations for the solution to the problem given by eqns.

4.1 and 4.2 using Green's identity

$$u_1(\mathbf{r}', \kappa) = - \int_{z_2} u_1(\mathbf{r}_0) \frac{\partial}{\partial \mathbf{n}_{\mathbf{r}_0}} G_0(\mathbf{r}', \mathbf{r}_0, \kappa) \, d\,l_{\mathbf{r}_0} - \int_{Q_1} G(\mathbf{r}', \mathbf{r}_0, \kappa) f(\mathbf{r}_0) \, d\,\mathbf{r}_0 \tag{4.21}$$

$$u_2(\mathbf{r}', \kappa) = - \int_{z_1} u_2(\mathbf{r}_0) \frac{\partial}{\partial \mathbf{n}_{\mathbf{r}_0}} G_1(\mathbf{r}', \mathbf{r}_0, \kappa) \, d\,l_{\mathbf{r}_0}, \quad \mathbf{r}' \in Q_1 \cup Q_2 \tag{4.22}$$

where $G_1(\mathbf{r}', \mathbf{r}_0, \kappa)$ is Green's function of the first boundary-value problem in Q_2 for the Helmholtz equation. Calculating the limits for $\mathbf{r}' \to \mathbf{r} \in z_1$ in eqn. 4.21 and for $\mathbf{r}' \to \mathbf{r} \in z_2$ in eqn. 4.22 with taking into account that $u_1 = u_2$ in $Q_1 \cap Q_2$ we obtain the system boundary integral equation equivalent to the problem given by eqns. 4.1 and 4.2

$$\begin{cases} u_1(\mathbf{r}, \kappa) = - \int_{z_2} u_2(\mathbf{r}_0) \frac{\partial}{\partial \mathbf{n}_{\mathbf{r}_0}} G_0(\mathbf{r}, \mathbf{r}_0, \kappa) \, d\,l_{\mathbf{r}_0} \\ \quad - \int_{Q_1} G(\mathbf{r}, \mathbf{r}_0, \kappa) f(\mathbf{r}_0) \, d\,\mathbf{r}_0, & \mathbf{r} \in z_1 \\ u_2(\mathbf{r}, \kappa) = - \int_{z_1} u_1(\mathbf{r}_0) \frac{\partial}{\partial \mathbf{n}_{\mathbf{r}_0}} G_1(\mathbf{r}, \mathbf{r}_0, \kappa) \, d\,l_{\mathbf{r}_0}, & \mathbf{r} \in z_2 \end{cases} \tag{4.23}$$

Eqn. 4.23 may be considered as an operator equation $\mathbf{u} = T\mathbf{u} + \mathbf{F}$ with respect to unknowns $\mathbf{u} = \{u_1, u_2\}$ in the space $L_2(z_1) \times L_2(z_2)$. Here

$$\mathbf{F} = \left\{ - \int_Q G(\mathbf{r}, \mathbf{r}_0, \kappa)|_{\mathbf{r} \in z_1} f(\mathbf{r}_0) \, d\,\mathbf{r}_0, \ 0 \right\}$$

and $T = \{T_{il}\}_{i,l=1}^2$ is the matrix operator with the operator elements

$$T_{11} = T_{22} = 0$$

$$T_{11}[-u] = - \int_{z_2} u(\mathbf{r}_0) \frac{\partial}{\partial \mathbf{n}_{\mathbf{r}_0}} G_0(\mathbf{r}, \mathbf{r}_0, \kappa) \, d\,l_{\mathbf{r}_0}$$

$$T_{22}[u] = - \int_{z_1} u(\mathbf{r}_0) \frac{\partial}{\partial \mathbf{n}_{\mathbf{r}_0}} G_1(\mathbf{r}, \mathbf{r}_0, \kappa) \, d\,l_{\mathbf{r}_0}$$

Due to the properties of Green's functions $G_0(\mathbf{r}, \mathbf{r}_0, \kappa)$ and $G_1(\mathbf{r}, \mathbf{r}_0, \kappa)$ $T(\kappa)$ is a compact finite-meromorphic operator-valued function of $\kappa \in H$ in $L_2(z_1) \times L_2(z_2)$. Hence $I - T(\kappa)$ is a canonical Fredholm finite-meromorphic operator-valued function and this assertion completes the proof of lemma 4.4.

4.4 Existence of fundamental frequencies

The problem stated in the title may be solved by comparatively simple methods only for a limited number of canonical structures but may form the basis for consideration of rather general families of electrodynamic objects.

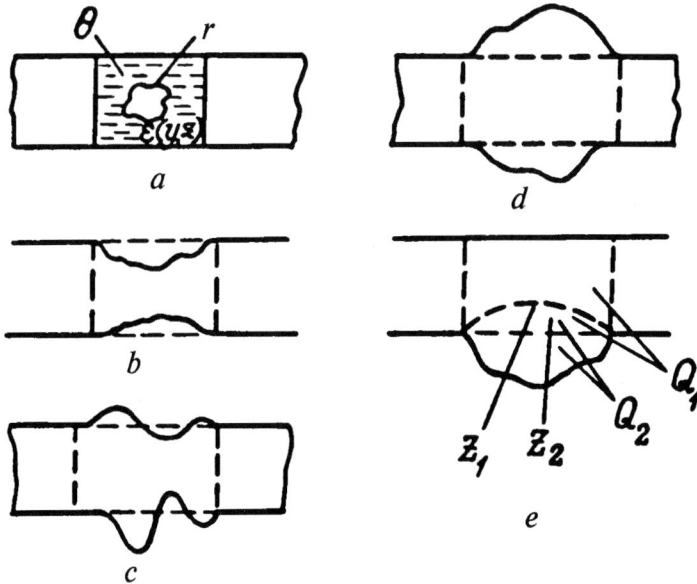

Figure 4.3 *Two-dimensional plane waveguide resonators*

Let us point out some canonical types of open waveguide resonators beginning with the resonator depicted in Figure 4.3. The analytical continuation of the corresponding Green's function G_0 with respect to the complex spectral parameter $\kappa \in H$ given by eqn. 4.3 has simple poles in the points $\tau_n = \sqrt{\kappa - \kappa_n}$ $(n = 0, 1, 2, \ldots)$ and

$$\operatorname{res} G_0 \big|_{\tau_n = 0} = \frac{j}{4\pi} \phi_n(x_2) \phi_n(x_{20}) \frac{1}{2\kappa_n}$$

are the residues of G_0 in these poles. $v_n = \phi_n(x_2)$ are the non-trivial solutions of the problem given by eqns. 4.6 and 4.7 which do not depend on the longitudinal co-ordinate parallel to waveguide's axis and determine the eigenoscillations having the form of "standing waves". This example indirectly confirms the necessity to consider analytical continuations of Green's functions and spectral problems on the whole Riemann surface H.

One can prove in a similar way that there exist non-trivial solutions of the problem given by eqns. 4.6 and 4.7 in open waveguide resonators with perfectly conducting "knife" inhomogeneities and in other types of perturbed parallel-plane waveguide resonators (some of them are presented in Figure 4.4).

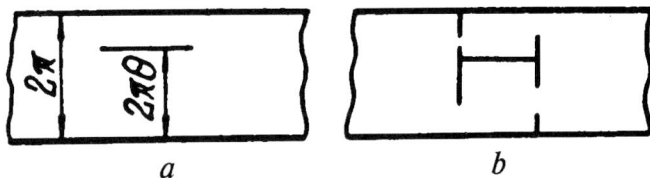

Figure 4.4 *Knife-type open waveguide resonators in a plane waveguide*

Let us consider a parallel-plane waveguide resonator with a dielectric plate (see Figure 4.1, $\epsilon =$const). Here the spectral problem in the TE case is reduced to the dispersion equation

$$\omega_{n\epsilon}[1 \mp \exp(4j\pi\delta\omega_{n\epsilon})] + \omega_{n\epsilon}[1 \pm \exp(4j\pi\delta\omega_{n\epsilon})] = 0$$

whose roots determine the fundamental frequencies of the eigenoscillations with

$$E_{x_1}(\mathbf{r}) = \begin{cases} b_{n1}\exp[j\omega_n(x_3 - 2\pi\delta)]\sin\frac{1}{2}nx_2, & x_3 > 2\pi\delta \\ b_{n2}\exp[-j\omega_n(x_3 + 2\pi\delta)]\sin\frac{1}{2}nx_2, & x_3 < -2\pi\delta \end{cases}$$

where the signs $\pm$ in $b_{n1} = \pm b_{n2}$ correspond to the oscillations which are symmetric (anti-symmetric) with respect to the plane $x_2 = 0$ and the signs of $\omega_{n\epsilon}(\kappa)$ are co-ordinated with the values of $\omega_n(\kappa)$ on the first sheet of H.

Now we will proceed from explicit consideration of important particular examples where one can easily find fundamental frequencies to a general approach based on a "small" perturbation of canonical problems. We will prove the existence of fundamental frequencies for the problem given by eqns. 4.1 and 4.2 in a parallel-plane waveguide resonator with an arbitrary perfectly conducting inhomogeneity situated in the domain Q (see Figure 4.4a where

$\epsilon(\mathbf{r})$ =const). Let us denote by θ (Γ_θ) the characteristic size (the contour of the transversal cross-section) of the latter and by $\kappa_n \in H$ the branch points of the analytical continuation of canonical Green's function.

Lemma 4.5: Let the domain $D_\gamma \subset H$ be bounded by a simple closed curve γ which surrounds $\kappa_{n'}$ and does not contain any other point $\kappa_n \in H$ $(n \neq n')$. Then there exists a sufficiently small $\theta > 0$ such that the perturbed problem given by eqns. 4.1 and 4.2 has at least one fundamental frequency $\tilde{\kappa}_{n'}$ in a two-sheet circle $D \subset D_\gamma$.

The proof of this lemma is based on the methods of the perturbation theory summarised by Sanchez-Palencia (1980). In fact, $\kappa = \kappa_{n'}$ is the only singularity in D of the solution $u = u(\mathbf{r}, \kappa, \theta)$ of the canonical problem given by eqns. 4.1 and 4.2 with the right-hand side $f = f(\mathbf{r}) \in L_2(Q)$, supp $f \subset Q$ and a perfectly conducting inhomogeneity bounded by Γ_θ. If it is assumed that beginning from a certain positive θ D contains only regular points, one can find a sequence $\theta_m \to 0$ $(m \to \infty)$ such that $u = u(\mathbf{r}, \kappa, \theta_m)$ are the holomorphic solutions of $\kappa \in D$ to the diffraction problems given by eqns. 4.1 and 4.2. Hence

$$\|u(\mathbf{r}, \kappa, \theta_m)\| \leq C \tag{4.24}$$

with respect to the L_2-norm uniformly for all $\kappa \in D_\gamma$ and

$$\|u(\mathbf{r}, \kappa, \theta) - u(\mathbf{r}, \kappa, \theta_m)\| \to 0, \quad m \to \infty \tag{4.25}$$

for a fixed $\kappa \in \gamma$. But

$$\int_\gamma u(\mathbf{r}, \kappa, \theta) \, d\tau_n = w(\mathbf{r}) \not\equiv 0, \quad w(\mathbf{r}) \in L_2(Q) \quad (\mathbf{r} \in Q) \tag{4.26}$$

since the Laurent series in D of the analytical continuation of canonical Green's function contains terms proportional to $\tau_{n'}^{-1}$ where $\tau_{n'} = \sqrt{\kappa - \kappa_{n'}}$ is the local variable in D. In addition to this,

$$\int_\gamma u(\mathbf{r}, \kappa, \theta_m) \, d\tau_n \equiv 0 \quad (\mathbf{r} \in Q \setminus \text{int} \, \Gamma_{\theta_m}) \tag{4.27}$$

since $u(\mathbf{r}, \kappa, \theta_m)$ has no singularities in D (here intΓ_{θ_m} is the domain occupied by the inhomogeneity).

Taking the limit $m \to \infty$ in eqn. 4.27 under the integral sign (which is possible due to eqns. 4.24 and 4.25) we come to the result which contradicts to eqn. 4.26. The last assertion completes the proof of lemma 4.5.

This lemma may be used for solving spectral problems considered as "continuous modifications" of the canonical problems with respect to an appropriate (non-spectral) parameter. Justification of the corresponding approach is based on the spectrum continuous dependence on problem's parameters.

Lemma 4.6: Let D and $T(\kappa, \mu)$ be a simply-connected domain of the Riemann surface H and a family of compact operators acting in a Banach space X defined for $\kappa \in D$, $\mu \in [\bar{\mu}_1, \bar{\mu}_2]$, respectively. Assume that

(i) $T(\kappa, \mu)$ is a holomorphic function of κ for every $\mu \in [\bar{\mu}_1, \bar{\mu}_2]$;

(ii) $T(\kappa, \mu)$ is a continuous function on $D \times [\bar{\mu}_1, \bar{\mu}_2]$ with respect to the operator norm;

(iii) a bounded operator $R_T(\kappa, \mu) = [I - T(\kappa, \mu)]^{-1}$ (the resolvent of T) exists at least in one point $\kappa \in D$ for all $\mu \in [\bar{\mu}_1, \bar{\mu}_2]$. Then the poles of R_T continuously depend on μ, i.e. they may emerge (or disappear) only on the boundary of D.

Let us present the scheme of the proof belonging to Sanchez-Palencia (1980). Let $\mu_m \to \mu_\infty \in [\bar{\mu}_1, \bar{\mu}_2]$ $(m \to \infty)$ and the poles κ_m of $R_T(\kappa, \mu_m)$ tend to $\kappa_\infty \in D$. Then κ_∞ is a pole of $R_T(\kappa, \mu)$. If it is assumed that κ_∞ is a regular point of $R_T(\kappa, \mu_\infty)$, then $R_T(\kappa, \mu)$ is a continuous function in a vicinity of $(\kappa_\infty, \mu_\infty)$ with respect to the operator norm, not possible since κ_∞ is the accumulation point of the poles of $R_T(\kappa, \mu_m)$ for $\mu_m \to \mu_\infty$. If κ_∞ is a pole of $R_T(\kappa, \mu_\infty)$, then there exists at least one pole of $R_T(\kappa, \mu_m)$ in every (sufficiently small) vicinity of κ_∞. Assume, on the contrary, that there exists the vicinity P of κ_∞ and the sequences $\mu_m \to \mu_\infty$ such that $R_T(\kappa, \mu_m)$ have no poles in P for every m. Then $R_T(\kappa, \mu_m)$ are holomorphic operator-functions in P due to the analytic Fredholm theorem and it is easy to get a contradiction similar to the one obtained in the previous case.

In order to show the existence of fundamental frequencies in parallel-plane waveguide resonators with arbitrary dielectric inhomogeneities it is sufficient, in accordance with the last lemma, to take an appropriate canonical problem (for example, a parallel-plane waveguide resonator without inclusions or containing a dielectric plate with the constant permittivity $\epsilon(\mathbf{r}) =$const) and to prove the continuity (with respect to the L_2-norm) of the operator-valued function $A(\kappa)[u]$ introduced in lemma 4.2 for $\kappa \in H$, $\kappa \neq \kappa_n$ when $\epsilon(\mathbf{r})$ varies as an element of $L_2(Q)$. The required property of $A(\kappa)[u]$ follows from the inequality

$$\|A(\kappa, \epsilon_1)[u] - A(\kappa, \epsilon_2)[u]\| \leq \text{const} \, \|u\| \|\epsilon_1 - \epsilon_2\|$$

which is obtained after estimating the difference of two expressions given by eqn. 4.14 (for $\epsilon = \epsilon_1$ and $\epsilon = \epsilon_2$) with the help of the Schwartz inequality.

The same procedure is applied for the proof of the existence of the fundamental frequencies in parallel-plane waveguide resonators with arbitrary perfectly conducting inhomogeneities. Here it is convenient to introduce the specially constructed "perturbing" operator which describes variations of the boundary Γ of an inhomogeneity. Let us rewrite eqn. 4.20 in the form

$$B(\Gamma)[\mu] = \int_Q \mu(\mathbf{r}_0) P(\mathbf{r}, \Gamma) P(\mathbf{r}_0, \Gamma) \frac{\partial}{\partial \mathbf{n}} G(\mathbf{r}, \mathbf{r}_0, \kappa) \, \mathrm{d}\mathbf{r}_0, \quad \Gamma \subset Q \qquad (4.28)$$

and consider $B(\Gamma)[\mu]$ as an operator acting in $L_2(Q)$. Here

$$P(\mathbf{r}, \Gamma) = \begin{cases} 1, & \mathbf{r} \in \Gamma \\ 0, & \mathbf{r} \notin \Gamma \end{cases}$$

Forming the difference of two operators $B(\Gamma_1)[\mu]$ and $B(\Gamma_2)[\mu]$ defined in eqn. 4.14 for two different contours Γ_1 and Γ_2 we obtain after integrating over Q and applying the Schwartz inequality that

$$\|B(\Gamma_1)[\mu] - B(\Gamma_2)[\mu]\| \leq \text{const} \, \|\mu\| \sqrt{B_{12}} \qquad (4.29)$$

where

$$B_{12} = \int_Q \int_Q \big| P(\mathbf{r}, \Gamma_1) P(\mathbf{r}_0, \Gamma_1) \frac{\partial}{\partial \mathbf{n}} G(\mathbf{r}, \mathbf{r}_0, \kappa)$$
$$- P(\mathbf{r}, \Gamma_2) P(\mathbf{r}_0, \Gamma_2) \frac{\partial}{\partial \mathbf{n}} G(\mathbf{r}, \mathbf{r}_0, \kappa) \big| \, \mathrm{d}\mathbf{r} \, \mathrm{d}\mathbf{r}_0$$

It follows from eqn. 4.29 that the operator $B(\Gamma)[\mu]$ continuously depends on variations $\Delta\Gamma$ of the contour Γ if $\Delta\Gamma = O(B_{12})$ which obviously holds if small increments of the length of Γ yields small variations of the directions of normal vectors (for example, when Γ is a Lyapunov contour).

The results of lemmas 4.5 and 4.6 (which give sufficient conditions for the existence of eigenoscillations) may be generalised for the cases of complicated open waveguide resonators presented in Figure 4.1a. The corresponding proof is not difficult but rather bulky so we will limit ourselves to the formulation of the main theorem having in mind that this theorem concerns the types of resonators shown in Figures 4.1a and 5.2a.

Theorem 4.9: The eigenoscillations spectrum of open waveguide resonators is not empty.

In conclusion one can find evident exceptions for which it is not possible to apply the pertubation technique described. For example, there are no TE eigenoscillations in a short-circuit (semi-infinite) parallel-plane waveguide.

4.5 The local expansion theorem

The properties of the analytical continuation of Green's function $G(\mathbf{r}, \mathbf{r}_0, \kappa)$ for the problem given by eqn. 4.1 and 4.2 with respect to the complex spectral parameter $\kappa \in H$ allow us to assert that for every finite point $\bar{\kappa} \in H$ there exists a vicinity $D \subset H$ which does not contain any singularities of G except for poles and the real second-order algebraic branch points $\kappa_{bn} = \pm \frac{1}{2}n$ (see eqn. 4.2) ($\bar{\kappa}$ may coincide with one of poles). Let us introduce the local variable in H

$$\tau = \begin{cases} \kappa - \bar{\kappa}, & \bar{\kappa} \neq \kappa_{bn} \\ \sqrt{\kappa - \bar{\kappa}}, & \bar{\kappa} = \kappa_{bn} \end{cases} \tag{4.30}$$

Then one can represent $G(\mathbf{r}, \mathbf{r}_0 \kappa)$ in D in the form of the Laurent series

$$G(\mathbf{r}, \mathbf{r}_0 \kappa) = \sum_{n=-N}^{\infty} c_n(\mathbf{r}, \mathbf{r}_0) \tau^n, \quad \kappa \in D \tag{4.31}$$

Eqn. 4.31 may be re-written in the form

$$G(\mathbf{r}, \mathbf{r}_0, \kappa) = \begin{cases} \sum\limits_{n=-N}^{\infty} c_n(\mathbf{r}, \mathbf{r}_0)(\kappa - \bar{\kappa})^n, & \bar{\kappa} \neq \kappa_{bn} \\ \sum\limits_{n=-N}^{\infty} c_n(\mathbf{r}, \mathbf{r}_0)(\kappa - \bar{\kappa})^{n/2}, & \bar{\kappa} = \kappa_{bn} \end{cases} \tag{4.32}$$

The same series may be obtained for any characteristic of the problem linearly depending on G. For example, the solution admits the representations

$$u(\mathbf{r}, \kappa) = - \int_Q G(\mathbf{r}, \mathbf{r}_0 \kappa) f_0(\mathbf{r}_0) \, d\mathbf{r}_0 \tag{4.33}$$

when an open waveguide resonator is excited by concentrated sources (the source function $f_0(r_0)$ enters the right-hand side of eqn. 4.1) or

$$u^{sc}(\mathbf{r}, \kappa) = -\int_Q G(\mathbf{r}, \mathbf{r}_0 \kappa)[1 - \epsilon(\mathbf{r}_0)]\kappa^2 u(\mathbf{r}_0)\, d\mathbf{r}_0 - \int_{\partial S} \frac{\partial}{\partial \mathbf{n}} G(\mathbf{r}, \mathbf{r}_0, \kappa) u(\mathbf{r}_0)\, d\mathbf{r}_0$$

(4.34)

in the case of excitation by a E-polarised plane wave. The total field is determined by the sum of the scattered and incident fields $u^{sc}(\mathbf{r}) + u_0(\mathbf{r})$ where $u_0(\mathbf{r}) \neq 0$ in the whole volume S (see Figure 4.1a). Finally one can formulate the following result.

Theorem 4.10: Any diffraction characteristic $f(\kappa)$ of the problem which linearly depends on G admits the representations

$$f(\kappa) = \sum_{n=-N}^{\infty} f_n \tau^n \quad \kappa \in D$$

(4.35)

where τ is the local variable given by eqn. 4.30.

In the case when $\bar{\kappa}$ is a simple pole ($N = 1$) one has

$$f(\kappa) = \frac{f_{-1}}{\kappa - \bar{\kappa}} + \sum_{n=0}^{\infty} f_n(\kappa - \bar{\kappa})^n = R(\kappa) + F(\kappa)$$

(4.36)

separating thus the "resonant" and "background" parts of the diffraction characteristic. Here $f_{-1} = \mathrm{res} f(\bar{\kappa})$ is determined by the eigenfield at the resonant frequency $\bar{\kappa}$. The "resonant" part does not always prevail and the ratio of R and F depends on the distance d_p from κ to the nearest singular point $\bar{\kappa}_1 \neq \bar{\kappa}$ and on the relation between d_p and $|\kappa - \bar{\kappa}|$. The separation given by eqn. 4.36 looks most naturally when $1 \gg |\kappa - \bar{\kappa}|$ and $|\kappa - \bar{\kappa}_1| \gg |\kappa - \bar{\kappa}|$.

The contributions made by several (simple) poles spaced at comparable distances from κ ($|\kappa - \bar{\kappa}_1| \sim |\kappa - \bar{\kappa}_2| \sim \cdots \sim |\kappa - \bar{\kappa}_m|$) may be determined with the use of the residue theorem:

$$f(\kappa) = \sum_{k=1}^{m} \frac{f_{-k}}{\kappa - \bar{\kappa}_k} = \sum_{k=1}^{m} R_k(\kappa) + F(\kappa), \quad \kappa \in D,\ \bar{\kappa}_k \in D\ (k = 1, 2, \ldots m)$$

(4.37)

Here one can use the usual metrics in the complex plane since D is chosen in such a way that the branch points are situated out of D and the projections of every two points $\kappa \in D$ on the first sheet of H do not coincide.

The expansions given by eqns. 4.36 and 4.37 allow us to estimate to a first approximation the contributions to diffraction characteristics caused by excitation of oscillations which are close to the eigenoscillations of open waveguide resonators. In particular, the main contribution is made by poles corresponding to the fundamental frequencies $\bar{\kappa}$ from the first sheet of H with small $|\operatorname{Im}\bar{\kappa}|$. For these values of κ ($\operatorname{Im}\kappa = 0$, $\operatorname{Re}\kappa > 0$) the processes of stationary wave scattering are observed. The poles situated on other sheets of H may essentially influence the resonant part of a diffraction characteristic if κ sides with one of the branch points and D is a two-sheet circle containing a pole from a neighbour sheet. The corresponding example is considered in Chapter 6.

For practical computations several first terms of eqn. 4.36 may be taken into account and the unknown coefficients are determined by the general properties of diffraction fields. For example, in order to clear up the mechanism of forming a definite response of an open waveguide resonator to the stationary resonant one-mode excitation (in a vicinity of a pole $\bar{\kappa}$) one can use the approximate formula

$$a = \frac{c}{\kappa - \bar{\kappa}} + d, \quad b = \frac{\alpha}{\kappa - \bar{\kappa}} + \beta \quad (\operatorname{Im}\kappa = 0) \tag{4.38}$$

for the amplitudes a and b of the principal waves propagating in the radiation channels, where the unknown coefficients c, d, α and β do not depend on κ and may be found with the help of the additional information concerning the scattering matrix and symmetry of the corresponding eigenoscillations. The presence of inhomogeneities in open waveguide resonators may cause both the total transmission ($|b| = 1$, $|a| = 0$) and total reflection ($|b| = 0$, $|a| = 1$) of energy carried by the incident wave and these effects may be analysed already to a first approximation using formulas similar to the one given by eqn. 4.38. This approach is considered in detail in Chapter 6.

Theorem 4.10 enables us to describe in a mathematically correct way various threshold phenomena in open waveguide resonators.

Let us note that the threshold points coincide with the branch points κ_{bn} : $\omega_n(\kappa_{bn}) = 0$ situated on the positive half-axis $\operatorname{Re}\kappa > 0$ of the first sheet of H. In these points the damping normal modes of the radiation channels are transformed into propagating modes. Using the notations introduced in this Section (see theorem 4.10) and in Section 4.4 one can formulate a generalisation of theorem 4.10:

Theorem 4.11: Any anomalous phenomena reflected in the diffraction characteristics $f(\kappa)$ of open waveguide resonators admits the representations

$$f(\kappa) = \sum_{m=0}^{\infty} f_m^n (\kappa - \kappa_{bn})^{m/2} \tag{4.39}$$

in a vicinity of a threshold point κ_{bn} (in the two-sheet circle $D = D(\kappa_{bn})$ with the origin in κ_{bn}) if it is not a pole of the analytical continuation in H of the resolvent for the problem given by eqns. 4.1 and 4.2. If κ_{bn} coincides with resolvent's pole (of the order N) then

$$f(\kappa) = \sum_{m=-N}^{\infty} f_m^n (\kappa - \kappa_{bn})^{m/2} \tag{4.40}$$

The first-order poles ($N = 1$) are characteristic for resonators with simple geometry (like the regular waveguide when Green's function G_0 has simple poles in κ_{bn}).

Eqns. 4.39 and 4.40 should be used in sufficiently small vicinities of the threshold points where the influence of neighbouring singularities is negligible, otherwise the closest poles may essentially contribute to the formation of an anomalous scattering characteristic. Such phenomena in the theory of dielectric waveguides are called double surface resonances.

Some threshold points may open new communication channels for waves propagating between the zones of transmission and reflection (as an example one may take waveguide modes of the waveguide section forming the communication domain Q). Since the shape of Q does not influence the choice of Riemann surface H, the passage of a threshold point must not cause anomalies of the scattering characteristics given by eqns. 4.39 and 4.40. Every possible variation of $f(\kappa)$ is described here by eqn. 4.36.

The intensity of anomalous behaviour caused by the singularities of the first derivative of $f(\kappa)$ depends on the geometry of an open waveguide resonator, field's polarisation, type of diffraction characteristic, number of a threshold point, etc. The threshold resonance described by eqn. 4.39 may be smoothed out due to the influence of resolvent's poles situated close to the threshold point. Most of these situations are considered in Part 3 for dielectric waveguides.

Chapter 5

Solution of spectral problems for open waveguide resonators

In this chapter the mathematically justified algorithms for the numerical analysis of the eigenoscillations spectra in basic types of open waveguide resonators with semi-infinite regular communication channels are constructed.

The approach developed is based on the principle that all singularities of the analytical continuation to the complex domain of the frequency parameter κ for the resolvents of the wave diffraction problems may be found as a result of the solution to the homogeneous (spectral) problems whose eigenfunctions determine the complete set of the quasi-stationary free oscillations of electromagnetic field. This approach enables us to study mathematical objects (operator-valued functions of different nature) which admit direct analysis of their spectra. Here one should not understand the word direct literally. In fact, homogeneous boundary-value problems for partial differential equations are useless, as a rule, for carrying out global numerical experiments and their equivalent reduction to operator equations are required which admits application of convergent (regularising) calculation procedures. Such re-formulations of the problems may be carried out by means of the integral equations method (Zakharov and Pimenov, 1982, Ilyinsky *et al.*, 1993), the methods of partial domains (Nikolsky, 1977), and the semi-inversion method (Shestopalov, 1971).

The efficiency of the final calculation algorithms constructed for the diffraction problems within the frame of classical approaches is determined by the "lucky" choice of operators, functional spaces and forms of solutions connected with separating the identity operator and equivalent reduction to the second-kind operator equations. In the spectral problems, on the contrary, efforts are concentrated on the studies (and calculations) of not the regular but singular (spectral) points of resolvent's analytical continuation. The analysis of operators is replaced by the analysis of operator-valued functions which can give researchers dealing with applied problems an opportunity to use various methods based, for example, on the analytical Fredholm theorem (Reed and Simon, 1980). Among the problems arising here is the necessity to prove the existence of infinite determinants considered as functions of the complex spectral parameter and the convergence of the sequences of approximately determined characteristic numbers (of finite-dimensional operators) to zeros of these infinite determinants (i.e. to the exact characteristic numbers). Operator equations with kernel operators or Koch-type operators described in the previous Sections enable one to solve these problems.

In this chapter we will treat E-polarisation and TE oscillations, assuming that the H-case may be considered in a similar way. The analysis of the spectra of open waveguide resonators will repeat the main stages considered in Chapter 4 (in particular, the proof of the equivalence principle and justification of the truncation method) so we will describe the basic steps in detail for only one typical problem, dispersing them between different sections.

5.1 Thin apertures in a regular parallel-plane waveguide

The solution to the spectral problem given by eqns. 4.6 and 4.7 in the open waveguide resonator formed by a thin perfectly conducting aperture in a regular parallel-plane waveguide (Figure 5.1) may be represented as

$$v(\mathbf{r}) = \sum_{n=1}^{\infty} b_n \exp(j\omega_n |x_3|) \sin \frac{1}{2} n x_2 \qquad (5.1)$$

where $v(\mathbf{r}) = E_{x_1}(\mathbf{r})$, $\omega_n = \omega_n(\kappa)$ are determined in Section 4.1 and

$$\mathbf{b} = (b_n)_{n=1}^{\infty} \in \tilde{l}_2 = \{\mathbf{a} = (a_n)_{n=1}^{\infty} : \sum_{n=1}^{\infty} |a_n|^2 n < \infty\} \qquad (5.2)$$

which yields the fulfillment of Meixner's condition. The left-hand side of eqn. 5.1 satisfies also the homogeneous Helmholtz equation, the boundary-value conditions on the waveguide's walls and the conditions at infinity (see Section 4.2).

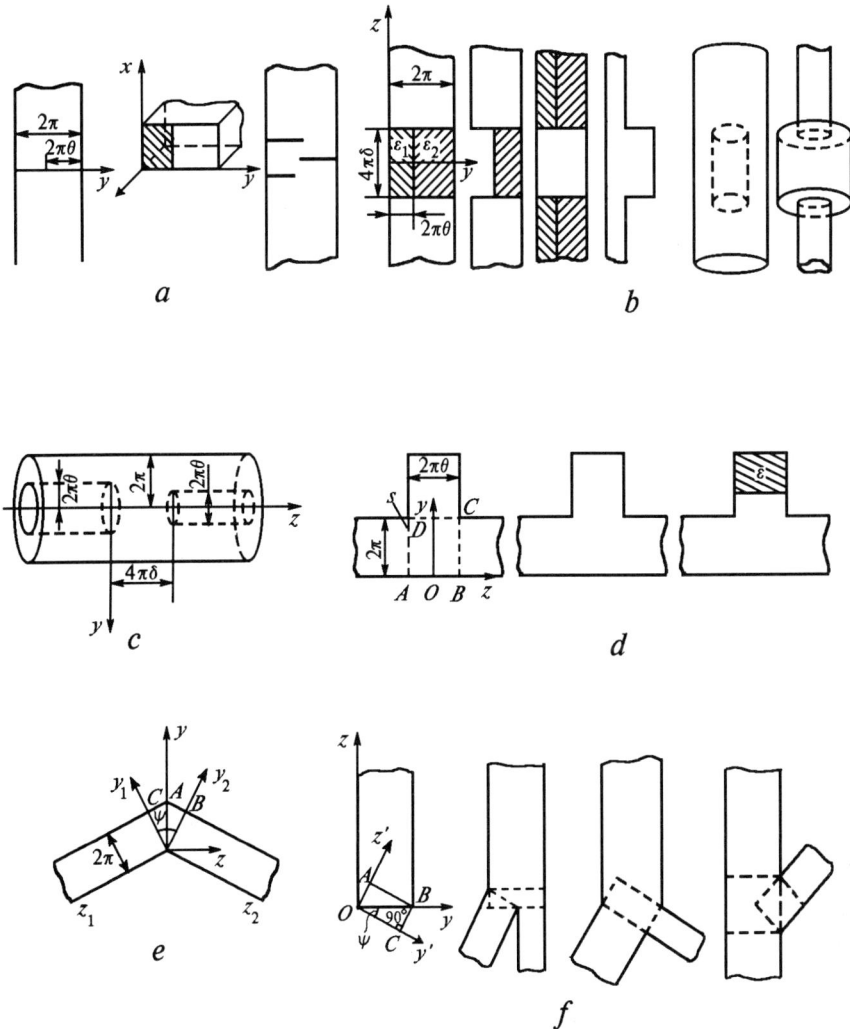

Figure 5.1 *Modifications of open waveguide resonators*

The problem is to determine the values $\kappa \in H$ (where H is the Riemann surface of the analytical continuation of canonical Green's function introduced in Section 4.1) where there exist non-trivial solutions of the spectral problem given by eqns. 4.6 and 4.7 having the form given by eqn. 5.1 and describing free-field oscillations in the considered open waveguide resonator. We will call these κ the spectral points. It is easy to see that these solutions define the free oscillations in a rectangular waveguide with a capacity aperture which do not vary along the direction x (see Figure 4.1b).

In accordance with theorem 4.5, the spectral points may exist only on the first sheet of H, coinciding with the real threshold points $\kappa = \kappa_n$ where $\omega_n(\kappa) = 0$.

The field matching in the plane $z = 0$ applied to eqn. 5.1 leads to the equivalent infinite system of linear algebraic equations

$$[I - A(\kappa)]\mathbf{b} = \mathbf{o} \tag{5.3}$$

obtained after the regularising (semi-inversion) procedure (Shestopalov, 1971) where

$$A(\kappa) = \|a_{mn}(\kappa)\|_{m,n=1}^{\infty}, \quad a_{mn}(\kappa) = \frac{[n + 2i\omega_n(\kappa)](V_{m-1}^{n-1} - V_{m-1}^{-n-1})}{m}$$

is the matrix operator-valued function acting in $\tilde{l}_2$ and the quantities V_m^n are defined in the book by Shestopalov (1971).

Lemma 5.1: $A(\kappa) : l_2 \to l_2$, $\kappa \in H$ is the kernel operator-valued function acting in $l_2 = \{\mathbf{a} = (a_n)_{n=1}^{\infty} : \sum_{n=1}^{\infty} |a_n|^2 < \infty\}$ and holomorphic in H.

In fact, the series $\sum_{m,n} |a_{mn}| < \infty$ due to the estimates

$$\operatorname{Im}\omega_n = \frac{n}{2} + O(n^{-1}), \quad V_{m-1}^{n-1} < \text{const} \frac{\sqrt{m}}{|m - n|\sqrt{n}}$$

(Sirenko *et al.*, 1985a). Hence the kernel property of $A(\kappa)$ follows from the fact that $\sum_n A(f_n, f_n)$ converges absolutely for every orthogonal and normalised basis $\mathbf{f} = (f_n)_{n=1}^{\infty}$ in l_2.

The series

$$\sum_n [n + 2j\omega_n(\kappa)]f_n t_n \quad (\mathbf{f}, \mathbf{t} \in l_2) \tag{5.5}$$

uniformly converges in every compactum $D \subset H$ and $A(\kappa)$ admits the representation

$$A(\kappa) = BC(\kappa), \quad C(\kappa) = \|c_{mn}(\kappa)\|_{m,n=1}^{\infty}, \quad c_{mn} = \frac{1}{2}\delta_{mn}[n + 2j\omega_n(\kappa)] \tag{5.6}$$

Hence, in accordance with the results of Sanchez-Palencia (1980), $A(\kappa)$ is holomorphic in H (in the weak sense) due to the corollary of the Riesz theorem on functionals representation (Kantorovich and Akilov, 1977).

One can prove that the non-trivial solutions $\mathbf{b} \in l_2$ of eqn. 5.3 belong to $\tilde{l}_2$ [due to the estimate $|b_n| = O(n^{-32})$ obtained by Podolsky (1965)]. Hence it follows from the uniqueness theorem (see Chapter 4) and the equivalence principle that the resolvent set of the operator-valued function $I - A(\kappa)$ is not empty and, consequently, its characteristic numbers (finite-order poles of $[I - A(\kappa)]^{-1}$), which coincide with fundamental frequencies of the considered open waveguide resonator, form not more than a countable set $\Omega \in H$ without finite accumulation points.

Characteristic numbers of $A(\kappa)$ (forming the set Ω) may be approximated by the ones of the truncated problem

$$\mathbf{b}_N = P_N A(\kappa) P_N \mathbf{b}_N \quad (N = 1, 2, \ldots) \tag{5.7}$$

(forming the sets $\Omega_N \in H$, $N = 1, 2, \ldots$) where $P_N : P_N(\mathbf{t}) = \mathbf{t}_N = (t_n)_{n=1}^N$ is the orthogonal projector onto the finite-dimensional sub-space of l_2. The following theorem is the justification of the corresponding numerical algorithm (similar to the one constructed in Section 1.11) for calculating the fundamental frequencies of the open waveguide resonator formed by a thin perfectly conducting aperture in a regular parallel-plane waveguide.

Theorem 5.1: If $\kappa \in \Omega$, then there exists a sequence $\{\kappa_N\}$ such that $\kappa_N \in \Omega_N$ and $|\kappa_N - \kappa| \to 0$, $N \to \infty$. On the other hand, if there exists a sequence κ_N, $\kappa_N \in \Omega_N$ such that $|\kappa_N - \kappa| \to 0$, $N \to \infty$, then $\kappa \in \Omega$.

This theorem may be proved by the methods developed by Vainikko and Karma (1974) (similar to the proof of theorem 1.3 from Section 1.11).

But we will prove the assertion of theorem 5.1 in a different way using the perturbation theory and considering $A(\kappa) = A(\kappa, N) : l_2 \to l_2$ as a function of two variables where N plays the role of a real parameter. Since $A(\kappa)$ is the kernel operator-valued function,

$$\lim_{N \to \infty} \|A(\kappa, \infty) - A(\kappa, N)\| = 0 \tag{5.8}$$

where $A(\kappa, \infty)$ denotes a limiting operator (with respect to the operator norm

$\| \cdot \| : l_2 \to l_2)$ and

$$\lim_{N \to \infty} \|[I - A(\kappa, \infty)]^{-1} - [I - A(\kappa, N)]^{-1}\| = 0 \tag{5.9}$$

in the vicinity of (κ_∞, ∞) if (κ_∞ is a regular point of $[I - A(\kappa, \infty)]^{-1}$. In addition to this, there exists the function

$$F(\kappa) = \det[I - A(\kappa)] \tag{5.10}$$

holomorphic in H together with every function

$$F_N(\kappa) = \det[I - A(\kappa, N)] \quad (N = 1, 2, \ldots) \tag{5.11}$$

and the following inequalities hold (see Shestopalov *et al.*, 1973)

$$|\det[I - A(\kappa, \infty)] - \det[I - A(\kappa, N)]|$$
$$\leq \|A(\kappa, \infty) - A(\kappa, N)\|_1 \exp[\|A(\kappa, \infty)\| + \|A(\kappa, N)\| + 1] \tag{5.12}$$

Assume that $\kappa_N \in \Omega_N$ is a characteristic number of $A(\kappa, N)$, i.e.

$$F_N(\kappa_N) = 0 \tag{5.13}$$

Then it follows from eqn. 5.12 that

$$|F(\kappa_N)| \leq \|A(\kappa_N) - A(\kappa_N, N)\|_1 \exp[2\|A(\kappa_N)\|_1 + 1] \tag{5.14}$$

(here $\| \cdot \|_1$ denotes the kernel norm which is estimated by the finite quantity $\sum_{m,n} |a_{mn}|$ for a matrix operator $A = \|a_{mn}\|_{m,n=1}^\infty$).

Since $\lim_{N \to \infty} F(\kappa_N) = 0$ ($\kappa_N \in \Omega_N$), the zeros of $F(\kappa)$ are the sought for characteristic numbers.

Let us verify the rate of convergence for the sequence of approximate characteristic numbers κ_N introduced in lemma 5.2. Assume that $F(\kappa_\infty) = 0$ and $F'(\kappa_\infty) \neq 0$ (i.e. the characteristic number κ_∞ has the unit multiplicity). Then the Taylor series for $F(\kappa)$ in the vicinity of κ_∞ may be represented in the form

$$F(\kappa_N) = F'(\kappa_\infty)(\kappa_N - \kappa_\infty) \left\{ \sum_{n=2}^\infty \frac{1}{n!} \frac{F^{(n)}(\kappa_\infty)}{F'(\kappa_\infty)} (\kappa_N - \kappa_\infty)^{n-1} + 1 \right\} \tag{5.15}$$

The second factor in the right-hand side of eqn. 5.15 does not vanish when $N \to \infty$ and, consequently,

$$|\kappa_N - \kappa_\infty| < \text{const} \left| \frac{F(\kappa)}{F'(\kappa_\infty)} \right| \qquad (5.16)$$

for sufficiently large N. Substituting eqn. 5.14 to eqn. 5.16 we get the required estimate for the rate of convergence

$$|\kappa_N - \kappa_\infty| < \text{const} \, \|A(\kappa_N) - A(\kappa_N, N)\|_1 \qquad (5.17)$$

The kernel norm in eqn. 5.17 may be replaced by the finite quantity

$$\sum_{n=1}^{\infty} \sum_{m=N+1}^{\infty} |a_{mn}(\kappa_N)| + \sum_{n=N+1}^{\infty} \sum_{m=1}^{\infty} |a_{mn}(\kappa_N)|$$

which enables us to obtain the estimate

$$|\kappa_N - \kappa_\infty| < \text{const} \, N^{-12} \qquad (5.17)$$

using the properties of V_n^m and c_{mn} presented above.

The same technique may be applied for determination of the fundamental frequencies of waveguide resonators formed by circular and co-axial waveguides containing symmetrical infinitely thin apertures.

5.2 Finite bifurcation and thick apertures in a regular parallel-plane waveguide

The problem of the fundamental frequencies of the open waveguide resonator depicted in Figure 5.1a may be considered as a key one for the following reasons:

(i) the operator equations of the problem admit correct asymptotic treatment for large values of Im ϵ_1 (or Im ϵ_2) and, consequently, enable one to correctly analyse open waveguide resonator with thick apertures;

(ii) application of the operator-valued functions method and the numerical algorithms for calculating fundamental frequencies of the key problem may be used without essential changes for the waveguide resonators shown in Figure 5.1b (including axial-symmetric resonators).

Consider the key problem using the semi-inversion method with separation of the solution which corresponds to a semi-infinite bifurcation in a parallel-plane waveguide proposed by Sirenko (1978) and Kirilenko *et al.* (1978).

The solution to the spectral problem given by eqns. 4.6 and 4.7 in the open waveguide resonator formed by a thick perfectly conducting aperture in a parallel-plane waveguide (Figure 5.1*b*) may be represented in the form proposed by Masalov and Sirenko (1978)

$$
v(\mathbf{r}) = \begin{cases}
\displaystyle\sum_{n=1}^{\infty} b_{n,1}\exp[j\omega_n(x_3 - 2\pi\delta)]\sin\frac{1}{2}nx_2, & x_3 > 2\pi\delta \\[2ex]
\displaystyle\sum_{n=1}^{\infty}\{c_{n,1}\exp[-j\beta'_{n,1}(x_3 - 2\pi\delta)] \\
\quad + d_{n,1}\exp[j\beta'_{n,1}(x_3 + 2\pi\delta)]\}\sin\frac{1}{2\theta_1}nx_2, & 0 < x_2 < 2\pi\theta \\[2ex]
\displaystyle\sum_{n=1}^{\infty}\{c_{n,2}\exp[-j\beta'_{n,2}(x_3 - 2\pi\delta)] \\
\quad + d_{n,2}\exp[j\beta'_{n,2}(x_3 + 2\pi\delta)]\}\sin\frac{1}{2\theta_2}nx_2, & 2\pi\theta < x_2 < 2\pi \\[2ex]
\displaystyle\sum_{n=1}^{\infty} b_{n,2}\exp[j\omega_n(x_3 + 2\pi\delta)]\sin\frac{1}{2}nx_2, & x_3 < 2\pi\delta
\end{cases}
\tag{5.19}
$$

where $v(\mathbf{r}) = E_{x_1}(\mathbf{r})$,

$$
\omega_n = \sqrt{\kappa^2\epsilon_l - \frac{n^2}{4}}, \qquad\qquad \beta_{m,l} = \sqrt{\kappa^2 - \frac{m^2}{4\theta_l^2}}
$$

$$
\beta'_{m,l} = \sqrt{\kappa^2\epsilon_l - \frac{m^2}{4\theta_l^2}}, \qquad\qquad \theta_l = \begin{cases} \theta, & l = 1 \\ 1 - \theta, & l = 2 \end{cases}
$$

The branches of the functions $\beta_{m,l}$ and $\beta'_{m,l}$ are chosen in accordance with eqn. 4.4 and assumed to be fixed on all sheets of the Riemann surface H since the signs of their real and imaginary parts do not influence the form of the equivalent operator equation given by eqn. 5.20.

The left-hand side of eqn. 5.19 satisfies the homogeneous Helmholtz equation, the boundary-value conditions on waveguide's walls and the conditions at infinity (see Section 4.2). The conjugation conditions in the planes $|x_3| = 2\pi\delta$ applied to eqn. 5.19 lead to the equivalent infinite system of linear algebraic equations

$$\sum_{n=1}^{\infty}(b_{n,1} \pm b_{n,2})\frac{\sin n\pi\theta}{\omega_n - \beta_{m,l}} = \mu_{m,l}^{\pm}\sum_{n=1}^{\infty}(b_{n,1} \pm b_{n,2})\frac{\sin n\pi\theta}{\omega_n + \beta'_{m,l}} \qquad (5.20)$$

Here

$$\mu_{m,l}^{\pm} = \frac{\lambda_{m,l} \mp e_{m,l}}{1 \mp \lambda_{m,l}e_{m,l}}, \quad \lambda_{m,l} = \frac{\beta'_{m,l} - \beta_{m,l}}{\beta'_{m,l} + \beta_{m,l}}, \quad e_{m,l} = \exp(4j\pi\delta\beta_{m,l})$$

The signs $+$ and $-$ (omitted below) correspond to the symmetrical and anti-symmetrical eigenoscillations (with respect to the plane $x_3 = 0$) where they coincide with the eigenoscillations in a resonator with one radiation channel when the plane $x_3 = 0$ is replaced by a perfectly-conducting plug.

Every equation of the system given by eqn. 5.20 may be identified for a fixed pair of indices m, l with the meromorphic function

$$Q(w) = \sum_{s,l}\phi_{s,l}(w)\tau_{s,l} \quad (s = 1, 2, \ldots)$$

vanishing at infinity as a function of the complex variable w. Here $\tau_{s,l}$ are the right-hand sides of eqn. 5.20,

$$\phi_{s,l}(w) = -\exp[j(w - \beta_{s,l})\psi(\theta)]\prod_{\substack{n=1\\n\neq s}}^{\infty}\frac{(\omega_n - \beta_{s,l})(w - \beta_{n,1})(w - \beta_{n,2})}{(\omega_n - w)(\beta_{n,1} - \beta_{s,l})(\beta_{n,2} - \beta_{s,l})}$$

$$\psi(\theta) = \theta\ln\theta - (1 - \theta)\ln(1 - \theta)$$

and the following conditions hold (Highman, 1966)

$$\text{res } Q(\omega_n) = t_n \quad (n = 1, 2, \ldots) \quad \text{res } Q(\beta_{m,l}) = -\tau_{m,l} \quad (m = 1, 2, \ldots)$$

where $t_n = (b_{n,1} \pm b_{n,2})\sin n\pi\theta$.

The first condition leads to the second-kind infinite system of linear algebraic equations with respect to the new unknowns

$$\tau_{m,l} = -\mu_{m,l}\sum_{n,l}\phi_{n,l}(-\beta_{m,l})\tau_{n,l}, \quad \tau_l = (\tau_{m,l})_{m=1}^{\infty} \in l_2 \qquad (5.21)$$

such that

$$t_n = \sum_{m,l}\text{res }\phi_{m,l}(\omega_n)\tau_{m,l} \qquad (5.22)$$

Lemma 5.2: The problem on characteristic numbers given by eqns. 5.21 and 5.22 is equivalent to eqn. 5.20.

Assume that $\kappa \in H$ (where H is the Riemann surface of the analytical continuation of canonical Green's function introduced in Section 4.1) is the point where there exist a non-trivial solution $\eta \in \tilde{l}_2$ of eqn. 5.21. Since

$$\phi_{m,l} = O\left[\frac{\sqrt{m}}{(w - \beta_{m,l})\sqrt{w}}\right], \quad \operatorname{res} \phi_{m,l}(\omega_n) = O\left[\frac{\sqrt{m}}{(n-m)\sqrt{n}}\right] \qquad (5.23)$$

eqn. 5.22 defines a (bounded) Hilbert-Schmidt operator in $\tilde{l}_2$. Let us show that $\mathbf{t} \in \tilde{l}_2$ given by eqn. 5.22 is a solution of eqn. 5.20. Multiplying eqn. 5.22 by $(\omega_n - \beta_{s,l})^{-1}$ and calculating the sum with respect to $n = 1, 2, \ldots$ we get

$$\sum_n \frac{t_n}{\omega_n - \beta_{s,l}} = \sum_{m,l} \tau_{m,l} \sum_n \frac{\operatorname{res} \phi_{m,l}(\omega_n)}{\omega_n - \beta_{s,l}}$$

$$= \sum_{m,l} \tau_{m,l} \delta_{ms} \delta_{il} = \tau_{s,i} = \mu_{s,i} \sum_n \frac{t_n}{\omega_n + \beta_{s,l}} \qquad (5.24)$$

(the changes of the order of summation are possible because $\eta \in \tilde{l}_2$ and $\mathbf{t} \in \tilde{l}_2$) and hence $\mathbf{t}$ satisfies eqn. 5.20.

Assuming that $\mathbf{t}$ is a non-trivial solution of eqn. 5.20, then

$$\lim_{R \to \infty} \int_{C_R} \left[\frac{\sum_{m,l} \phi_{m,l}(w)\tau_{m,l}}{w - \beta_{s,l}}\right] d\,w = \sum_n \frac{\sum_{m,l} \operatorname{res} \phi_{m,l}(\omega_n)\tau_{m,l}}{\omega_n - \beta_{s,l}}$$

$$+ \sum_{m,l} \phi_{m,l}(\beta_{s,i})\tau_{m,l} = 0 \qquad (5.25)$$

where $\eta \in \tilde{l}_2$, which yields uniform convergence of the series $\sum_{m,l} \phi_{m,l}(w)\tau_{m,l}$ with respect to w out of the poles of $\phi_{m,l}(w)$. In accordance with eqn. 5.25, $\mathbf{t}$ may be represented in the form given by eqn. 5.22 if

$$\tau_{m,l} = \mu_{m,l} \sum_{m=1}^{\infty} \frac{t_n}{\omega_n + \beta_{m,l}} \qquad (5.26)$$

Substituting the expression given by eqn. 5.22 to eqn. 5.26 enables us to make the required conclusion that $\eta \in \tilde{l}_2$ satisfies eqn. 5.21.

Let us rewrite eqn. 5.21 with respect to the new unknowns

$$\tau = R(\kappa)\tau \tag{5.27}$$

where

$$\tau = \left\|\begin{array}{c} \tau_1' \\ \tau_2' \end{array}\right\| \quad \tau \in l_2^2 = \left\{ a = \left\|\begin{array}{c} a_1 \\ a_2 \end{array}\right\| \, a_l \in l_2 \, l = 1, 2 \right\} \quad \tau_l' = (m^{32}\tau_{m,l})_{m=1}^{\infty} \in l_2$$

$$R(\kappa) = \left(\begin{array}{cc} R_{11} & R_{12} \\ R_{21} & R_{22} \end{array}\right), \quad (R_{il})_{m,n} = -\mu_{m,l}\phi_{n,l}(-\beta_{m,l})m^{3/2}n^{-3/2} \tag{5.28}$$

Lemma 5.3: $R(\kappa) : l_2 \to l_2$ is a kernel finite-meromorphic operator-valued function of the complex variable $\kappa \in H$.

The proof of the lemma is based on the estimate

$$(R_{il})_{m,n} = O[m^{-1}n^{-1}(m + n)^{-1}] \quad (m, n \to \infty) \tag{5.29}$$

which is easily obtained with the help of eqn. 5.23 from the explicit representation of the elements of $R(\kappa)$ given in eqn. 5.28. The kernel property of $R(\kappa)$ follows from the convergence of the series $\sum_{m,n} |(R_{il})_{m,n}|$. In order to prove that $R(\kappa)$ is a finite-meromorphic operator-valued function with respect to $\kappa \in H$, it is sufficient to check that the series

$$\sum_{m,n}(R_{il})_{m,n}(\kappa)t_m^1 t_n^2 \quad (t^l = (t_n^l)_{n=1}^{\infty} \in l_2 \, l = 1, 2)$$

converges uniformly in every compactum $D \in H$ which does not contain the poles of $(R_{il})_{m,n}(\kappa)$.

Now one can use the analytical Fredholm theorem and state that the fundamental frequencies of the open waveguide resonator considered are the isolated points in H where the operator-valued function $[I - R(\kappa)]^{-1}$ has finite-order poles.

The point spectrum Ω of the problem given by eqn. 5.27 may be approximated by the characteristic numbers of the truncated problem

$$\tau_N = P_N R(\kappa) P_N \tau_N \quad (N = 1, 2, \ldots) \tag{5.30}$$

(forming the sets $\Omega_N \in H$, $N = 1, 2, \ldots$) where

$$P_N(l_2^2) = \left\{ \mathbf{a} = \left\| \begin{matrix} \mathbf{a}_1 \\ \mathbf{a}_2 \end{matrix} \right\|, \quad \mathbf{a}_l = (a_n^l)_{n=1}^\infty \in l_2, \ a_n^l = 0, \ n > N, \ l = 1, 2 \right\}$$

is the orthogonal projector onto the finite-dimensional sub-space of l_2^2. The proof of the following theorem is similar to the one of theorem 5.1 carried out in Section 5.1.

Theorem 5.2: The points $\kappa_\infty \in \Omega$ may be approximated by the sequences $\kappa_N \in \Omega_N$ of the characteristic numbers of the operator-valued functions $I - P_N R(\kappa) P_N$, coinciding with the roots of the equations $\det[I - P_N R(\kappa) P_N] = 0$, and the following estimate holds for the rate of convergence $|\kappa_N - \kappa_\infty| <$ const $N^{-1+\alpha}$, $N \to \infty$.

We will prove only the last estimate. Due to eqns. 5.27 and 5.30, the corresponding inequality may be represented in the "operator" terms

$$|\kappa_N - \kappa_\infty| \le \text{const} \, \|R(\kappa_N) - P_N R(\kappa_N) P_N\|_1$$

(compare with eqn. 5.17) where the kernel norm is estimated by the quantities

$$\sum_{n=1}^\infty \sum_{m=N+1}^\infty |(R_{il})_{m,n}(\kappa_N)| + \sum_{n=N+1}^\infty \sum_{m=1}^\infty |(R_{il})_{m,n}(\kappa_N)| \quad (i, l = 1, 2)$$

Now the explicit estimate for $(R_{il})_{m,n}(\kappa)$ given by eqn. 5.29 enables us to apply the integral summation formulae for $\alpha \in (0, 1)$ and to obtain the required inequality $|\kappa_N - \kappa_\infty| <$ const $N^{-1+\alpha}$ $(N \to \infty)$.

Eqn. 5.31 gives only a rough estimate for the rate of convergence. In the next Sections we will present the results of computer modelling which demonstrate the real efficiency of the numerical algorithms constructed on the basis of this truncation method.

Most of the results obtained here may be applied within the frame of the same mathematical model in the case $\text{Im}\,\epsilon_1 \gg 1$ for open waveguide resonators with lossy thick apertures shown in Figure 5.1b, considering lossy metal walls as the boundaries of lossy dielectric layers.

5.3 Coaxial waveguide resonator

Let us consider the problem on fundamental frequencies of the open waveguide resonator formed by a section of a circular waveguide bounded by coaxial-waveguide bifurcations (see Figure 5.1c). Unlike the waveguide resonators considered in Sections 5.1 and 5.2, this open waveguide has four radiation channels characterised by different parameters.

Among numerous works devoted to the solution of the diffraction problem in this open waveguide, only Koshparenok *et al.* (1980, 1983) treated the free oscillations but limiting themselves to the determination of the real fundamental frequencies under the assumption that all the radiation channels are over cut-off. Rigorous mathematical models justifying analytical and numerical analysis of the spectral properties of coaxial waveguide resonators have not been elaborated until recently. The technique developed in this Section is partially based on the works of Goldstein (1969) and Roger and Maystre (1977).

The solution to the spectral problem given by eqns. 4.6 and 4.7 in the coaxial waveguide resonator (Figure 5.1c) may be represented in the form similar to the one given by eqn. 5.19

$$
u(r, x_3) = \begin{cases}
\sum_{n=1}^{\infty} b_{n,1} J_1\left(\nu_{1n}\dfrac{r}{2\pi\theta_1}\right) \exp(-j\omega_{n,1} x_3) \\
\quad 0 \le r \le 2\pi\theta_1 \\[2mm]
\sum_{n=1}^{\infty} d_{n,1} \left[J_1\left(\mu_{n,1}\dfrac{r}{2\pi}\right) - N_1\left(\mu_{n,1}\dfrac{r}{2\pi}\right) \dfrac{J_1(\mu_{n,1}\theta_1)}{N_1(\mu_{n,1}\theta_1)} \right] \exp(-j\phi_{n,1} x_3) \\
\quad 2\pi\theta_1 < r < 2\pi \\[2mm]
\sum_{n=1}^{\infty} b_{n,2} J_1\left(\nu_{1n}\dfrac{r}{2\pi\theta_2}\right) \exp[j\omega_{n,2}(x_3 - 4\pi\delta)] \\
\quad 0 \le r \le 2\pi\theta_2,\ x_3 > 4\pi\delta \\[2mm]
\sum_{n=1}^{\infty} d_{n,2} \left\{ J_1\left(\mu_{n,2}\dfrac{r}{2\pi}\right) - N_1\left(\mu_{n,2}\dfrac{r}{2\pi}\right) \dfrac{J_1(\mu_{n,2}\theta_2)}{N_1(\mu_{n,2}\theta_2)} \right\} \\
\quad \times \exp[j\phi_{n,2}(x_3 - 4\pi\delta)], \quad 2\pi\theta_1 < r < 2\pi,\ x_3 > 4\pi\delta \\[2mm]
\sum_{n=1}^{\infty} J_1\left(\nu_{1n}\dfrac{r}{2\pi}\right) \left[a_n^+ \exp(j\omega_n) \mp a_n^- \exp[-j\omega_n(x_3 - 4\pi\delta)] \right] \\
\quad 0 < x_3 < 4\pi\delta
\end{cases}
$$

$$(5.32)$$

where $u(r, x_3) = E_\phi(r, x_3)$,

$$\omega_n = \sqrt{\kappa^2 - \frac{\nu_{1n}^2}{4\pi^2}}, \quad \omega_{n,l} = \sqrt{\kappa^2 - \frac{\nu_{1n}^2}{4\pi^2\theta_l^2}}, \quad \phi_{n,l} = \sqrt{\kappa^2 - \frac{\mu_{n,l}^2}{4\pi^2}}$$

ν_{1n} and $\mu_{n,l}$ are the nth roots of the equations $J_1(x) = 0$ and $J_1(x\theta)N_1(x) - J_1(x)N_1(x\theta_l) = 0$, respectively; $\mathbf{b}_i = (b_{n,l})_{n=1}^\infty \in l_2$, $\mathbf{d}_i = (d_{n,l})_{n=1}^\infty \in l_2$ and $\mathbf{a}^\pm = (a_n^\pm)_{n=1}^\infty \in l_2$ are the complex amplitudes of the partial components of the axial-symmetric free oscillations ($i = 1, 2$). We will consider the functions given by eqn. 5.32 on the Riemann surface H_4 of the analytical continuation from the real axis κ of four canonical Green's functions of the semi-infinite waveguides forming the radiation channels. For $\text{Im}\,\kappa = 0$ these Green's functions satisfy the non-incoming radiation condition and, therefore, H_4 has four times as many branch points (zeros of $\omega_{n,l}$ and $\phi_{n,l}$) as the Riemann surface H described in the previous Chapter. Let us note that the inequalities $\omega_{n,l}(\kappa < 0$ and $\phi_{n,l}(\kappa) < 0$ hold in every fixed point $\kappa \in H_4$, as well as on H, for not more than a finite number of n.

Imposition of the conjugation conditions in the planes $x_3 = 0$ and $x_3 = 4\pi\delta$ together with application of the standard re-expansion procedure result in the equivalent infinite system of linear algebraic equations

$$\begin{cases} \sum_{n=1}^\infty \dfrac{a_n^+ J_1(\nu_n\theta_1)}{\omega_n - \phi_{m,1}} = \sum_{n=1}^\infty \dfrac{a_n^- J_1(\nu_n\theta_1)}{\omega_n + \phi_{m,1}} e_n \\[3mm] \sum_{n=1}^\infty \dfrac{a_n^+ J_1(\nu_n\theta_1)}{\omega_n - \omega_{m,1}} = \sum_{n=1}^\infty \dfrac{a_n^- J_1(\nu_n\theta_1)}{\omega_n + \omega_{m,1}} e_n \end{cases} \tag{5.33}$$

$$\begin{cases} \sum_{n=1}^\infty \dfrac{a_n^- J_1(\nu_n\theta_2)}{\omega_n - \phi_{m,2}} = \sum_{n=1}^\infty \dfrac{a_n^+ J_1(\nu_n\theta_2)}{\omega_n + \phi_{m,2}} e_n \\[3mm] \sum_{n=1}^\infty \dfrac{a_n^- J_1(\nu_n\theta_2)}{\omega_n - \omega_{m,2}} = \sum_{n=1}^\infty \dfrac{a_n^+ J_1(\nu_n\theta_2)}{\omega_n + \omega_{m,2}} e_n \end{cases} \tag{5.34}$$

where $m = 1, 2, \ldots$ and $e_n = \exp(4j\pi\delta)$. The matrices in eqns. 5.33 and 5.34 define compact operators in l_2 and, hence, in order to construct numerical procedures similar to the algorithms presented in Sections 5.1 and 5.2 for the canonical Fredholm operator equations we will regularise the operators of these systems.

Lemma 5.4: The problem given by eqns. 5.33 and 5.34 is equivalent to the system of coupled operator equations

$$\begin{cases} \mathbf{a}^+ = A^+(\kappa)\mathbf{a}^- \\ \mathbf{a}^- = A^-(\kappa)\mathbf{a}^+ \end{cases} \tag{5.35}$$

where $A^\pm$ are the compact finite-meromorphic operator-valued functions of $\kappa \in H_4$ given by the infinite matrices $\|A^\pm_{m,n}\|^\infty_{m,n=1}$ with the elements

$$A^\pm_{m,n} = \frac{J_1(\nu_{1n}\theta_i)\operatorname{res}Q_{nl}(\omega_n)}{J_1(\nu_{1m}\theta_i)}$$

where

$$Q_{nl}(w) = \frac{\exp[j(w+\omega_n)\psi(\theta_i)]}{w+\omega_n}\prod_{s=1}^\infty \frac{(\omega_n+\omega_s)(w-\phi_{s,l})(w-\omega_{s,l})}{(\omega_s-w)(\phi_{s,l}+\omega_n)(\omega_{s,l}+\omega_n)} \tag{5.36}$$

are the meromorphic functions of the complex variable w such that

$$Q_{nl}(w) = O[n^{12}|w|^{-12}(w+\omega_n)^{-1}] \quad n, |w| \to \infty \tag{5.37}$$

Here the values $i = 1$ and $i = 2$ correspond to the signs $+$ and $-$, respectively.

The proof of this lemma is again based on the possibility to identify every equation of the systems given by eqns. 5.33 and 5.34 with the Laurent expansions of the meromorphic functions

$$Q_i(w) = \sum_{n=1}^\infty Q_{ni}(w)a_n \mp J_1(\nu_n\theta_i)e_n \quad (i = 1, 2)$$

having the simple poles in the points $w = \pm\omega_n$ where

$$\operatorname{res}Q_1(\pm\omega_n) = a_n^\pm J_1(\nu_n\theta_1)\tilde{e}_n^\pm$$
$$\operatorname{res}Q_2(\pm\omega_n) = a_n^\mp J_1(\nu_n\theta_2)\tilde{e}_n^\pm$$

where

$$\tilde{e}_n^\pm = \begin{cases} 1 & \text{for the sign } + \\ e_n & \text{for the sign } - \end{cases}$$

vanishing at infinity as a function of the complex variable w on a certain correct system of contours C_R $(R \to \infty)$ and satisfying the "normalising" conditions $Q_i(\phi_{m,i}) = 0$, $Q_i(\omega_{m,i}) = 0$ $(m = 1, 2, \ldots, i = 1, 2)$. It follows from the definition of $Q_i(w)$ that

$$a_n + J_1(\nu_n \theta_1) = \sum_{n=1}^{\infty} \operatorname{res} Q_{m1}(\omega_n) a_m - J_1(\nu_m \theta_1) e_m$$
$$\hspace{5cm} (n = 1, 2, \ldots) \quad (5.38)$$
$$a_n - J_1(\nu_n \theta_2) = \sum_{n=1}^{\infty} \operatorname{res} Q_{m2}(\omega_n) a_m + J_1(\nu_m \theta_2) e_m$$

and eqn. 5.38 is the element-by-element representation of the coupled system given by eqn. 5.35.

The equivalence between eqn. 5.35 and the problem given by eqns. 5.33 and 5.34 may be easily proved if to use the scheme presented in Section 5.2.

The compact property of the operators $A^{\pm}(\kappa) : l_2 \to l_2$ follows from the estimate

$$A_{m,n}^{\pm} = O[(m + n)^{-1} \exp(-4\pi \delta n)] \quad (n \to \infty) \qquad (5.39)$$

which is obtained with the help of eqn. 5.37 and the known asymptotic behaviour $\nu_{1n} = O(n)$ of the roots ν_{1n} of Bessel's function $J_1(x)$ when $n \to \infty$.

Lemma 5.4 enables us to assert that due to the analytical Fredholm theorem the fundamental frequencies of the considered open waveguide resonator are the isolated points in H_4 and they coincide with finite-order poles of the resolvent of the operator defined by eqn. 5.35

Characteristic numbers of the problem given by eqn. 5.35 may be approximated by the characteristic numbers κ_N of the truncated problem

$$\begin{cases} \mathbf{a}_N^+ = P_N A^+(\kappa) P_N \mathbf{a}_N^- \\ \mathbf{a}_N^- = P_N A^-(\kappa) P_N \mathbf{a}_N^+ \end{cases} \quad N = 1, 2, \ldots \qquad (5.40)$$

But in order to estimate the rate of convergence we cannot apply the operator approach developed in Sections 5.1 and 5.2 since both $A^{\pm}(\kappa)$ and

$$A(\kappa) = \begin{pmatrix} 0 & A^+(\kappa) \\ A^-(\kappa) & 0 \end{pmatrix}$$

are not kernel operators and, consequently, the function $F(\kappa) = \det I - A(\kappa)$ may not exist. In accordance with the general definition given by Gohberg and Krein (1965) and the estimates given by eqn. 5.39, one can prove that $A^{\pm}(\kappa)$

are the Koch (operator) matrices, i.e. the elements of the corresponding Banach space with the norm

$$\|A^{\pm}(\kappa)\|_K = \sum_m |(A^{\pm}_{m,n}| + \sqrt{\sum_{m,n} |(A^{\pm}_{m,n}|^2}$$

For such operators it is possible to apply the results of Reed and Simon (1978) and to prove the estimate

$$|F(\kappa_N)| \le \|A(\kappa_N) - A_N(\kappa_N)\|_K \exp[2\|A(\kappa_N)\|_K + 1] \qquad (5.41)$$

where

$$A_N(\kappa) = \begin{pmatrix} 0 & P_N A^+ P_N \\ P_N A^- P_N & 0 \end{pmatrix}$$

Substitution of the inequality given by eqn. 5.41 to eqn. 5.16 yields the estimate $|\kappa_N - \kappa_\infty| < \text{const} N^{-12}$ $(N \to \infty)$. This result may be considered as a weak one because the elements of $A^{\pm}(\kappa)$ exponentially decrease along the line (but their rate of decrease is non-uniform with respect to the index m as it follows from the estimate given by eqn. 5.39). In this way we will apply the technique that enables one to essentially improve the rate of convergence and reduce eqn. 5.35 to the equivalent system of coupled operator equations

$$\begin{cases} \mathbf{b}^+ = B^+(\kappa)\mathbf{b}^- \\ \mathbf{b}^- = B^-(\kappa)\mathbf{b}^+ \end{cases} \qquad (5.42)$$

The characteristic numbers of the corresponding operator-valued function $B(\kappa)$ defined by eqn. 5.42 coincide with the characteristic numbers $\kappa_\infty \in \Omega$ of $A(\kappa)$ and, consequently, with the point spectrum of the considered spectral problem in a coaxial waveguide resonator. Here $\mathbf{b}^{\pm} = (b^{\pm}_n)^{\infty}_{n=1}$, $b^{\pm}_n = a^{\pm}_n \sqrt{e_n}$ (and $\mathbf{b}^{\pm} \in l_2$ if $\mathbf{a}^{\pm} = (a^{\pm}_n)^{\infty}_{n=1} \in l_2$ is a solution of eqn. 5.35), $B^{\pm} : l_2 \to l_2$ are the compact finite-meromorphic operator-valued functions of $\kappa \in H_4$ given by the infinite matrices $\|B^{\pm}_{m,n}\|^{\infty}_{m,n=1}$ with the elements

$$B^{\pm}_{m,n} = A^{\pm}_{m,n} \sqrt{e_m} \sqrt{e_n}$$

which have the following asymptotic behaviour

$$B^{\pm}_{m,n} = O[(m+n)^{-1} \exp(-4\pi\delta(m+n))] \quad (m, n \to \infty) \qquad (5.43)$$

Unlike eqn. 5.39, this estimate is uniform with respect to both indices what allows us to prove that $B^\pm$ are the kernel finite-meromorphic operator-valued functions acting in l_2.

Theorem 5.3: The points $\kappa_\infty \in \Omega$ may be approximated by the sequences $\kappa_N \in \Omega_N$ of the characteristic numbers of the finite-dimensional operator-valued functions produced by the system

$$\begin{cases} \mathbf{b}_N^+ = P_N B^+(\kappa) P_N \mathbf{b}_N^- \\ \mathbf{b}_N^- = P_N B^-(\kappa) P_N \mathbf{b}_N^+ \end{cases} \quad N = 1, 2, \ldots \quad (5.44)$$

and the exponential estimate holds for the rate of convergence

$$|\kappa_N - \kappa_\infty| < \text{const} \cdot \exp(-2\pi N \delta), \quad N \to \infty$$

The effective scheme for computing the infinite determinants $F(\kappa) = \det I - B(\kappa)$ (which exist for the kernel operators $B(\kappa)$) was proposed by Kirilenko and Masalov (1982). Calculations of the approximate characteristic numbers were carried out on the basis of the modified Newton method and initial approximations were chosen with the use of different limiting situations (described in Chapter 4).

As a result of numerous numerical simulations, it has been established that in order to provide the 10^{-4} accuracy of the κ_N calculations it is sufficient to take the truncation order $N \sim 2d + 5$, where d is the characteristic diameter of waveguide's inhomogeneity.

5.4 T-junction of rectangular waveguides

Let us consider determination of the fundamental frequencies of the TE oscillations (whose field is constant along the x_1-axis) in the T-junction of rectangular waveguides (see Figure 5.1d). We will use a variant of the mode matching method proposed by Kevanishvili (1978)

The solution to the spectral problem given by eqns. 4.6 and 4.7 may be represented here in the form similar to eqns. 5.1 and 5.19

$$
v(\mathbf{r}) = \begin{cases}
\displaystyle\sum_{n=1}^{\infty} b_n \exp[j\beta_n(x_2 - 2\pi)] \sin \frac{n}{2\theta}(x_3 + \pi\theta), & x_2 > 2\pi \\[2ex]
\displaystyle\sum_{n=1}^{\infty} a_{n,1} \exp[-j\omega_n(x_3 + 2\pi)] \sin \frac{1}{2}nx_2, & x_3 < \pi\theta \\[2ex]
\displaystyle\sum_{n=1}^{\infty} a_{n,2} \exp[j\omega_n(x_3 - 2\pi)] \sin \frac{1}{2}nx_2, & x_3 > \pi\theta
\end{cases}
\tag{5.45}
$$

where

$$
v(\mathbf{r}) = E_{x_1}(\mathbf{r}), \quad \omega_n = \omega_n(\kappa) = \sqrt{\kappa^2 - \frac{n^2}{4}}, \quad \beta_n = \beta_n(\kappa) = \sqrt{\kappa^2 - \frac{n^2}{4\theta^2}}
$$

$$
\mathbf{b} = (b_n)_{n=1}^{\infty} \in \tilde{l}_2, \quad \mathbf{a}_l = (a_{n,l})_{n=1}^{\infty} \in \tilde{l}_2 \quad (l = 1, 2)
$$

are the complex amplitudes of the partial components of the TE free oscillations. Thus the function defined by eqn. 5.45 satisfies the Homogeneous Helmholtz equation and Meixner's conditions. All the functions of κ are considered on the Riemann surface H_2 of the analytical continuation from the real axis κ of two canonical Green's functions of the parallel-plane waveguides having the width 2π and $2\pi\theta$. H_2 coincides with H for $\theta_2 = 1$. For $\text{Im}\,\kappa = 0$ these Green's functions satisfy the radiation condition of the absence of waves coming from infinity what provides the fulfillment of radiation conditions for $v(\mathbf{r})$.

Imposition of the conjugation conditions in the planes $|x_3| = \pi\theta$, $x_2 = 2\pi$ will result in the equivalent infinite system of linear algebraic equations.

In order to carry out this reduction let us apply the second Green's formula

$$
\int_S \left(V \frac{\partial U}{\partial \mathbf{n}} - U \frac{\partial V}{\partial \mathbf{n}} \right) ds = 0
\tag{5.46}
$$

in the domain bounded by the contour $S = ABCD$ (see Figure 5.1d) where the functions $V = v(\mathbf{r})$ and U equal to each of the eigenfunctions $v_{n,k}$, $k = 1, 2, 3$ of the main (of the width 2π) and lateral waveguides

$$
v_{n,1} = \sin \frac{1}{2}nx_2 \exp[j\omega_n(x_3 + \pi\theta)]
$$

$$
v_{n,2} = \sin \frac{1}{2}nx_2 \exp[-j\omega_n(x_3 - \pi\theta)]
\tag{5.47}
$$

$$
v_{n,3} = \sin \frac{n}{2\theta}(x_3 + \pi\theta) \sin \beta_n x_2
$$

Integration in eqn. 5.46 and elimination of the unknowns a_l lead to the required equivalent infinite system of linear algebraic equations

$$\mathbf{b} = A(\kappa)\mathbf{b}, \quad \mathbf{b} \in \tilde{l}_2 \tag{5.48}$$

where $A(\kappa) = B(\kappa)D(\kappa) + \bar{B}(\kappa)\bar{D}(\kappa)$ are the operator-valued functions given by the infinite matrices $\|B_{m,n}^{\pm}\|_{m,n=1}^{\infty}$ and $\|D_{m,n}^{\pm}\|_{m,n=1}^{\infty}$ with the elements

$$
\begin{aligned}
B_{m,n} &= \frac{(-1)^n mn[1 - \exp(4j\pi\beta_m)]}{2\pi\theta\beta_m(n^2 - 4\beta_m^2)} \\
D_{m,n} &= \frac{(-1)^m imn[1 - (-1)^n \exp(2j\pi\omega_m\theta)]}{2\pi\theta\omega_m(m^2 - 4\beta_m^2)}
\end{aligned}
\tag{5.49}
$$

such that $\bar{B}_{m,n}(\kappa) = (-1)^m B_{m,n}(\kappa)$, $\bar{D}_{m,n}(\kappa) = (-1)^m D_{m,n}(\kappa)$.

The equivalence is provided here by the possibility to represent every solution $u(\mathbf{r})$ of the Helmholtz equation in the domain $ABCD$ satisfying the boundary-value condition $u|_S = F$ as a sum $u(\mathbf{r}) = u_1(\mathbf{r}) + u_2(\mathbf{r})$, where $u_{1,2}(\mathbf{r})$ satisfy the Helmholtz equation in $ABCD$ and the conditions $u_1(x_2, -\pi\theta) = u_1(x_2, \pi\theta) = 0$ $(0 \leq x_2 \leq 2\pi)$ and $u_2(0, x_3) = u_2(2\pi, x_3) = 0$ $(x_3 < 2\pi)$. Further, $u_1(0, x_3)$, $u_1(2\pi, x_3)$ $(|x_3| < \pi\theta_2)$ and $u_2(x_2, -\pi\theta)$, $u_2(x_2, \pi\theta) = 0$ $(0 \leq x_2 \leq 2\pi)$ coincide with the corresponding values of F on S and each of the functions $u_{1,2}(\mathbf{r})$ may be uniquely represented with the help of the complete systems $\{v_{n,k}\}_{n=1}^{\infty}$ $(k = 1, 2, 3)$.

In order to construct operator equations of the problem with kernel operator-valued functions we will change the variables using the formula $\mathbf{c} = (\sqrt{n}b_n)_{n=1}^{\infty}$ and rewrite eqn. 5.48 in the form

$$\mathbf{c} = K(\kappa)\mathbf{c}, \quad \kappa \in H_2 \tag{5.50}$$

where $K(\kappa) = S(\kappa)T(\kappa) + \bar{S}(\kappa)\bar{T}(\kappa)$ and S, T, $\bar{S}$, $\bar{T}$ are the operator-valued functions given by the infinite matrices $\|S_{m,n}^{\pm}\|_{m,n=1}^{\infty}$, $\|T_{m,n}^{\pm}\|_{m,n=1}^{\infty}$, $\|\bar{S}_{m,n}^{\pm}\|_{m,n=1}^{\infty}$ and $\|\bar{T}_{m,n}^{\pm}\|_{m,n=1}^{\infty}$ with elements

$$
S_{m,n} = B_{m,n}\sqrt{\frac{m}{n}}, \qquad\qquad \bar{S}_{m,n} = \bar{B}_{m,n}\sqrt{\frac{m}{n}}
$$

$$
T_{m,n} = D_{m,n}\sqrt{\frac{m}{n}}, \qquad\qquad \bar{T}_{m,n} = \bar{D}_{m,n}\sqrt{\frac{m}{n}}
$$

Eqn. 5.49 yield the estimates

$$S_{m,n} = O[\sqrt{mn}(m^2 + n^2\theta^{-2})^{-1}]$$
$$T_{m,n} = O[\sqrt{mn}(m^2 + n^2\theta^{-2})^{-1}] \quad (m, n \to \infty) \tag{5.51}$$

Following Gribanov (1963), we will call an operator K produced by an infinite matrix $\|K_{m,n}\|$ ω-completely continuous if $\|K - P_N K P_N\| \to 0$ for $N \to \infty$.

Lemma 5.5: $S(\kappa)$, $T(\kappa)$, $\bar{S}(\kappa)$ and $\bar{T}(\kappa)$ are the finite-meromorphic ω-completely continuous operator-valued functions of $\kappa \in H_2$ in the space $l_1 = \{\mathbf{a} = (a_n)_{n=1}^\infty : \sum_n |a_n| < \infty\}$.

The proof of this lemma is based on the estimate

$$\sum_{m=1}^\infty \frac{\sqrt{m}}{m^2 + n^2} \leq \frac{\text{const}}{\sqrt{n}} \left| \ln \frac{1 + n + \sqrt{n}}{\sqrt{n^2 + 1}} - \arctan \frac{\sqrt{n}}{n - 1} \right|$$

which implies fulfillment of the condition

$$\lim_{k \to \infty} \sup_{k \leq m < \infty} \sum_{n=1}^\infty |a_{m,n}| = 0 \tag{5.52}$$

(Gribanov, 1963) providing the complete ω-continuity (here $a_{m,n}$ denotes the element of each of the operators S, T, $\bar{S}$ and $\bar{T}$).

Completely continuous operators contain the set of ω-completely continuous operators. Therefore, $I - K(\kappa) : l_1 \to l_1$ is a canonical Fredholm operator-valued function and, consequently, the fundamental frequencies of the considered open waveguide resonator are the isolated points in H_2 (finite-order poles of the resolvent $[I - K(\kappa)]^{-1}$)

As well as in the previous cases, the characteristic numbers κ_∞ of the problem given by eqn. 5.50 may be approximated by the characteristic numbers κ_N of the truncated problem

$$\mathbf{c}_N = P_N K(\kappa) P_N \mathbf{c}_N, \quad \mathbf{c}_N \in l_1 \quad (N = 1, 2, \ldots)$$

where $\det[I - P_N K(\kappa_N) P_N] = 0$ and $|\kappa_\infty - \kappa_N| \to 0$, $N \to \infty$. Here one cannot prove the existence of the infinite determinant $\det[I - K(\kappa)]$ and, hence it is not possible to directly estimate the rate of convergence.

Table 5.1

N	$\kappa = \kappa' + j0$	$\kappa = \kappa' + j\kappa''$	
		κ'	$-\kappa''$
1	0.483497	0.923919	0.090132
2	0.482927	0.923151	0.089794
3	0.482647	0.922674	0.089583
4	0.482482	0.922349	0.089439
5	0.482376	0.922116	0.089335
6	0.482302	0.921940	0.089257
7	0.482249	0.921804	0.089196
8	0.482208	0.921696	0.089147
9	0.482176	0.921608	0.089108
10	0.482151	0.921536	0.089076
11	0.482130	0.921475	0.089048
12	0.482113	0.921423	0.089025
13	0.482098	0.921378	0.089005
14	0.482086	0.921340	0.088988
15	0.482076	0.921306	0.088973
16	0.482067	0.921277	0.088959

In this way we will present some numerical examples which will demonstrate the efficiency of this truncation procedure. The *a priori* estimate $|\kappa_\infty - \kappa_N| < |\eta|^{N-\tau}$ together with the algorithm for the choice of the parameters η and τ are proposed by Shestopalov *et al.* (1984). The data in Table 5.1 presented for the T-junction with $\theta = 2$ enable one to estimate the convergence. Here, in accordance with the results of Shestopalov *et al.* (1984) $\eta = -0.006$, $\tau = 1.53$ for the real values of κ and $\eta = 0.006$, $\tau = 0.85$ for $|\kappa_N| = |\kappa_N' + j\kappa_N''|$ when $\operatorname{Im}\kappa \neq 0$.

One of the serious problems influencing the convergence is taking into account the rate of decrease of the rest of the truncated sums

$$\Delta_{m,n}^Q = \sum_{q=Q+1}^{\infty} d_{m,n}^q \tag{5.53}$$

where $d^q_{m,n} = S_{m,q} T_{q,n} = O(q)^{-3}$. In order to explicitly estimate the right-hand side of eqn. 5.53, let us take $Q > N$ such that $Q^2 \exp(-2\pi \operatorname{Im}\omega_Q\theta) < 1$ and $4\beta_l^2 < Q^2$ $(l = m, n)$. Then eqn. 5.53 yields

$$\Delta_{m,n}^Q = \frac{i\sqrt{mn}[1 - \exp(4i\pi\beta_m)]}{2\pi^2\theta^3\beta_m} \sum_{s=3,5,\dots} \hat{d}_s[\zeta(s) - \sum_{q=1}^{Q} q^{-s}] \tag{5.54}$$

Limiting oneself to the first k terms in the series given by eqn. 5.54, it is easy to show that $\Delta_{m,n}^Q \sim O(\hat{d}_{k+1} Q^{-(k+1)})$ where $\zeta(s)$ is the Riemann zeta-function and $\hat{d}_s$ depend on the indices m, n and the parameters κ, θ. In particular, $d_3 = 1$, $d_5 = 10\kappa^2 - \theta^2(m^2 + n^2)$, $d_7 = 70\kappa^4 - 14\kappa^2\theta^{-2}(m^2 + n^2) + \theta^{-4}(m^4 + n^4 + m^2 n^2)$.

5.5 Free oscillations in irregular domains

The semi-inversion technique developed by Kirilenko *et al.* (1979), Kirilenko and Rud' (1980) and Shestopalov *et al.* (1984) will form the basis for determination of forced and free oscillations in irregular domains depicted in Figure 5.1*d*. The operators in the operator equations of the second kind are easily identified there with the operators describing scattering on elementary inhomogeneities (when the corresponding diffraction problems may be solved explicitly). Masalov and Sirenko (1978) have proved that the spectra of these operators are localised inside the unit disc on the complex plane of the spectral parameter κ (except for not more than a finite number of real points) which turns out to be sufficient for the proof of the unique solvability of the diffraction problems and convergence of numerical procedures based on the truncation of infinite matrix operators.

Let us consider first the problem of free oscillations in the open resonator formed by a symmetric waveguide break with an arbitrary break angle ψ (note that $\psi = 45°$ in Figure 5.1*d*). The solution to the spectral problem given by eqns. 4.6 and 4.7 satisfying the homogeneous Helmholtz equation, the boundary-value conditions and radiation conditions at infinity may be represented as

$$u(\mathbf{r}_i) = \sum_{m=1}^{\infty} A_{ml} \exp(j\omega_m x_{3i}) \sin \frac{1}{2} m x_{2i} \qquad (5.55)$$

where $\mathbf{r}_i = (x_{2i}, x_{3i})$, $u(\mathbf{r}_i) = E_{x_1}(\mathbf{r}_i)$, $\omega_m = \omega_m(\kappa) = \sqrt{\kappa^2 - \frac{m^2}{4}}$ ($i = 1, 2$). The problem is to determine the (spectral) values $\kappa \in H$ (H is the Riemann surface of the analytical continuation of canonical Green's function introduced in Section 4.1) where there exist non-trivial solutions of the spectral problem given by eqns. 4.6 and 4.7 in the form given by eqn. 5.55. The fields in the irregular domain $OBAC$ (see Figure 5.1e) are connected with the fields in the regular waveguides given by eqn. 5.55 by means of Green's formula (eqn. 5.46) where U are taken to be equal to the eigenfunctions of the parallel-plane waveguides of the width 2π

$$v_{qi}^{(1)} = \sin(qx_{2i})\exp(j\omega_q x_{3i}), \quad v_{qi}^{(2)} = \sin(qx_{2i})\exp(-j\omega_q x_{3i})$$

Successive integration in eqn. 5.46 along the contours OBA and OAC leads to the infinite system of linear algebraic equations with respect to the unknowns $\alpha_n^\pm$

$$\sum_{n=-\infty}^{\infty} \alpha_n^+[(\Gamma_n^+ - \omega_q)^{-1} \mp (\Gamma_n^- + \omega_q)^{-1}] = 0 \quad (q = 1, 2, \ldots) \qquad (5.56)$$

and corresponding expressions for the coefficients

$$v_q A_q^\pm = -\sum_{n=-\infty}^{\infty} \alpha_n^\pm[(\Gamma_n^+ + \omega_q)^{-1} \mp (\Gamma_n^- - \omega_q)^{-1}] = 0 \quad (q = 1, 2, \ldots) \quad (5.57)$$

Here

$$u(x_2, 0) = \sum_{n=-\infty}^{\infty} \alpha_n^+ \exp\left(j2\pi n\frac{x_2}{l}\right) \sin\frac{1}{2}mx_{2j}$$

or

$$\frac{\partial u}{\partial x_3}(x_2, 0) = j\frac{2\pi}{l} \sum_{n=-\infty}^{\infty} \Gamma_n \alpha_n^- \exp\left(j2\pi n\frac{x_2}{l}\right) \sin\frac{1}{2}mx_{2j}$$

for the magnetic or electric plug, respectively, where

$$\Gamma_n = \sqrt{\frac{\kappa^2}{\cos^2\psi} - m^2}$$

$$\Gamma_n^\pm = \Gamma_n \cos\psi \pm n\sin\psi$$

$$v_q = 2q\hat{\omega}_q \cos^2\psi\{\exp[j(\hat{\omega}_q \sin\psi + \pi q)] - 1\}$$

$$\hat{\omega}_q = \omega_q(\cos\psi)^{-1}$$

and the signs $+$ $(-)$ correspond to the symmetric (anti-symmetric) oscillations (with respect to break's axis).

Applying the inversion of the matrix operators with the elements $\Gamma_n^+ - \omega_q)^{-1}$ and using the semi-inversion proposed by Kirilenko *et al.* (1979)) and

Shestopalov *et al.* (1984) transform eqns. 5.55–5.57 to the equivalent system

$$\alpha_n^\pm \mp \sum_{s=-\infty}^{\infty} \gamma_{ns}\alpha_s^\pm = 0 \quad (n = 0, \pm1, \pm2, \ldots) \tag{5.58}$$

$$A_q^\pm = \sum_{s=-\infty}^{\infty} f_s(-\hat\omega_q)\alpha_s^\pm \quad (q = 1, 2, \ldots) \tag{5.59}$$

where

$$\gamma_{ns} = -\frac{\operatorname{res} f(\Gamma_n^+)}{f(-\Gamma_s^-)(\Gamma_n^+ + \Gamma_s^-)}, \quad f_s(-\hat\omega_q) = \frac{f(-\hat\omega_q)}{v_q f(-\Gamma_s^-)(\hat\omega_q + \Gamma_s^-)}$$

$$f(w) = \frac{\exp[jw(t(\psi)]}{\Gamma_0^+ - w} \prod_{m=1}^{\infty} \frac{2jm\cos\psi(\hat\omega_m - w))}{(\Gamma_m^+ - w)(\Gamma_{-m}^+ - w)}$$

$$t(\psi) = 2[\cos\psi \ln(2\cos\psi) + \psi\cos\psi]$$

The estimate

$$\gamma_{|ns|} \leq \frac{1}{\sqrt{n}} \frac{\operatorname{const}(\kappa, \psi)\sqrt{s}}{|n| + s\exp\{j\psi[\operatorname{sign}(n) + \operatorname{sign}(s)]\}} \quad (|n|, |s| \to \infty) \tag{5.60}$$

enables one both to prove that eqns. 5.58 and 5.59 define a canonical Fredholm operator in l_2 and to justify the truncation algorithm described in the previous Sections. The data in Table 5.2 illustrate the convergence of the characteristic numbers κ_N approximating the fundamental frequencies of the irregular domain of the symmetric right-angle waveguide's break (and being close to the fundamental frequencies of the anti-symmetric oscillations in the T-junction with $\theta = 2$ presented in Table 5.1 what indirectly confirms the quality of the model and numerical algorithms). The truncation errors are $\Delta\kappa_N \approx -0.0017N^{-1.4}$ and $\Delta|\kappa_N| \approx -0.0074N^{-1.4}$ for the real and complex fundamental frequencies, respectively.

Consider application of this semi-inversion method based on the use of the "residue" series (given by eqns. 5.56 and 5.57) for the analysis of the free oscillations in the irregular domain of waveguide's plug (Figure 5.1f). The corresponding spectral problem is equal to the determination of the fundamental frequencies (of anti-symmetric oscillations) in the open resonator formed by waveguide's break with a mirror. Both waveguide resonators contain a rectangular domain with partially (or completely) "free" (non-earthed)

Table 5.2

N	$\kappa = \kappa' + j0$	$\kappa = \kappa' + j\kappa''$	
		κ'	$-\kappa''$
3	0.485506	0.936795	0.095240
5	0.483683	0.927709	0.091578
7	0.483043	0.925190	0.090594
9	0.482728	0.923967	0.090101
11	0.482544	0.923248	0.089805
13	0.482425	0.922780	0.089608
15	0.482343	0.922452	0.089470
17	0.482283	0.922212	0.089367
19	0.482237	0.922029	0.089288
21	0.482202	0.921886	0.089226
23	0.482173	0.921772	0.089176
25	0.482150	0.921678	0.089135
27	0.482131	0.921600	0.089101
29	0.482115	0.921534	0.089072
31	0.482101	0.921478	0.089048
33	0.482090	0.921430	0.089026

boundaries (denoted in Figure 5.1d by dashed lines) which enables one to apply the technique developed in this Section. A similar approach has been used by Krutin and Sirenko (1983) for solving some problems of waves diffraction on lossy periodical surfaces.

The solution to the spectral problem given by eqns. 4.6 and 4.7 satisfying the homogeneous Helmholtz equation in the irregular domain of the plug (see Figure 5.1f), boundary-value conditions (on the walls $\theta = \sin\psi$, $\beta = \cos\psi$) and radiation conditions at infinity may be represented as

$$
u(\mathbf{r}) = \begin{cases}
\displaystyle\sum_{n=1}^{\infty} a_n \exp(j\omega_n x_3) \sin\frac{1}{2}nx_2, & x_3 \geq 0 \\[2em]
\displaystyle\sum_{n=1}^{\infty} c_n [\exp(j\gamma_n x_3') - \exp(-j\gamma_n x_3')] \sin\frac{1}{2}n\delta x_2' \\[1em]
\quad + \displaystyle\sum_{n=1}^{\infty} \bar{c}_n \{\exp[j\beta_n(x_2' - 2\pi\delta)] \\[1em]
\quad - \exp[-j\beta_n(x_2' - 2\pi\delta)]\} \sin\frac{1}{2}n\theta x_3', & x_3', x_2' \in OABC
\end{cases}
$$

$$\tag{5.61}$$

where $u(\mathbf{x}) = E_{x_1}(\mathbf{x})$, the local co-ordinates x_2', x_3' satisfy the relations

$$\begin{cases} x_2 = x_2'\delta + x_3'\theta \\ x_3 = x_3'\delta - x_2'\theta \end{cases}$$

$$\gamma_n = \omega_n(\kappa) = \sqrt{\kappa^2 - \frac{n^2}{4\delta^2}}, \quad \beta_n = \beta_n(\kappa) = \sqrt{\kappa^2 - \frac{n^2}{4\theta^2}}$$

are the main branches of the corresponding multi-valued functions of $\kappa \in H$ co-ordinated with the choice of the branch of the function $\omega_n(\kappa)$ (and the choice of the Riemann surface H is determined here, as well as in the previous problem considered in this Section, by the presence of a single radiation channel).

Fields matching on the boundaries OA and OB results after the series of bulky transformations in the matrix operator equation equivalent to the initial spectral problem

$$\mathbf{b} = T(\kappa)\mathbf{b}, \quad T(\kappa) = A(\kappa)B(\kappa) + \bar{A}(\kappa)\bar{B}(\kappa)$$
$$\mathbf{b} = (b_n)_{n=1}^{\infty} \in \tilde{m} = \{\mathbf{c} = (c_n)_{n=1}^{\infty} : \sup|c_n| < \infty\} \qquad (5.62)$$
$$b_n = \sqrt{n}a_n \quad (n = 1, 2, \ldots)$$

where $A = \|A_{mn}\|_{m,n=1}^{\infty}$, $B = \|A_{mn}\|_{m,n=1}^{\infty}$,

$$A_{mn} = \frac{\gamma_{2n}^{-}[\Gamma_m^1(\gamma_{1n}^{+}) - \Gamma_m^1(-\gamma_{1n}^{+})] - \gamma_{2n}^{+}[\Gamma_m^1(\gamma_{1n}^{-}) - \Gamma_m^1(-\gamma_{1n}^{-})]}{2\sin 2\pi\gamma_n\theta\omega_m}\sqrt{\frac{m}{n}}$$

$$B_{mn} = \Gamma_m^{\delta}(-\omega_{1n}^{-})\exp(2j\pi\theta\omega_{2n}^{+}) - \Gamma_m^{\delta}(-\omega_{1n}^{+})\exp(2j\pi\theta\omega_{2n}^{-})\sqrt{\frac{m}{n}}$$

$$\gamma_{1n}^{\pm} = \pm\frac{n}{2} + \gamma_n\theta, \quad \gamma_{2n}^{\pm} = \pm\frac{n}{2}\frac{\theta}{\delta} + \gamma_n\delta$$
$$\omega_{1n}^{\pm} = \pm\frac{n}{2}\delta + \omega_n\theta, \quad \omega_{2n}^{\pm} = \pm\frac{n}{2}\theta + \omega_n\delta$$

Lemma 5.6: $A(\kappa)B(\kappa)$ and $\bar{A}(\kappa)\bar{B}(\kappa)$ are the finite-meromorphic ω-completely continuous operator-valued functions of $\kappa \in H_2$ in the space $\tilde{m}$.

The proof of this lemma follows from the limiting relations

$$\lim_{k \to \infty} \sup_{k \leq m < \infty} \sum_n |A_{mn}| = 0 \qquad \sup_{1 \leq m < \infty} \sum_n |B_{mn}| < \infty \qquad (5.63)$$

which are based on the estimate

$$A_{mn} = O\{\sqrt{mn}[m^2 + k_1(\psi)n^2 + ik_2(\psi)n^2]^{-1}\}$$
$$B_{mn} = O\{\sqrt{m^3 n^{-1}}[m^2 + k_3(\psi)n^2 + jk_4(\psi)n^2]^{-1}\}$$

(5.64)

where $k_i(\psi)$ $(i = 1, 2, 3, 4)$ are the real-valued positive functions uniformly bounded with respect to all the non-spectral parameters.

One can conclude that $I - T(\kappa) : \tilde{m} \to \tilde{m}$ in eqn. 5.62 is a canonical Fredholm operator-valued function and, consequently, the fundamental frequencies of the considered open waveguide resonator are the isolated points in H (finite-order poles of the resolvent $[I - T(\kappa)]^{-1}$). Due to the limiting relation $\lim_{N \to \infty} \|T - P_N T P_N\|_{\tilde{m}} = 0$ which follows from the ω-continuity of the operator-valued function T, the characteristic numbers κ_∞ of the problem given by eqn. 5.62 may be approximated by the characteristic numbers κ_N of the truncated problem $\mathbf{b}_N = P_N T(\kappa) P_N \mathbf{b}_N$ $(N = 1, 2, \ldots)$ where $\det[I - P_N K(\kappa_N) P_N] = 0$ and $|\kappa_\infty - \kappa_N| \to 0$, $N \to \infty$. The proof of these assertions is similar to the one of theorem 5.1.

5.6 Generalised scattering matrices

The method of generalised scattering matrices proposed and developed by Shestopalov and Shcherbak (1968) and Nikolsky (1978) is a powerful tool which enables one to generalise the technique applied in the modelling of waveguides and resonators. We will demonstrate the capabilities of this method taking the problem of fundamental frequencies of the open waveguide resonator formed by two elementary (basic) inhomogeneities (see Figure 5.2) as a typical example. In accordance with the results of the present Chapter, the fundamental frequencies of the elementary resonators (which are assumed to be known) form a set of isolated points in the Riemann surface H and the complex amplitude vectors $\mathbf{a}$, $\mathbf{b}$, $\mathbf{c}$ and $\mathbf{d}$ of the partial free fields components obey the operator relations

$$\mathbf{a} = T_1 E^\delta \mathbf{d}, \quad \mathbf{d} = R_2 E^\delta \mathbf{c}, \quad \mathbf{c} = R_1 E^\delta \mathbf{d}, \quad \mathbf{b} = T_2 E^\delta \mathbf{c} \qquad (5.65)$$

where $R_{1,2}$ and $T_{1,2}$ are produced by the corresponding generalised scattering matrices and define (matrix) bounded finite-meromorphic operator-valued

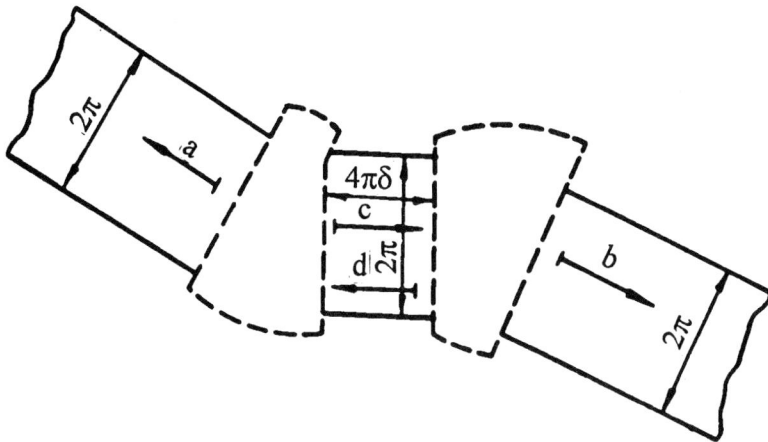

Figure 5.2   *Basic types of inhomogeneities in open waveguide resonators*

functions of $\kappa \in H$ acting in $\tilde{l}_2$ (for the definition of $\tilde{l}_2$ see Section 5.1). The
diagonal matrix

$$E^\delta = ||\delta_{mn} \exp[2j\pi\delta\sqrt{4\kappa^2 - n^2}]||_{m,n=1}^\infty$$

defines a compact operator-valued function $E^\delta(\kappa)$ in $\tilde{l}_2$. The system of operator relations given by eqn. 5.6 may be equivalently reduced to one equation

$$\mathbf{d} = S(\kappa)\mathbf{d}, \quad \mathbf{d} \in \tilde{l}_2 \tag{5.65}$$

with the canonical Fredholm operator-valued function $I - S(\kappa)$ where $S(\kappa) = R_2(\kappa)E^\delta R_1(\kappa)E^\delta$. Hence, as well as in the cases considered in Sections 5.1–
5.5, the fundamental frequencies of the open waveguide resonator formed
by two basic inhomogeneities are the isolated points in H and may be
approximated by the sequences of roots κ_N of the "truncated" equations
$\det[I - P_N S(\kappa_N)P_N] = 0$.

We may conclude that the knowledge of the spectra of several basic structures allows one to effectively treat by means of the generalised scattering
matrices method a wide class of open waveguide resonators whose communication domains are formed by a combination of elementary resonators.

Chapter 6

Free oscillations and resonance excitation of open waveguide resonators

In this Chapter the results of calculations of the qualitative characteristics of the free oscillations spectra in two groups of open waveguide resonators are presented. Among them are waveguide junctions with communication domains formed by the sections of regular waveguides whose cross-sections differ from semi-infinite radiation channels (e.g. finite waveguide discontinuities) and open waveguide resonators with irregular communication domains (like T-junctions or waveguide breaks).

In accordance with the uniqueness theorems proved in Chapter 4, the numerical data obtained concern the variation of the complex spectral parameter $\kappa = \kappa' + j\kappa''$ on the first ("physical") sheet of the Riemann surface H (i.e. the spectral points situated at the shortest distance from the real axis with respect to the natural "physical" metric and, consequently, from those real frequencies where the real stationary scattering processes are observed) with cuts along the curves

$$d_{kl} = (\kappa')^2 - (\kappa'')^2 - [\alpha_{kl}(\theta_l)]^2, \quad \frac{7}{2}\pi \le \arg \kappa \le 2\pi \tag{6.1}$$

The real value $\alpha_{kl}(\theta_l)$ in eqn. 6.1 is the transversal wavenumber of the lth rectangular or circular waveguide. In this case the oscillations attenuate in time and the radiation conditions for the lth semi-infinite waveguide of the relative width θ_l (situated in the plane $x_{3l} = 0$)

$$u_l(\eta, x_{3l}) = \sum_s a_{sl} \Phi_s(\eta) \exp(j\omega_{sl} x_{3l}) \quad [\eta = x_{1l}(x_{2l}) \text{ or } \phi_l] \qquad (6.2)$$

contain a finite number of the functions $\omega_{nl} = \sqrt{(\kappa\theta_l)^2 - \alpha_{nl}^2}$ with $\mathrm{Re}\,\omega_{nl} \geq 0$, $\mathrm{Im}\,\omega_{nl} \leq 0$ (the corresponding terms in eqn. 6.2 exponentially grow for $x_{3l} \to \infty$) and an infinite number with $\mathrm{Re}\,\omega_{nl} < 0$, $\mathrm{Im}\,\omega_{nl} > 0$. Here $\Phi_s(\eta)$ is the eigenfunctions of the lth waveguide.

Some examples are related to the higher order sheets of H. The passage to them is connected with the change of signs of $\mathrm{Re}\,\omega_{nl}$ and $\mathrm{Im}\,\omega_{nl}$, where l and n are the numbers of the (lth) waveguide and the (nth) cut, respectively.

The approximate values of the fundamental frequencies κ_N (zeros of the determinants $F_N(\kappa) = \det\|a_{mn}\|_{m,n=1}^N$ of the matrices of the approximating finite-dimensional operators) are calculated by means of Newton's method and the iterations stop when either $F_N(\kappa_{N*}) < 10^{-4}$ or $|\kappa_N - \kappa_{N-1}| < 10^{-5}$.

6.1 Open waveguide cylindrical resonators

Let us consider open waveguide resonators containing radiation channels in the form of coaxial semi-infinite circular waveguides with different cross-sections.

The radiation problems for such waveguides are more complicated than that for rectangular ones since here the diffraction communication is carried out between a waveguide resonator and two different radiation channels, a circular waveguide (with a simple connected cross-section) and a coaxial waveguide (with a multi-connected cross-section). Let us successively describe all symmetrical oscillations for these three types of open waveguide resonators which are good models of real devices widely applied in the super-high frequency technology.

We will denote by $k = k(M)$ the number of waves carrying energy out of a resonator to a circular (coaxial) waveguide, where (k, N), (M, N) and $\left(\frac{M}{k}, N\right)$ correspond to the coaxial semi-infinite circular waveguides, to the coaxial waveguides or to the circular and coaxial waveguides, respectively. The value of N is determined by the condition $\mathrm{Re}\,\omega_n \geq 0$, $\mathrm{Im}\,\omega_n \leq 0$, $n = 1, 2, \ldots, N$. Assuming that the frequency imaginary part is negligibly small, one can connect the boundaries of the indicated regimes with the critical wave frequencies in the narrow (k), coaxial (M) and internal wide (N) waveguides. $\kappa = \frac{ka}{2\pi}$ is chosen as a frequency spectral parameter, where a is the radius of the internal waveguide forming the cavity in an open waveguide resonator.

The isolines $E_\phi = \text{const}$, $H_\phi = \text{const}$ constructed at eigenfrequencies (corresponding thus to the fields of certain free oscillations) were used, in addition to various a-priori assumptions, to determine the oscillations types.

As a result of numerical experiments, we may separate two groups of oscillations differing by Q-factors, fields and the behaviour of their spectral curves with respect to the geometrical parameters. The properties of the oscillations belonging to the first group (H_{0mn} and E_{0mn} where n, m are the number of field variations with respect to the radius and resonator's length, respectively) enable us to assert that they are close to the oscillations of the initial cylindrical volume resonators. These oscillations have the high Q-factors $Q = \frac{\kappa'}{-\kappa''} \sim 10^2 - 10^3$ and their eigenfields are concentrated in the resonant volume. The oscillations of the second group ($\hat{H}_{0mn}$ and $\hat{E}_{0mn}$) have a strong communication with the external domain and, therefore, they have the low Q-factor. Their eigenfields are concentrated in semi-infinite waveguides.

Let us consider the cylindrical resonator symmetrically loaded by the circular waveguides. Let a, l and b be the radius and the length of the internal resonant domain and the radius of the semi-infinite waveguides, respectively, such that $\theta = \frac{b}{a} < 1$.

The qualitative changes in a closed cylindrical resonator with diffraction losses adjusted to two semi-infinite waveguides may be followed considering its eigenfrequencies as functions of the communication hole diameter $\theta - \kappa_{0nm}$ where θ depends on the cuts positions given by eqn. 6.1 (determined by the relation $\alpha_k = \nu_{ks}(2\pi\theta)^{-1}$ on the Riemann surface $\mathcal{H}$). On the complex surface κ these cuts may be defined as

$$\kappa'_n = \sqrt{(\kappa''_n)^2 + \frac{\nu_{ns}}{(2\pi\theta)^2}} \qquad (6.3)$$

where ν_{ns} is the nth root of the equation $J_s(\nu) = 0$, $s = 0$ and $s = 1$ correspond to the E and H oscillations, respectively.

Let us discuss firstly the properties of the H_{0mn} oscillations whose fundamental frequencies vary on the first sheet of $\mathcal{H}$ considering $\kappa_{0nm} = \kappa_{0nm}(\theta)$

for $L = la^{-1} = 1$ ($\theta = 0$ and $\theta = 1$ correspond to the closed resonator and open waveguide, respectively). Here two situations take place:

(i) there are no waves in the radiation channels which carry out the field energy ($k = 0$), the fundamental frequencies are real and lie in the domain

$$\frac{\nu_{11}}{2\pi} < \kappa' < \frac{\nu_{11}}{2\pi}, \quad \kappa'' = 0 \tag{6.4}$$

(ii) the resonator's energy is radiated to the waveguides ($k > 0$), the fundamental frequencies are complex and the domain of their existence are determined by the inequality

$$\kappa' > \sqrt{(\kappa'')^2 + (\frac{\nu_{11}}{2\pi})^2} \tag{6.5}$$

The currents of the given type of oscillation has a definite structure ($j_{x_3} = 0$, $j_\rho = 0$, $j_\phi \neq 0$) and, hence the symmetrical holes at the resonator's edges cause weak perturbations of its eigenfrequencies in the $(0, N)$ regime. The latter practically do not depend on the communication hole diameter when κ' varies in the domain defined by eqn. 6.4 and κ_{0nm} coincides with the closed resonator eigenfrequencies

$$\nu_{0nm}^c = \sqrt{\left(\frac{\nu_{n1}}{2\pi}\right)^2 + \left(\frac{m}{2L}\right)^2}$$

Increasing θ yields emergence of a wave (in narrow waveguides) carrying the energy out of the resonator, i.e. diffraction losses and κ_{0nm}'' appear.

The transfer from the regime $(0, N)$ to the regime $(1, N)$ is accompanied by the absence of changes in the field's structure and by the breaks of the branches of the function $\kappa_{0nm}(\theta)$ for $n = 1$ when crossing the cut defined by eqn. 6.3. These breaks are different for different oscillations and they cause minimal addition terms for θ for the low longitudinal index oscillations. With growing longitudinal index the breaks increase and the field structure is violated. For example, when $\kappa_{022}(\theta)$ transfers to the regime defined by eqns. 6.1 and 6.2, the eigenfield is concentrated out of the open waveguide resonator that is characteristic for the $\hat{H}_{022}$ oscillations and such a configuration is preserved while θ varies until the value when $|\kappa_{022}''|$ reaches its maximum. The field's concentration in the resonator grows with increasing θ in that part of the spectrum where $|\kappa''_{022}|$ falls. In particular, the isolines $E_\phi(r, x_3) = $ const are identical for $\theta = 0.34$ and for $\theta = 0.74$.

The value of θ essentially influences eigenfrequencies (considered as functions of resonator's length). In particular, one can separate the domains on the first sheet of the Riemann surface κ where there exist only the H_{0nm} oscillations:

$$\sqrt{(\kappa''_{0nm})^2 + \left(\frac{\nu_{ns}}{2\pi}\right)^2} < \kappa'_{0nm} < \sqrt{(\kappa''_{0nm})^2 + \left(\frac{\nu_{ns}}{2\pi\theta}\right)^2} \qquad (6.6)$$

(where the number k of waves carrying off the radiation energy is always less than the number N of the direct and inverse H_{0n} waves forming a definite oscillation in resonator's volume) or $\hat{H}_{0nm}$ oscillations:

$$\sqrt{(\hat{\kappa}''_{0nm})^2 + \left(\frac{\nu_{ns}}{2\pi\theta}\right)^2} < \hat{\kappa}'_{0nm} < \sqrt{(\hat{\kappa}''_{0nm})^2 + \left(\frac{\nu_{n+1,s}}{2\pi}\right)^2} \qquad (6.7)$$

(where $k = N$ what causes higher diffraction losses and low Q-factors of the $\hat{H}_{0nm}$ oscillations). The H_{01m} oscillations have real fundamental frequencies

$$\kappa = \kappa'_{01m} : \frac{\nu_{11}}{2\pi} < \kappa'_{01m} < \frac{\nu_{11}}{2\pi\theta}$$

and the cut-off frequencies coincide with the critical values of the first higher order H_{011} oscillation of the narrow waveguide. When $L \to \infty$, the dispersion curves asymptotically tend to the critical value $\frac{\nu_{11}}{2\pi}$ of the first higher order oscillation of the wide waveguide. At a fixed frequency the distance between different branches κ_{01m} is close to the constant value $\Delta L = (2\omega_1)^{-1}$ where $\omega_1 = \sqrt{\kappa^2 + \left(\frac{\nu_{11}}{2\pi}\right)^2}$. For $n > 1$ the H_{0nm} oscillations have complex fundamental frequencies and the curves κ''_{0nm} oscillate. For $L \to \infty$ the spectrum broadens, the Q-factors increase and may become greater than an arbitrary positive number due to the growth of the stored energy in the increasing resonator's volume.

The curves $\kappa_{0nm}(L)$ continuously pass to other sheets of the Riemann surface through the points slightly exceeding the threshold values of the H_{0n} waves of semi-infinite waveguides and the shape of the curves κ''_{0nm} remains practically the same.

The structure of the isolines $|E_\phi| = $ const (see Figure 6.1) enables one to conclude that the eigenoscillation's type is preserved along the whole curve

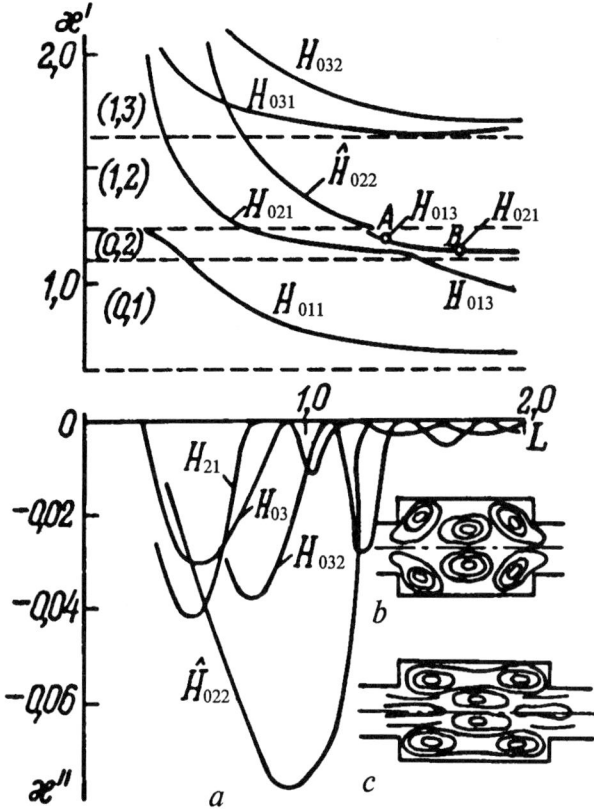

Figure 6.1   *Spectrum of eigenoscillations and fields' distributions of a symmetric open waveguide resonator*

$\kappa_{0nm}(L)$ at least for sufficiently wide radiation channels. The sequence of the (k, N) regimes presented in Figure 6.1 for $\theta = 0.5$ differs from that considered, for $\theta = 0.8$ which may lead to the intertype interaction of oscillations having both even or both odd longitudinal indices (for example, the interaction of the H_{021} or H_{013} oscillations) formed by the $k + 1$ and $k + 2$ standing waves $(k = 0, 1, 2, \ldots)$. Intertype interaction is not observed for oscillations with even and odd longitudinal indices although the real parts of their fundamental frequencies may be degenerated (see the branches κ_{022} and κ_{031} in Figure 6.1a for $L \approx 0.64$).

Figures 6.1b and 6.1c illustrate transformation of eigenfields, whose fundamental frequencies lie on one of the branches of $\kappa(L)$, before and after the interaction region (points A and B in Figure 6.1a). In the considered case there are no diffraction losses in the process of interaction and the latter is caused by the mutual transformation of the fields of two oscillations on resonator's boundary (Yashina, 1979). The amplitude distribution of the H_{013} oscillation for $L = 1.34$, $\kappa = 1.201$ (point A in Figure 6.1a) and the H_{013} oscillation for $L = 1.34$, $\kappa = 1.201$ (point B in Figure 6.1a) are presented in Figure 6.1b.

As follows from the analysis of the data presented in Figure 6.1a the H_{01m} oscillations with the smallest longitudinal index have purely real fundamental frequencies for average values of θ. The fundamental frequency of the H_{021} oscillation [which tends to $\left(\frac{\nu_{21}}{2\pi}\right)$ when $L \to \infty$] sharply grows with decreasing resonator's length and the curve κ_{021} does not "feel" the change of the (k, N) regimes (as well as in the case $\theta = 0.8$). The same behaviour is characteristic for the branch $\hat{\kappa}_{022}$ which has no asymptotic value for $L \to \infty$.

The axial-symmetric E_{0nm} oscillations should be treated separately because their fields have specific structure caused by the discontinuity of the j_ρ component (due to the presence of communication holes in resonator's butt-ends). We will not present here their detailed analysis and indicate only the resemblance of the curves $\kappa_{0nm}(\theta)$ of the H_{0nm} and E_{0nm} oscillations for small values of θ and the fact that the diffraction Q-factor of the latter is greater than that of the former.

Let us consider the second type of open waveguide resonators with the radiation channels formed by semi-infinite coaxial waveguides. The corresponding spectral problem becomes more complicated because here a cylindrical resonator (a simply-connected domain) interact with coaxial waveguides (a doubly-connected domain). The cuts in eqn. 6.1 are given by the equality

$$\kappa' = \sqrt{(\operatorname{Im} \kappa_n)^2 - \left(\frac{\mu_{ns}}{2\pi}\right)^2} \tag{6.8}$$

where μ_{ns} is a root of the equation $J_1(\mu_n)N_s(\mu_n\theta) - J_s(\mu_n\theta)N_1(\mu_n) = 0$.

The properties of the H-type oscillations are similar to that of the resonator loaded with circular waveguides (see the present Section) because here the "electrical" widths of the radiating circular and coaxial waveguides are close.

The qualitative picture essentially differs for the E-type oscillations when the lowest TEM mode (of the coaxial waveguide) with the wavenumber $\omega_{31} = \kappa$ carries away the energy along the coaxial waveguide. Therefore, the free oscillations cannot have real fundamental frequencies and the (k, N) regime with $k > N$ may be realised (similar to the situation in the waveguide with finite apertures).

The behaviour of the fundamental frequencies as functions of the diameter of the communication hole is similar to that of the E_{0nm} oscillations although an infinitely thin conductor strongly influences the E_ρ and E_{x_3} components (being at the same time the reason for generating the TEM mode in this domain). Therefore, it is not possible to carry out the correct limiting transition $\theta \to 0$ describing the transformation to a regular circular waveguide.

The structure of the doubly-connected radiation channels essentially influences the functions $\kappa_{0nm}(L)$. The first group of the E-type oscillations excited at $\theta = 0.25$ is situated in the θ, κ domain where the condition $M = N$ holds and the second group is characterised by the condition $M > N$. The curves $\kappa_{0nm}(L)$ and $\hat{\kappa}_{0nm}(L)$ have similar qualitative properties: $\kappa'(L)$ and $\hat{\kappa}'(L)$ monotonically decrease with growing L and tend to the threshold E-type frequencies of the internal circular waveguide, $2\pi\kappa'_{0nm} \to \nu_{1n}$. $\kappa''(L)$ and $\hat{\kappa}''(L)$ decrease when $L \to 0$, $\kappa''_{0nm} \to 0$ while the curves $\hat{\kappa}''_{0nm}$ break at the finite values of L corresponding to the points of intersection with the cuts given by eqn. 6.8.

The of principle E_{0n0} oscillation from the first group should be distinguished. The corresponding spectral curves on the first sheet of H intersect in the break points not with the nth but with the $(n-1)$th cut. On the second sheet of H the curves $\kappa_{0n0}(L)$ come (with decreasing L) to certain negative points on the imaginary axis. The functions $\kappa_{0nm}(L)$ corresponding to the higher order E_{0nm} oscillations $(m > 0)$ have a similar behaviour. In particular, such a "turn" is observed for them on the successive sheets of H and all the spectral curves "arrive" to the points $\kappa'' < 0$.

Let us consider the third type of open waveguide resonators with four radiation channels formed by circular and coaxial waveguides. Two infinite sets of cuts on the Riemann surface H are given here by eqns. 6.3 and 6.8.

We will limit ourselves to the case of the H_{0nm} oscillations. Here an infinitely thin conductor situated along waveguide's axis does not influence the radiated field because $E^{\mp}(0, x_3) = 0$ and the curves $\kappa_{0nm}(\theta)$ are identical when $\theta \to 0$ and $\theta \to 1$, i.e. when carrying out a transition to a circular waveguide, $2\pi\kappa'_{0nm} \to \nu_{1n}$. The value $\theta = 0.546$ corresponds to the case when the wall of the narrow waveguide is in the antinode of the H_{01} mode and in the knot of the H_{02} mode of the internal waveguide and in this situation the spectra $\kappa_{01m}(\theta)$ and $\kappa_{02m}(\theta)$ are symmetric. In fact, the incident H_{02} mode is practically not reflected by the bifurcation and its energy is uniformly separated between the narrow and coaxial waveguides, while the reflection coefficient of the H_{02} mode is rather high.

In the first region defined by the inequalities $\nu_{11}2\pi\kappa' < \sqrt{(\kappa'')^2 - m^2(2\pi\theta)^{-2}}$, the losses are caused by radiation to the coaxial waveguide because there are no waves in the narrow waveguide which carry away the energy $(k = 0)$. In the second region $\nu_{11}2\pi\kappa' < \sqrt{(\kappa'')^2 + \mu_{11}^2(2\pi)^{-2}}$, the losses are caused by radiation to the narrow waveguide $(M = 0)$ and $\kappa_{0nm}(\theta)$ are greater than in the first region due to essential diffraction communication between the open waveguide resonator and the circular or coaxial waveguides. In the third region which is the intersection of the first two the radiation losses are absent $(M = k = 0)$ and the open waveguide resonator has real fundamental frequencies.

Only the H_{0nm} oscillations satisfying the condition $N \geq k+M$ may exist in the considered open waveguide resonator. Eigenoscillations of the first group exist if $N > k + M$ in the region

$$\sqrt{(\kappa'')^2 + \left(\frac{\nu_{ns}}{2\pi}\right)^2} \geq \kappa' \geq \sqrt{(\kappa'')^2 - \left(\frac{\alpha_n}{2\pi}\right)^2} \qquad (6.9)$$

and of the second group if $N = k + M$ in the region

$$\sqrt{(\kappa'')^2 + \left(\frac{\alpha_n}{2\pi}\right)^2} \geq \hat{\kappa}' \geq \sqrt{(\kappa'')^2 - \left(\frac{\alpha_{n+1}}{2\pi}\right)^2} \qquad (6.10)$$

on the first sheet of H. Here $\alpha_n = \min(\nu_n\theta^{-1}, \mu_n)$. Hence the following sequence of regimes is realised for every θ with respect to the growing κ': $N < k+M$, $N = k+M$, $N < k+M$, and so on. Therefore, the fundamental frequencies of the H_{0nm}, $\hat{H}_{0nm}$ and H_{0n+m}, $\hat{H}_{0n+m}$ eigenoscillations are situated in different domains on the complex plane κ bounded by the cuts given by eqns. 6.3 and 6.8 and cannot interact.

The fundamental frequencies have been calculated for $\theta = 0.25$ (the narrow waveguide is below-cutoff), $\theta = 0.8$ (the coaxial waveguide is below-cutoff) and $\theta = 0.546$ (when the losses are caused by radiation to both waveguides). It is interesting that in the last case the H_{02m} and $\hat{H}_{0nm}$ eigenoscillations do not exist. The properties of the E_{0nm} eigenoscillations are similar to those in open waveguide resonators with the radiation channels formed by semi-infinite coaxial waveguides. For example, the curves $\kappa_{0nm}(\theta)$ and $\hat{\kappa}_{0nm}(\theta)$ are symmetric with respect to the point $\theta = 0.436 = \nu_{10}\nu_{20}^{-1}$. Here the wall of the narrow waveguide is in the antinode of the E_{02} mode and strong diffraction communication is observed with the coaxial waveguide.

The numerical data obtained enable us to follow the influence of an additional radiation channel in the form of a narrow waveguide. For $\theta > 0.436$ ($\theta < 0.436$) the radiation to the circular waveguide influences the lower order (higher order) oscillations because of stronger communication with the coaxial (for $\theta > 0.436$) or circular (for $\theta < 0.436$) waveguides. The E_{031} oscillation transforms to the E_{030} oscillation when $\theta = 0.8$.

6.2 Open waveguide resonators with irregular communication domains

The waveguide junctions whose communication domains are not sections of regular waveguides but have irregular boundaries form a special family of open waveguides. Among them, in particular, are angle and T-junctions of rectangular waveguides.

The irregular character of a junction excludes the possibility of considering free oscillations as a result of interaction of the direct and inverse waves of regular waveguides (studied in Section 5.6) and necessitates investigation in more detail the correspondence between spectral characteristics and eigen-fields in order to correctly classify the latter. This is especially important for waveguide resonators with angle-type discontinuities where one cannot usu-ally predict the one-to-one correspondence between a spectral curve and an eigenoscillation with definite field's structure.

Let us begin with the study of this type of angle junction, considering the symmetric break as an example. The communication domain here is formed by triangles and its volume is determined by the angle between the axes of intersecting waveguides. We will limit ourselves to TE oscillations with the only non-zero E_{x_1} field's component directed along break's edge parallel to the narrow walls of rectangular waveguides.

The angle symmetry of this waveguide junction allows us to separate the symmetric $\left(\frac{\partial E_{x_1}}{\partial x_3}\Big|_{x_3=0} = 0\right)$ and anti-symmetric $\left(E_{x_1}|_{x_3=0} = 0\right)$ oscillations. Analysis of the variations of the spectral parameter $\kappa = \frac{ka}{2\pi}$ (a is the width of the waveguide) on the first sheet of the Riemann surface $\kappa = \kappa' + j\kappa''$ enables one to conclude that for $\kappa' < 0.5$ the fundamental frequencies are real (in this frequency range the energy is not radiated out of the open waveguide resonator), situated in the interval $0.25 < \kappa' < 0.5$, and correspond to the symmetric oscillations. In addition to this, $\kappa' \to 0.25$ for $\psi \to 90°$ when the irregular domain is transformed to a short-circuit double-width waveguide. The propagation of the H_{10} waves running to meet each other is possible in the transformed communication domain beginning from $2\kappa' = 0.5$. Their superposition forms oscillations with a symmetric field's structure. An anti-symmetric oscillation in the "limiting" waveguide may be formed by the H_{20} waves. However, these waves cannot propagate in the communication domain when $2\kappa' < 1$. The curves $\kappa = \kappa(\psi)$ concentrate in the vicinity of the branch point $\kappa = 0.5$ of the function $\omega_1 = \sqrt{\kappa^2 - 0.25}$ when $\psi \to 0$. This point coincides with the H_{10} critical (cut-off) frequency.

The curves $\kappa = \kappa_n(\psi)$ ($n = 2, 3, 4, 5$) tend to the same point when ψ increases. The corresponding fundamental frequencies vary between the cuts d_1 and d_2 on the complex plane and each branch κ_n is characterised by the specific values $\psi = \psi_n$ when $\kappa_n = 0.5$. Here the curves $\kappa_n(\psi)$ pass to the domain $\kappa' < 0.5$ where the fundamental frequencies become real. The curves κ_n ($n = 6, 7, 8, 9$) lie between the cuts d_2 and d_3 and here only one of them tends to the H_{03} cut-off point of the regular waveguide when ψ decreases (to the point $\kappa = 1.5$ where $\omega_3 = 0$).

The behaviour of $\kappa_n(\psi)$ on other sheets of the Riemann surface H is shown in Figure 6.2 where $n = 1, 4, 6$ and the bold lines correspond to the first sheet, the one-dot lines to the second sheet and two-dot lines to the third sheet. The numeration of the sheets is purely conditional and their numbers are equal to the number of cuts intersecting the graphs $\kappa_n(\psi)$ when passing from one sheet to another. Each of the curves κ_n returns to the point $\kappa = 0.5$ (on the real axis of the corresponding sheet) after "running through" a finite number of sheets. Other extreme points of these curves also lie on the real axis (where $\omega_n = 0$, $n > 1$).

The qualitative features of the fundamental frequencies analysed above are

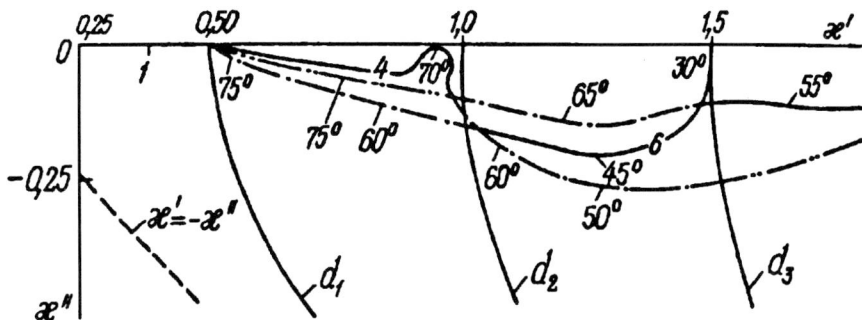

Figure 6.2 *Fundamental frequencies κ against ψ on different sheets of the Riemann surface*

in complete agreement with the principles of continuous spectrum's dependence on non-spectral parameters stated in Chapter 2 if the open waveguide resonator considered degenerates either to a regular waveguide ($\psi = 0$) or to a semi-infinite short-circuit bifurcation of a regular waveguide ($\psi = 90°$).

In addition to the domain of small angles, the super-high Q-factor (symmetric) oscillations may exist for $\psi \approx 58.5°$ (κ_3) and $\psi \approx 70°$ (κ_4). Increasing volume of the communication domain leads to the spectrum's condensation and the growth of diffraction Q-factor, while in the higher frequency range the Q- factor falls, an obvious result of the growing number of radiation channels caused by the generation of the higher order H_{m0} waves with propagation constants $\operatorname{Re}\omega_n > 0$.

The characteristic bends observed at the intersection points of κ_3'' and κ_4'' point to the intertype interaction of oscillations with comparable Q-factors inherent to closed resonators with perturbed boundaries. The Q-factors of the interacting oscillations essentially differ after the interaction point: $|\kappa_4''|10^{-6} - 10^{-7}$ is minimal in the interval where $|\kappa_3''|$ is maximal and the Q-factor of the former is maximal.

In order to calculate the fields in triangular regions (where it is inconvenient to represent them as a series of basis functions) we will apply Green's identity (Tikhonov and Samarsky, 1978) for the sought for function $E_{x_1}(\mathbf{r})$ and canonical Green's function $G_0(\mathbf{r}_0, \mathbf{r})$ of a parallel-plane waveguide (see

Section 2.1) in the communication domain bounded by the contour S:

$$E_{x_1}(\mathbf{r}_0) = \int_S \left[G_0(\mathbf{r}_0, \mathbf{r}) \frac{\partial E_{x_1}(\mathbf{r})}{\partial \mathbf{n}} - E_{x_1}(\mathbf{r}) \frac{\partial G_0(\mathbf{r}_0, \mathbf{r})}{\partial \mathbf{n}} \right] d s \qquad (6.11)$$

The isolines $|E_{x_1}(\mathbf{r})| = $ const in the plane parallel to the wide walls of the waveguide junction indicates maximal field concentration of the symmetric oscillations in the irregular domain and the point $|E_{x_1}(\mathbf{r})|\|_{max}$ lies on the bend line. The number of field variations (m) over the apertures of the regular waveguides corresponds to the H_{m0} waves ($\kappa_n(\psi)$ tends to their threshold frequencies when $\psi \to 0$) and, consequently, one may define the high Q-factor oscillations as the H_{m0} oscillations. The decrease of oscillation Q-factor caused by the increase in energy radiated out of the resonator forces the point $|E_{x_1}(\mathbf{r})|\|_{max}$ out of the irregular domain. The number of field variations in the regular waveguides is determined then by the field structure of the H_{m0} wave carrying the radiation energy.

Interacting oscillations have a similar structure of isolines $|E_{x_1}(\mathbf{r})| = $ const: at different branches $\kappa_n(\psi)$ ($\psi \approx 64°$).

Let us consider the H_{m0n} eigenoscillation spectrum of the waveguide junction having the communication domain in the form of a rectangular parallelepiped whose three narrow sides coincide with apertures of semi-infinite rectangular regular waveguides (the only non-zero electric field component is parallel to the narrow walls). Such a T-junction has an additional radiation channel of varying width $\theta = a_1/a$ and the number of radiation channels is greater than that in the coaxial waveguide resonator studied in Section 5.3.

A preliminary analysis of radiation conditions given by eqn. 6.2 in every shoulder yields existence of three groups of symmetric and anti-symmetric oscillations (with respect to the plane $z = 0$) whose fundamental frequencies lie on the first sheet of the Riemann surface H of the complex spectral parameter κ:

   (i) over cut-off regime for all waveguides;
  (ii) over cut-off regime for either the lateral shoulder or the direct shoulders;
 (iii) radiation to all shoulders.

The eigenoscillations of the first group have not more than two real fundamental frequencies κ_{11} and κ_{12} with $\kappa' < 0.5$ on the first sheet of H corresponding to the H_{101} and H_{102} symmetric and anti-symmetric oscillations existing in the intervals $0.5 < \kappa' < (2\theta)^{-1}$, $0.5 < \kappa' < (\theta)^{-1}$, respectively, and $\kappa_{11} \to 0.5$ when $\theta \to 0$. Violation of the junction's symmetry gives rise to the imaginary part κ''_{12} due to possible radiation to the lateral shoulder at the H_{10} wave or may even destroy this oscillation.

Further increase in width of a lateral shoulder in a symmetric junction with over cut-off direct shoulders forms the condition for the existence of high Q-factor oscillations with complex fundamental frequencies such that $|\kappa''| \ll 1$. The first is the H_{103} symmetric oscillation in the junction with $\theta = 3$ whose fundamental frequency varies between the cuts d_{11} and d_{32}. The field of this oscillation has three variations over the lateral aperture and the value of its final diffraction Q-factor is caused by the radiation to the lateral waveguide at the H_{01} wave since the H_{03} wave cannot propagate there and the H_{20} wave is not excited due to the symmetry of the H_{103} oscillation and the junction itself. H_{104} will be the next oscillation ($\theta \approx 4$) whose Q-factor is comparable with that of the H_{103} oscillation and all successive H_{10n} oscillations ($n > 4$) will have much lower Q-factors due to the lateral radiation at several H_{10n} waves. Let us note that the H_{10n} oscillations are "well separated" with respect to the parameter θ but not with respect to the frequency.

H_{201}, H_{202}, H_{301} and H_{302} are the eigenoscillations of the second group related with the lateral waveguide which is over cut-off for the H_{10} (H_{20}) wave in the case of symmetric (anti-symmetric) oscillations. They have rather high diffraction Q-factor (Q_{m0n}) reaching the level 10^2 - 10^3 and, consequently, an easily identified structure of the eigenfield. Q_{m01} essentially grows for $\theta \to 1$ and for $\theta = 0$ ($\kappa_{m01} = 0.5m, m = 2, 3, \ldots$) when the Q-factor becomes infinitely large. Hence a small slot in a thick wall creates the conditions for the existence of high Q-factor eigenoscillations whose fields are concentrated in a small inhomogeneity and fundamental frequencies are close to the threshold frequencies of the regular waveguide.

The eigenoscillations of the third group are supported by an open communication domain in the presence of the radiation to every shoulder at one or several waves. The spectral parameter varies between the cuts d_{m1} and d_{m2} for each of the branches $\kappa_{mn}(\theta)$ ($m, n = 2, 3, \ldots$) and the Q-factors of the corresponding H_{m0n} oscillations are two or three orders lower than that of the second group. The field structure is less regulated than in the previous case. The maximum of the $|E_{x_2}(\mathbf{r})| = \mathrm{const}$ distributions is situated not in the communication domain but out of it in the shoulder to which a greater part of energy is radiated. There are four such maxima for the anti-symmetric H_{204} oscillation for $\theta = 2$: two are in the lateral shoulder and the rest are situated in each of the direct shoulders at an equal distance from the boundary of the communication domain.

The identification of the low Q-factor oscillations is difficult because the field picture in semi-infinite waveguides is determined by the higher-order wave carrying the radiation energy, and with growing θ this wave in the communication domain may be destroyed even remaining at the same branch κ_{mn}. Anyway, in spite of "washing out" the field, the number of field's variations remain to be equal to two (along the axis Oy) and to one (along the axis Oz) and these oscillations are therefore denoted by the symbol $\hat{H}_{201}$ (the H_{201} oscillations differ from them by the fact that they exist in the regime when one of the shoulders is over cut-off). The $\hat{H}_{m0n}$ oscillations have different field structure in the communication domain and in regular waveguides at a certain distance from the boundary of the former. Hence their properties are similar to that of eigenoscillations in open waveguides.

Let us note that the spectral curves $\kappa_{mn}(\theta)$ of the T-junctions have no asymptotic values to which they approach with increasing θ as observed for the function $\kappa(L)$ in widening waveguides and coaxial waveguide resonators (see Section 5.3). In fact, the parameter determining the volume of communication domains in three-shoulder junctions enters radiation conditions for a lateral shoulder given by eqn. 6.2 which defines the geometry of the cuts d_{ml} on the complex plane κ and the cuts d_{m1} depend on the parameter θ.

6.3 Resonance excitation of open waveguide resonators

We will consider here the total resonant reflection and transmission caused by excitation of oscillations which are close to the eigenoscillations of a given open waveguide resonator. We will use the information concerning the eigenoscillations spectrum which enables one to predict resonator's diffraction properties while the knowledge of scattering matrices is not always sufficient for describing the resonance phenomena.

Let us begin with the study of symmetric open waveguide resonators excited by waveguide modes (H_{0p} and E_{0p}). We will denote by (k, N), (M, N) etc. the excitation regimes considered in Section 6.1 where k, M, N are the number of waves propagating in narrow, coaxial and wide waveguides.

Attempts to explain the mechanism of forming resonance oscillations in irregular open waveguide resonators were based on the notion of locked modes, namely, the higher-order waves propagating in the waveguide sections of enlarged cross-section and attenuating in the main waveguide channel (the regime with $N > k$).

In the case of a weak transformation of a locked mode to other types of waves the condition of total resonant reflection coincides with the longitudinal resonance conditions

$$2\pi L \omega_N + \arg R_{NN} = n\pi \quad (n = 1, 2, \ldots) \tag{6.2}$$

where R_{NN} is the reflection coefficient of the Nth the locked mode with the propagation constant ω_N, $L = l/b$ is the relative width of the widening (b and a are the radii of wide and narrow waveguides, respectively) $\kappa = b/\lambda$, $\theta = a/b$ and eqn. 6.12 may be applied if $1 - \theta \ll 1$. For $\theta > 0.8$ and $L > 0.5$ the total resonant reflection frequencies obtained from eqn. 6.12 and by means of a rigorous solution to the diffraction problem coincide. If $N = 1$, eqn. 6.12 describes the total transmission caused by the interference resonance at the lowest mode of the widening. These resonances correspond to the low Q-factor oscillations. When $\theta = 0.6$ the following regimes are observed with respect to the growing frequency (in the case of the H_{01} wave diffraction on the widening): $(0, 1), (1, 1), (1, 2), (1, 3), (2, 3)$ etc., i.e. two or three locked modes are excited providing the conditions for interaction of different oscillations formed by different modes in the widening. In this way it is interesting to consider the $(1, 3)$ regime when variation of κ' corresponds to the H_{02m} and H_{03q} oscillations which belong to the different classes of symmetry and do not interact.

The resonance frequencies of the forced H_{031} and H_{023} oscillations (interacting in a vicinity of the point $\kappa \approx 1.66$, $L \approx 1.14$) come together when the Q-factors of those oscillations increase the response leading to double extrema (such double extrema are not characteristic for open waveguide resonators satisfying the condition $|\kappa''_{0pt}| \ll 1$). The diffraction fields here practically coincide. The phenomena considered may be explained in terms of locked modes since the resonance at the higher order locked H_{03} (H_{02}) mode provides the total transmission (reflection). Such features of the resonances at locked modes are characteristic for open waveguide resonators with strongly transforming boundaries.

In the case of the E_{01} wave excitation the $(1,1)$ regime deserves special attention for small values of L when the higher order E_{02} wave attenuates (in the widening) and at the same time this wave generates the resonances of total reflection (when the locked modes are absent) due to its big excitation amplitude close to the cut-off frequency. When open waveguide resonators are excited by multi-mode waveguides at the boundaries between the regimes (e.g. when passing from the regime $(1,1)$ to $(2,2)$) such anomalous behaviour is not observed (at least at the boundary of the $(2,3)$ regime where the total reflection and transmission are not registered). When the frequency increases, the influence of eigenoscillations on the scattered fields decreases and the amplitude of transmitted waves takes practically constant values which may be determined by the methods of geometrical optics.

The spectral characteristics of the H_{0nm} oscillations of cylindrical resonators loaded with semi-infinite coaxial waveguides (excited by the H_{0p} waves) are similar to that of open waveguide resonators with step boundaries (see Section 6.2). It is especially interesting to consider the excitation by the TEM wave which has no cut-off frequency. Let us take an open waveguide resonator with $\theta = 0.25$ when the locked mode regime (and the corresponding resonances) cannot be supported for any κ. Here $\theta = c/b$, $L = l/b$ where b is the radius of the wide waveguide, l is its length and c is the radius of the coaxial waveguide. The total transmission of the TEM wave is carried out in the $(1,1)$ regime owing to the E_{010} oscillation. The value $\kappa_{\max}$ may be determined by eqn. 6.12 for the E_{01} wave of the circular waveguide facilitating the communication between two coaxial waveguides. The forced oscillation formed after the interaction of the direct and inverse E_{02} waves of the resonant volume (corresponding to the E_{020} free oscillation) causes the energy growth of the higher order E_{01} wave of the coaxial waveguide and its fall for the TEM wave close to the point $\kappa = \kappa'_{020}$ belonging to the frequency range of the $(2,2)$ regime.

Let us consider the properties of coaxial waveguide resonators. They have four feeding channels and the ratio of the narrow and coaxial waveguides

$\theta = b/a$ plays the crucial role in forming their diffraction characteristics. The strongest wave transformation at the boundary of coaxial waveguide resonators occurs when $0.5 < \theta \leq 0.7$ and the weakest when $\theta \leq 0.4$ and $\theta > 0.7$. There exist the values $\theta = \theta_s = \nu_{1s}/\nu_{2s}$ for which symmetric diffraction characteristics are observed. In fact, when the H_{02} or E_{02} wave falls on the bifurcation of a circular waveguide and $\theta = \theta_s$ ($\theta_s \approx 0.546$ for the H_{02} wave and $\theta_s \approx 0.436$ for the E_{02} wave) they are practically not reflected by a break since the latter is in the knot of the electric field of these waves.

The energy isolines of the H_{01} wave in the narrow waveguide are situated within the limits of the $(1,2)$ regime and equally spaced with respect to L ($\Delta L = 0.5\omega_{21}$), being close to the qualitative characteristics of the H_{02m} oscillations. Similar pictures observed in the $\left(\frac{0}{1}, 1\right)$ regime enable us to characterise the resonances as the $\hat{H}_{01m}$ oscillations with $\Delta L = 0.5\omega_{11}$. The same situations occur in the $\left(\frac{k}{0}, M\right)$ regimes for $\theta \approx 0.25$.

For $\theta = 0.546$ the following sequence of regimes are realised in coaxial waveguide resonators: $\left(\frac{0}{0}, 1\right)$, $\left(\frac{1}{1}, 2\right)$ $\left(\frac{1}{1}, 3\right)$, i.e. three energy channels are open, narrow and coaxial waveguides. A conditionally locked mode may exist here only if $N > M + k$ and in such regimes the resonant diffraction phenomena are connected with excitation of oscillations which are close to the first group of free oscillations. The low Q-factor H_{0n} eigenoscillations do not influence the energy characteristics of the excited oscillations, but already in the $\left(\frac{1}{1}, 3\right)$, regime the frequency dependence of the transmission coefficients has a resonant character. As well as in the case $\theta = 0.8$, the resonant field excited in a coaxial waveguide resonator has different types of diffraction communication with the H_{01} waves of narrow and coaxial waveguides.

The presence of a doubly-connected domain complicates the characteristics of coaxial waveguide resonators excited by the E_{0n} waves. The frequency domains where the locked modes may exist are absent because of the different propagation regimes in different parts of waveguides: the TEM wave propagates in a coaxial waveguide while the $N \leq k + m$ regimes are realised (for given field's polarisation) in coaxial waveguide resonators. Systematic calculations show that resonant splashes in the diffraction characteristics of the E_{0n} waves appear only when $N = k + M$ (for $N < k + M$ the diffraction characteristics vary monotonically with respect to increasing frequency).

The mechanism of forming the diffraction resonances for the E_{0n} and H_{0n}

waves is different, as well as their communication with the spectral data. In
the simplest $\left(\frac{1}{0}, 1\right)$ regime the total transmission points are connected with
the responses to the E_{01n} free oscillations from the first group (unlike the corresponding H_{0nm} oscillations). The $\hat{E}_{0nm}$ oscillations from the second group
have very low Q-factors ($Q \approx 1$) and, consequently, weak diffraction respons-
es. In the $\left(\frac{1}{1}, 2\right)$ regime the diffraction responses to the E_{02m} oscillations are
different for different m but their Q-factors are practically equal.

In the coaxial waveguide resonator with $\theta = 0.436$ the E_{02m} oscillations
are not excited (see Section 6.2) and, therefore, the eigenoscillation spectrum
is essentially rarefied. An unusual feature of this resonator is that the principle E_{011} and $\hat{E}_{011}$ oscillations from different groups (with equal indices) are
excited in one and the same domain situated between the cuts d_1 and d_2 on
the first sheet of the Riemann surface H of the complex parameter κ. They
have equal Q-factors and equally influence the diffraction characteristics of
the excited oscillations, which finally leads to the total transmission of the
TEM wave through a coaxial waveguide resonator.

Let us dwell on resonant phenomena in rectangular waveguides with breaks.
The numerical investigation of the H_{p0} waves diffraction on the H-plane symmetric break is carried out by Kirilento *et al.* (1974). In particular, it has
been established that the total reflection of the H_{10} wave coming to a break
from one of the shoulders is caused by excitation of oscillations in the irregular domain which are close to waveguide's eigenoscillation. The greater the
Q-factors, the closer are the resonant frequencies of the break and the fundamental frequencies of an open waveguide resonator (if the break angles ψ
are sufficiently small). Also characteristic of small angles are the high amplitudes of the non-propagating H_{20} waves (both in the reflected and transmitted
fields) which determine similar structures of the diffraction and free oscillation
fields in the vicinity of the break. The Q-factors fall with decreasing ψ which
essentially influences the diffraction fields.

Small breaks in two-mode waveguides also support the high Q-factor free
oscillations whose frequencies tend to the threshold values of the H_{30} waves
when $\psi \to 0$ and the fields of these non-propagating waves determines the
structure of the free oscillations. But the expected total reflection of the
H_{10} wave is not observed (Katzenelenbaum, 1961, Kirilento *et al.*, 1974).
In addition to this, the H_{30} wave amplitude sharply grows for $\kappa \to 1.5$ in

the reflected and transmitted fields which may be considered as a resonance response (without the total reflection of the incident H_{10} wave, as indicated in the case of a single-mode waveguide excited by the H_{20} wave when the amplitudes of both the incident and the higher order reflected waves increased in a resonant way). The resonant growth of the amplitude of the generated higher order wave weakens with increasing frequency.

The resonant splashes of the scattered amplitudes are connected with the interference of the H_{10} wave reflected from the irregular boundary and radiated to the waveguide by the resonance oscillation of the open waveguide resonator. The speed of the variation of the amplitude and phase of the secondary field depends on oscillation's Q-factor, hence the addition or subtraction of the primary and secondary fields occurs at frequencies which differ from the fundamental frequencies inversely proportional to the Q-factors of the oscillations excited in the communication domain. For large values of ψ the eigenoscillation spectrum broadens and the formation of resonance oscillations becomes complicated because several neighbouring oscillations may contribute to the secondary diffraction field.

Unlike the single-mode regime, the resonant reflection and transmission of the working wave in multi-mode waveguides is rarely total (Kirilenko *et al.*, 1974).

The equal-shoulder T-junction is considered by Feld and Sukharevsky (1969) and Rud' (1985) who model the total unbinding of all possible combinations of the shoulders close to the cut-off frequency of the H_{20} wave of a rectangular waveguide. The resonant behaviour of the diffraction amplitudes is displayed in the cleanest way when at least one of the shoulders is over cut-off.

The high Q-factor resonance of the total reflection for the principle wave of the direct waveguide is observed in the T-junction with $\theta = 0.4$ close to the H_{20} wave critical frequency ($\kappa = a/\lambda = 1$). When the T-junction is excited from the lateral waveguide by the non-propagating H_{10} wave, its own reflection coefficient in the resonance point reaches the value ~ 20. Comparing the resonance and fundamental frequencies of the T-junction, the structure of the diffraction and H_{201} fields, one can make a conclusion concerning the direct influence of the H_{201} oscillation on forming the resonant frequency dependence of this T-junction when κ is close to a fundamental frequency.

The equal-shoulder T-junction has two fundamental frequencies, $\kappa_{22} = 0.9982 - j0.0046$ (the H_{022} oscillation) and $\hat{\kappa}_{21} = 0.9449 - j0.1464$ (the $\hat{H}_{021}$ oscillation). The comparison of the diffraction fields calculated in two characteristic points with the eigenfields of the T-junction (carried out in Section 6.2 for $\theta = 1$) enables us to establish that, as well as in the cases described above, the excitation at the frequency which is close to the fundamental frequency of a low Q-factor oscillation (when the H_{01} wave comes from a lateral shoulder) leads to the total unbinding of a lateral and two other shoulders. If the excitation frequency is close to the fundamental frequency of a high Q-factor oscillation, the unbinding of the direct shoulders is observed (in the resonance points the diffraction fields are determined by the non-propagating H_{20} wave of the direct waveguide or of the direct and lateral waveguides).

6.4 The Morse critical points of open non-symmetric waveguide resonators

Spectral properties of open waveguide resonators determine the alternation of frequency regimes which depends on the radius (θ_2) of one of the radiation channels (for fixed θ_1; an example of the non-symmetric waveguide resonator with $\theta_1 = 0.5$ and $L = 1.0$ is depicted in Figure 6.3). In the non-symmetric case there are no limitations concerning the combinations of indices of interacting oscillations and, besides, one cannot distinguish between the E_{02m} and $\hat{E}_{0nm}$ eigenoscillations because of the different properties of feeding waveguides. For example, the following pairs of oscillations interact in the vicinity of the Morse critical points: E_{030} and E_{011}, E_{031} and E_{012}, E_{013} and E_{030}, E_{040} and E_{011}, E_{040} and E_{023} (see Figure 6.4). Violation of symmetry removes the degeneration of the Morse critical points where the even-odd or odd-even oscillations (with respect to the third index) interact. Variation of θ_2 changes the type of the Morse critical points and the character of the physical processes.

Let us consider the domains of spectrum thickening (see Figure 6.4) for the non-symmetric waveguide resonator depicted in Figure 6.3. The spectral curves $\operatorname{Re}\kappa_i(\theta_2)$ are similar to the Wien graphs for the E_{031} and E_{013} oscillations when $\theta_2 \in [0.7, 0.9]$ [as well as $\operatorname{Im}\kappa_i(\theta_2)\,(i = 1, 2)$ for the E_{021}

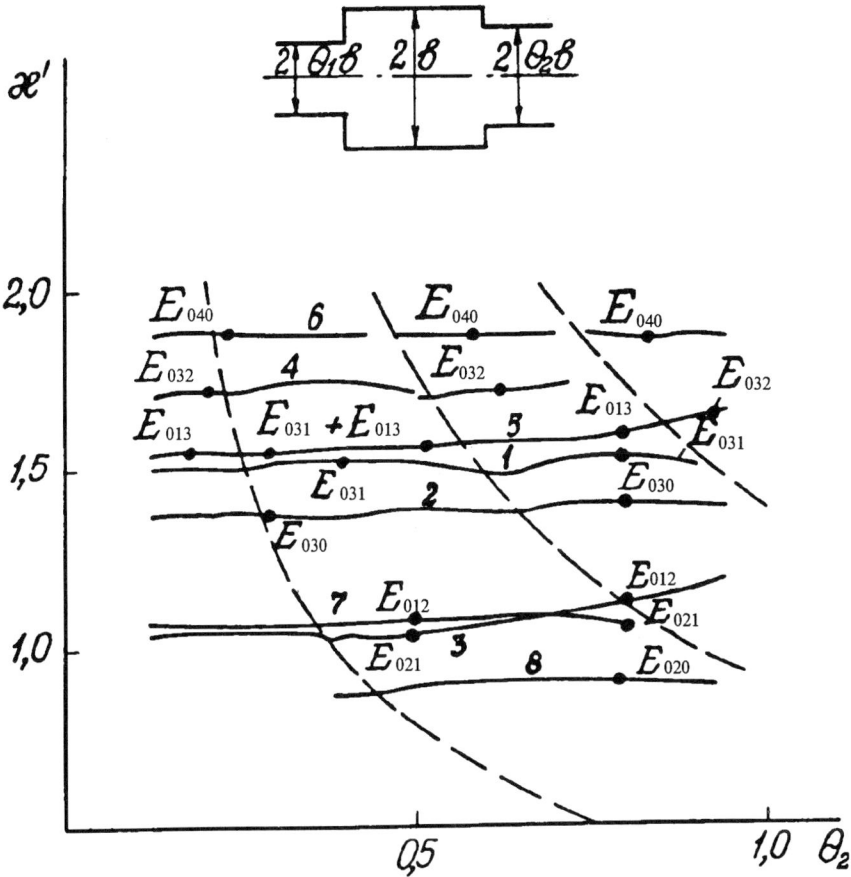

Figure 6.3   *Spectral curves and Morse critical point of a non-symmetric open waveguide resonator*

and E_{012} oscillations] and intersect for the E_{021} and E_{012} oscillations when $\theta_2 \in [0.65, 0.75]$. The curves $\mathrm{Im}\,\kappa_i(\theta_2)$, $(i = 1, 2)$ (denoted by 1 and 5 in Figure 6.3) have many intersection points. In the domains of their interaction the Q-factors become very close.

Let us follow the variation of one of the eigenoscillations along the curve 1 in Figure 6.3. It is clear that this oscillation becomes the hybrid $E_{031}+E_{013}$ oscillation when $\theta_2 \in [0.2, 0.84]$ and they have the non-degenerate Morse critical point close to $L = 1$. The "pure" oscillations are observed when $\theta_2 < 0.2$. The E_{021} and E_{012} oscillations have the degenerate Morse critical

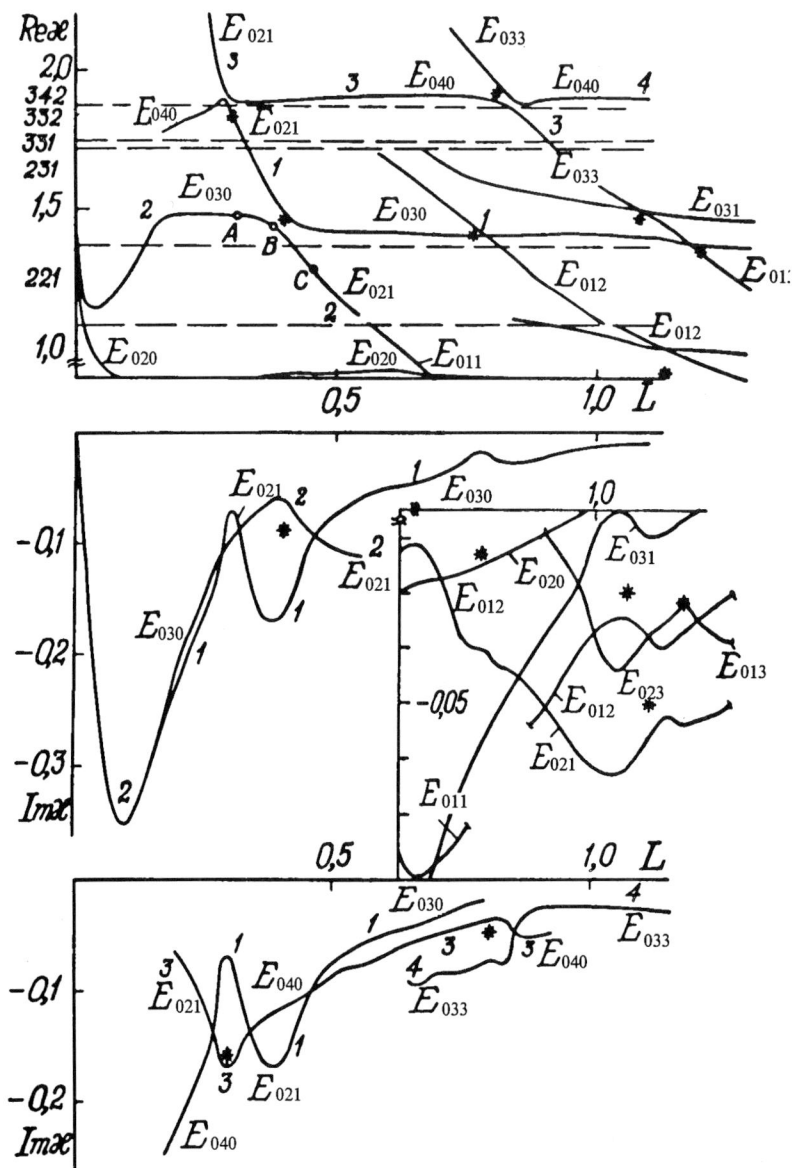

Figure 6.4 *Spectral curves for interacting oscillations of a non-symmetric open waveguide resonator*

point close to the same value $L = 1$ and the high Q-factor E_{031} oscillation suppresses the E_{013} oscillation (curves 3 and 4 in Figure 6.4). Here it is important that the curves $\operatorname{Re} \kappa_{1,2}(\theta_2)$ of both pairs of oscillations intersect and $\operatorname{Im} \kappa_{1,2}(\theta_2)$ are identical (curves 1,2 and 3 in Figure 6.4 which correspond to one of the communication diagrams presented in Figure 1.2).

The variety of forms of hyperplanes of the spectral sets (for the definition see Section 1.10) in the κ, θ, L co-ordinates is presented in Figure 6.3. The spectral curves $\operatorname{Re} \kappa_{1,2}(\theta_2)$ depicted in Figure 6.4 are separated by two sets of cuts on the first sheet of the Riemann surface H which start in the critical frequencies of each of the feeding waveguides.

One may conclude that intertype interaction of oscillations is a stable characteristic of this open waveguide resonator.

The spectral curves $\operatorname{Re} \kappa_i(\theta_2)$ $(i = 1, 2, \ldots)$ of open resonators formed by a closed resonator connected with different waveguide channels (Figure 6.5) are similar to the curves in Figure 6.3 but have a smoother character due to the absence of cuts on the Riemann surface. These curves are evidence of the thickening of the Morse critical points and successive interaction of oscillations (from one family E_{02m} which have no degeneration points in a closed resonator!) with respect to varying L. Variation of the H_ϕ field component (Figure 6.5b, c) along the arc AB of the spectral curve corresponding to the most intensive transformation of oscillations caused by this interaction leads to the $E_{022} \rightarrow E_{031} \rightarrow E_{030}$ transformation.

Deformation of the closed cylindrical resonator yields removal of degeneration and spectrum broadening. There are two degeneration points close to the non-degenerate $E_{021} \rightarrow E_{020}$ Morse critical point (see Figure 6.5) which have no analogues in the closed resonator.

An isolated Morse critical point is a stable singularity of a characteristic determinant with respect to the varying radius θ_1 of the circular waveguide as shown in Figure 6.6 where the co-ordinates of the degenerate $(E_{022} \rightarrow E_{030})$ and non-degenerate $(E_{030} \rightarrow E_{023})$ Morse critical points are presented as functions of θ_1.

The results of investigations of the eigenoscillations in the Morse critical points of open non-symmetric waveguide resonators allow us to make the following conclusions:

(i) the spectra broaden close to the degenerate and non-degenerate Morse

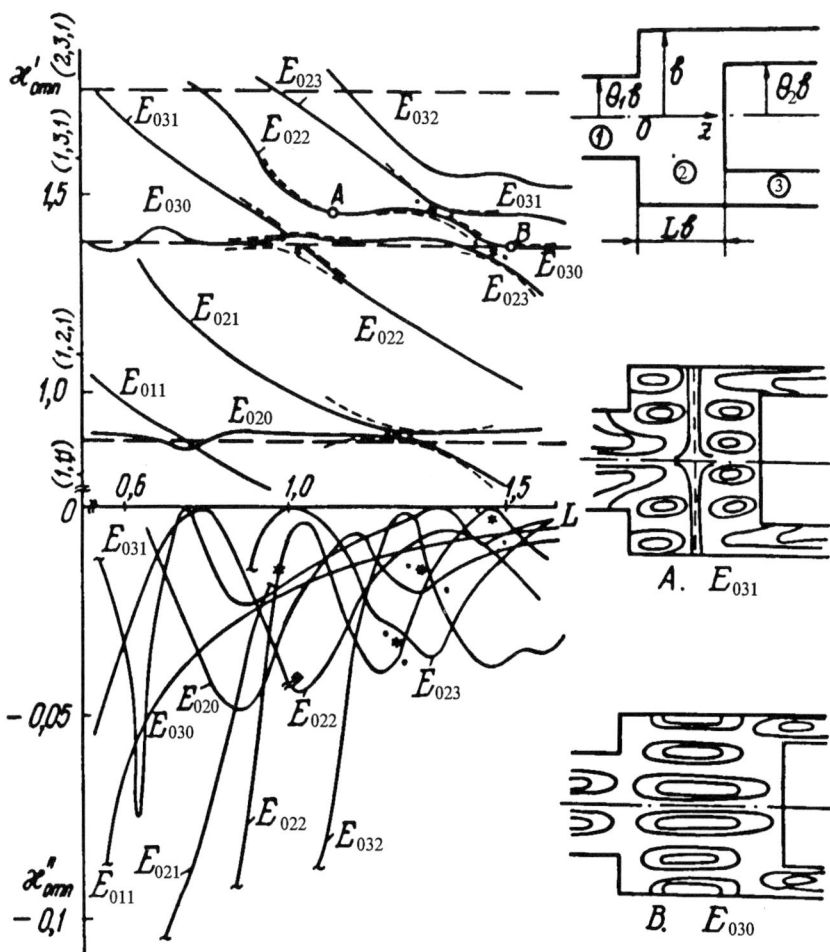

Figure 6.5   *Spectral curves of a non-symmetric open waveguide res-*
onator formed by a closed resonator connected with
waveguides of different cross-sections

critical points (in particular, of eigenoscillations belonging to the same
family);

(ii) some oscillations participate in several interactions successively which
leads to multiple variation of the type of these oscillations along the
spectral curve;

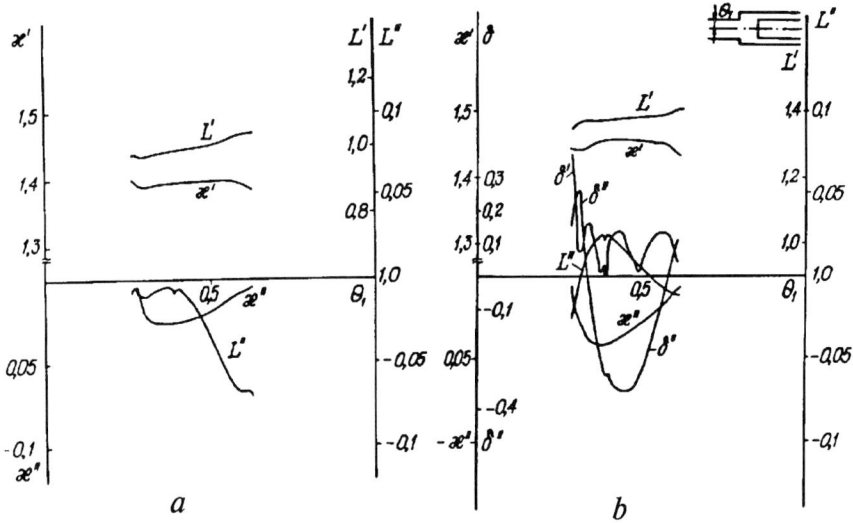

Figure 6.6   *Dynamics of the Morse critical point of a non-symmetric open waveguide resonator with respect to one of the feeding waveguides*

(iii) intertype interaction and the domains of spectrum broadening are also observed in the κ, θ co-ordinates;

(iv) violation of symmetry removes the Morse critical points degeneration;

(v) there is no prohibition concerning the even-odd and odd-even interaction (with respect to the third index);

(vi) resonator's deformation yields emergence of the Morse critical points which have no analogues in the closed resonator;

(vii) variation of θ_1 destroying the symmetry of an open waveguide resonator yields changes of the dispersion curve for the same pair of interacting oscillations;

(vii) some of the Morse critical points correspond to the degeneration points of the closed resonator.

One of the most interesting properties of interacting oscillations is the generation of hybrid eigenfields when the oscillations emerge which have no analogues in the closed resonator. Shteinshleiger (1955) represented a hybrid oscillation as a linear combination of two bearing "pure-type" oscillations whose interaction leads to the formation of the given intertype (hybrid) oscillation. The "bearing" eigenfunctions were calculated far from the interaction domain, i.e. far from the Morse critical point.

Let us demonstrate this approach taking the interaction of the high Q-factor E_{030} and E_{011} oscillations as an example. The $E_{011} \rightarrow E_{030}$ and $E_{030} \rightarrow E_{011}$ transformations occur with growing L along the curves $\kappa_1(L)$ and $\kappa_2(L)$, respectively, which produces the hybrid $E_{011}+E_{030}$ oscillation. The corresponding spectral curves together with the H_ϕ isolines (calculated in the points A, B and C on curve 2) are depicted in Figure 6.7.

In order to analyse in detail this transformation, let us choose the point $L = L_0 = 0.32$ situated far from the Morse critical point where the eigenoscillations (determined for $\kappa = \kappa_1(L_0)$, $\kappa = \kappa_2(L_0)$) have definite types and following Wilkinson (1970) calculate the normalized (with respect to the $\tilde{l}_2$ norm) eigenvectors $\mathbf{u}_l = \mathbf{u}_l[\kappa_l(L_0)]$ and $\mathbf{u}(L) = \mathbf{u}[\kappa_2(L)]$ of the matrices $A_l[\kappa_l(L_0)]$ and $A_l[\kappa_2(L)]$, respectively, corresponding to its zero eigenvalues $(l = 1, 2)$. Representing $\mathbf{u}(L)$ in the form

$$\mathbf{u}(L) = \alpha_1(L)\mathbf{u}_1 + \alpha_2(L)\mathbf{u}_2 + \mathbf{v}(L) \qquad (6.13)$$

where the scalar products $\alpha_l(L) = (\mathbf{u}(L), \mathbf{u}_l)$ $(l = 1, 2)$, $\|\mathbf{v}(L)\| \ll 1$ and $|(\mathbf{u}_1, \mathbf{u}_2)| \ll 1$, we thus obtain an expansion of $\mathbf{u}(L)$ with respect to the "almost" orthogonal basis $\mathbf{u}_1$, $\mathbf{u}_2$ with the coefficients $\alpha_l(L)$ describing the transformation dynamics. Their graphs are presented in Figure 6.7. The condition $|\alpha_1(\tilde{L})| = |\alpha_2(\tilde{L})|$ determines the existence of the hybrid oscillation $(\alpha_l(L)$ in Figure 6.7 are calculated for the spectral curve 2) where the E_{030} and E_{011} oscillations (corresponding to α_1 and α_2, respectively) make equal contributions to the hybrid oscillation (the contribution of other oscillations described by $\mathbf{v}(L)$ is negligibly small). When L deviates from $\tilde{L}$, one of the coefficients $\alpha_l(L)$ rapidly tends to zero and the other one approaches unit.

The same qualitative analysis may be carried out for the spectral curve 1 in Figure 6.7.

Let us formulate the results obtained by means of an alternative approach based on the Fourier analysis of the formation of hybrid oscillations and developed by Pochanina *et al.* (1989, 1991):

(i) the types of eigenoscillations are not preserved along the spectral curves which split into parts corresponding to equal types and interaction phenomena are observed close to the boundaries of these parts;

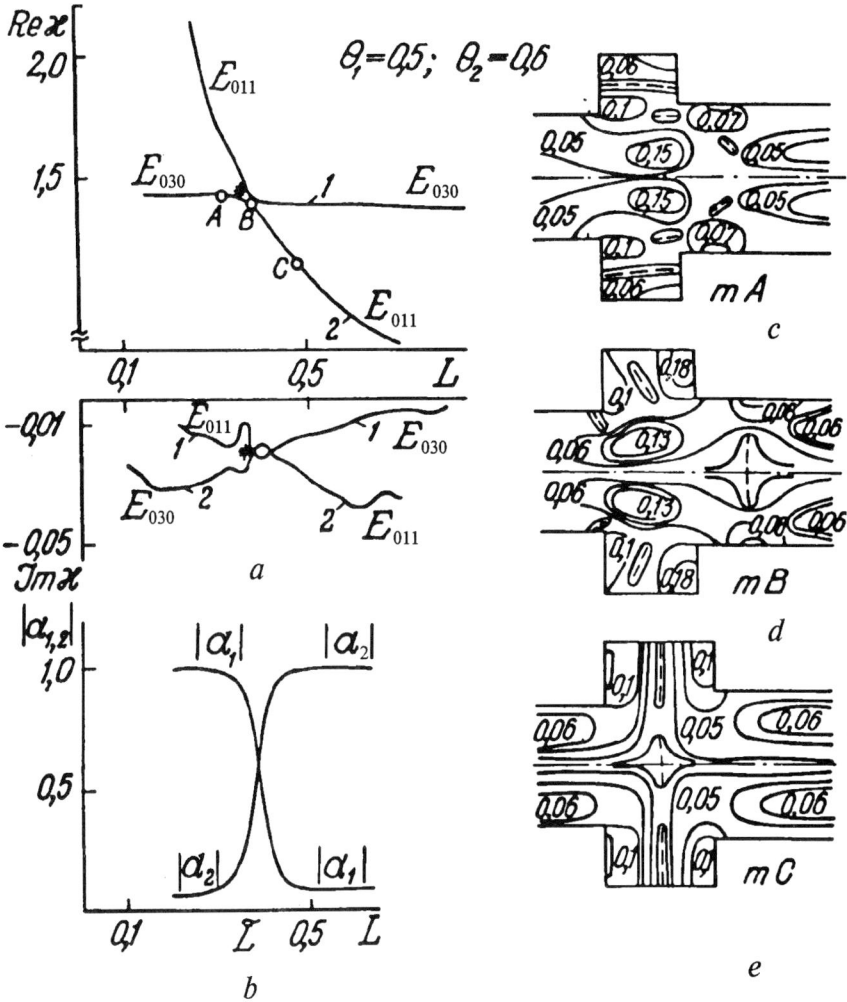

Figure 6.7 *Spectral curves and fields' distributions for the intertype*
E_{011}-E_{030}-*oscillations of an open waveguide resonator*

(ii) the parts of the spectral curves defined in (i) may be separated and the domains of the existence of hybrid oscillations may be described.

One can see that broadening of the spectra in open waveguide resonators is connected with the condenser-type E_{0m0} oscillations which have the largest "domains of existence" (localised close to the cut-off E_{m0} frequencies of the circular waveguide) and these oscillations mainly interact with other types of

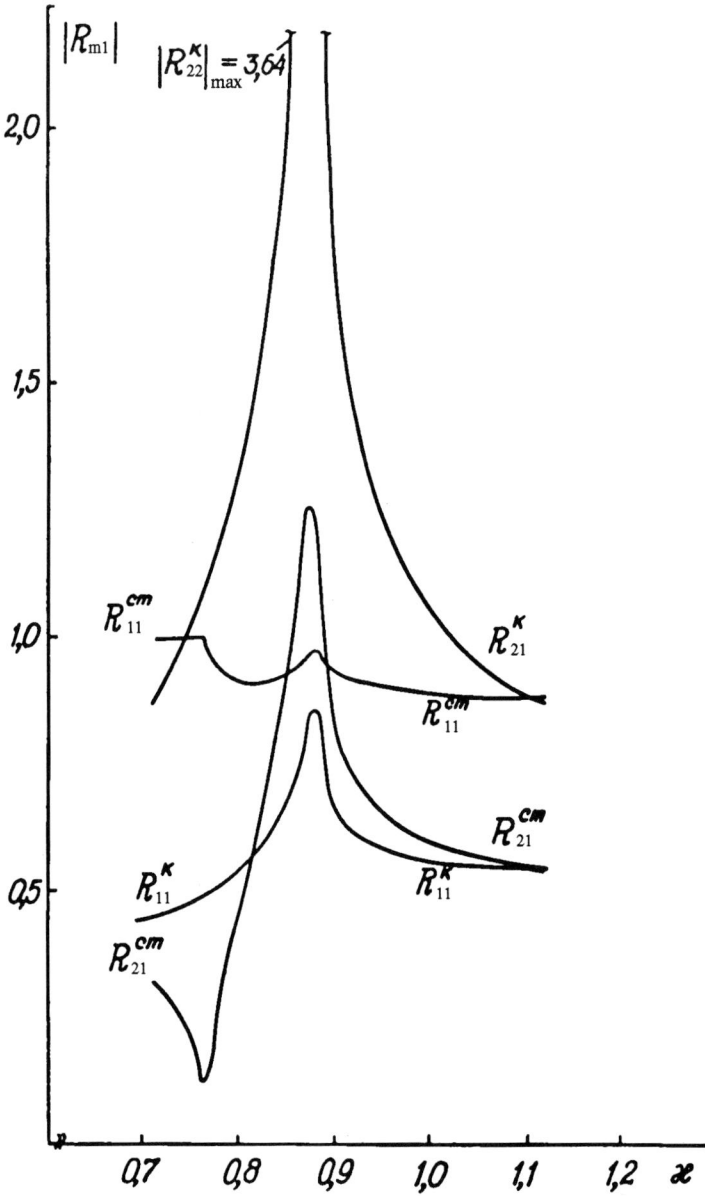

Figure 6.8 *Resonant reflection of waves of a circular waveguide by a step*

oscillations in the vicinity of the non-degenerate Morse critical points. The interaction mechanism may be determined by means of the solution to the problem of the E_{m0} wave diffraction on waveguide discontinuity [see Figure 6.8 where R_{mn} are the absolute values of the reflection coefficients of the nth mode of the circular waveguide $(n = 1)$ transforming to its mth mode as a result of the E_{m0} wave diffraction on a step (R_{mn}^s) or on a coaxial pivot (R_{mn}^c) calculated at the cut-off frequency $\kappa = 0.878$ of the E_{02} wave; the resonances of R_{mn} characterise the transformation properties of the discontinuities].

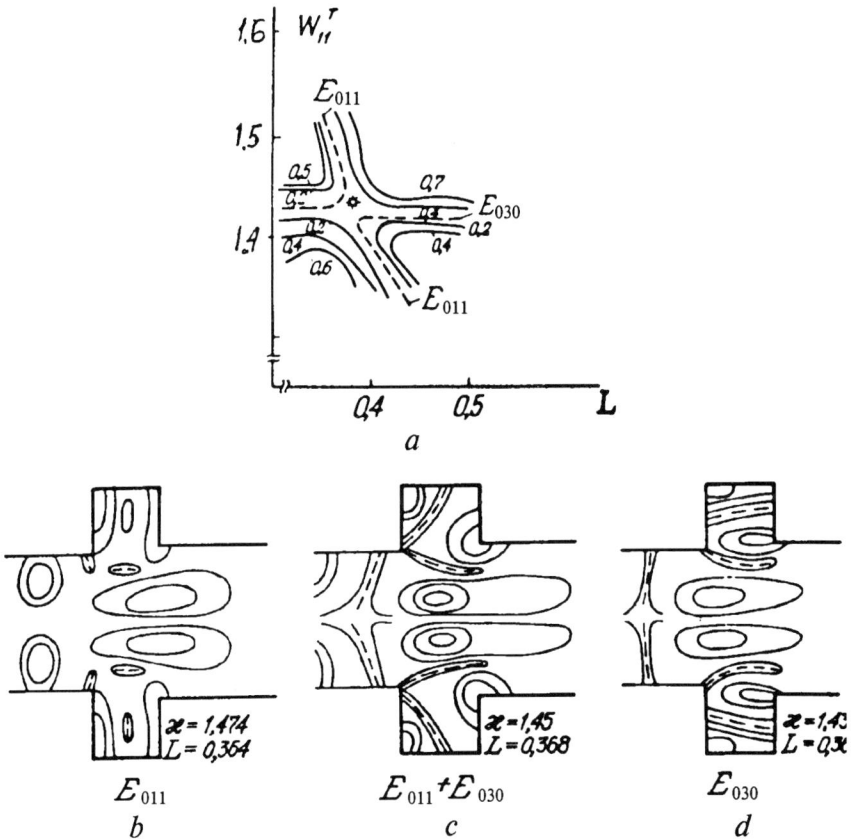

Figure 6.9   *Energy tramsmission coefficient and fields' distributions of the forced oscillations of a non-symmetric open wave-guide resonator*

The structure of the forced oscillations generated close to the Morse critical points are similar to that of the corresponding resonator's intertype oscilla-tions. The isolines $W_{11}(\kappa, L) = $ const of the energy transmission coefficients constructed close to the non-degenerate Morse critical point (Figure 6.9) are evidence of the strong influence of small deformations of a symmetric open waveguide resonator on its diffraction properties.

Part 3 Spectral theory and excitation of diffraction gratings

Diffraction gratings were discovered by Fraunhofer and Roulend [see historical notes presented by Bagbaya (1972), Born and Wolf (1973)] and from the very beginning they have been used as a spectroscopic tool that enables one to obtain spectra with a given dispersion and the highest sharpness of spectral curves.

The length l of the characteristic period of modern diffraction gratings varies from $1.5 \cdot 10^{-7}$ (optical gratings) to several metres (antenna gratings). Holographic gratings discovered by Denisuk in 1962 occupy a special place because they are manufactured without using the thread technology.

Experimental investigations carried out by Wood (1902, 1915) stimulated the theoretical studies of Lord Rayleigh (1907a, 1907b) who was the first to represent the scattered field close to a diffraction grating in the form of a plane wave series and to obtain an approximate solution to the diffraction problem (the Rayleigh approximation). Since the times of Lord Rayleigh the wave diffraction theory has been the most developed approach used for the modelling of diffraction gratings. But this theory cannot take into account all details of the real processes occurring in diffraction gratings.

The properties of diffraction gratings essentially depend on the parameter $\kappa = l\lambda^{-1}$ (λ is the excitation wavelength) and the long-wave ($\kappa \ll 1$), resonant ($\kappa \sim 1$) and short-wave ($\kappa \gg 1$) ranges are treated on the basis of different mathematical approaches.

The long-wave range has been investigated the most completely. Among the variety of results let us indicate the theory of multiperiodic gratings

(Moyzhes, 1955, Sivov, 1961, Marchenko and Khruslov, 1964) and the solution to the acoustic wave diffraction problem on a bar grating obtained by Astapenko and Maluzhinec (1970) and Astapenko (1971) within the frame of this theory, as well as the confirmation and generalisation of the Maluzhinec effect (Nefedov and Sivov, 1977) when the H-polarised wave passes without reflection through a thick grating made of metal bars (Sologub, 1967, Sologub and Shestopalov, 1968, Verbitsky, 1976, Masalov *et al.*, 1980, Borovsky and Khizhnyak, 1983).

The resonant range was considered from the very beginning as the most complicated region (Ignatovsky, 1938). Foundations of the mathematical methods for the studies of diffraction gratings in the resonant frequency range made by Ignatovsky (1938), Weinstein (1955), Katzenelenbaum (1955), Lisanov (1958), Deryugin (1952) formed preconditions for wide inculcation of mathematical and computer modelling. Further development of rigorous mathematical approaches was carried out on the basis of the method of partial domains and the mode matching method (Sologub, 1967, Weinstein, 1955, Deryugin, 1956), the integral equations method (Carlson and Heins, 1947, Baldwin and Heins, 1954, Twersky, 1956, Tarapov, 1965, Ilyinsky and Sveshnikov, 1977, Galishnikova and Ilyinsky, 1987), the modified Wiener-Hopf-Fock method (Avdeev and Voskresensky, 1969, Voskresensky and Zhukov, 1976), the modified residue method (Kirilenko and Yashina, 1980) and the semi-inversion of the matrix convolution-type equations (Masalov and Sologub, 1970, Shestopalov *et al.*, 1984). Following Deryugin (1954) and Deryugin and Fridman (1956) who were the first to analyse the surface and double resonances in reflection diffraction gratings, Hessel and Oliner (1965), Bolotovsky and Lebedev (1967), Sologub *et al.* (1967), Masalov *et al.* (1978a, 1978b), Shestopalov *et al.* (1987) investigated an anomalous scattering close to sliding points. Khruslov (1968), Galishnikova and Ilyinsky (1980) considered the total resonant transmission. Other resonant phenomena, in particular, the total resonant reflection of plane waves by semi-transparent diffraction gratings and total non-mirror reflection by reflection gratings were studied by Kapitza (1965), Kirilenko and Masalov (1972), Masalov (1977), Jull and Ebbeson (1977), Andrewarsha *et al.* (1979), Masalov and Sirenko (1980), Veliev (1980), Veliev *et al.* (1980), Rud (1984), Kusaikin and Sirenko (1985), Kirilenko and Senkevich (1985).

The work by Agranovich *et al.* (1962) occupies a special place in the theory of wave diffraction by gratings. The authors for the first time reduced the problem to a canonical Fredholm equation (in the case of a strip grating). They used the semi-inversion method which had been effectively applied later for various types of gratings (Shestopalov, 1971, 1983, Shestopalov *et al.*, 1973).

The method of the Riemann-Hilbert problem developed by Shestopalov (1971) was the basis for solving the matrix convolution-type equations (Sirenko, 1983a) arising in the problems of wave diffraction by gratings. The analytical solution to the problem of plane wave diffraction on a system of perfectly conducting planes (Carlson and Heins, 1947) formed the background for carrying out the justified semi-inversion procedure within the frames of this method.

The results obtained by the semi-inversion (semi-analytical) methods are permanently used as test data for the determination of the limits of application of heuristic approaches. Their efficiency depends on the quality of analytical (explicit) separation of an invertible operator and a part of the solution determined by free terms. In other words, it is important to know how "far" (with respect to a reasonable physical metric) the physical object (a grating) is situated from a canonical structure (for which the solution may be obtained explicitly). For example, the knife gratings may be effectively analysed when the height h of waveguide channels are sufficiently big (Shestopalov *et al.*, 1973). But when h tends to zero, the numerical algorithms diverge because for small h it is not possible to correctly separate an invertible part in the operator equation connected with analytical solution to the (canonical) diffraction problem on a system of perfectly conducting planes.

Diffraction gratings formed by arbitrary elements were investigated with the help of various direct methods (Kupradze, 1967, Ilyinsky, 1973, Ilyinsky and Repin, 1975, Nikolsky, 1977, Kevanishvili and Tzgareyshvili, 1977 Ilyinsky, 1978). Their obvious disadvantages are the slow convergence and instability of numerical procedures already in the resonant frequency range. The efficiency of direct methods falls sharply when electrodynamic objects have edges and when it is necessary to take into account the higher order harmonics. Combinations of semi-analytical and direct methods (Tuchkin and Shestopalov, 1985, Tuchkin, 1985) turn out to be very perspective in the modern diffraction theory. The universal character of the semi-inversion methods has essentially grown after working out the technique based on the use of generalised scattering matrices (Shestopalov and Shcherbak, 1968, Nikolsky and Nikolskaya, 1983).

One of the main problems arising in the short-wave range is estimating the lower bound for application of the physical optics (Brekhovskikh, 1952). Semi-inversion methods, which have been applied for the first time in this range by Podolsky (1965) for the analysis of a strip grating, enable one to penetrate into the region $\kappa > 10$ (Agranovich *et al.*, 1962, Kirilenko *et al.*, 1978). Weinstein (1966) and Sirenko (1978) constructed the asymptotically correct solutions of the diffraction problems on strip gratings and on gratings made of semi-infinite planes with complicated structure of the period. The former were also analysed by the method of separating a single element (Sologub, 1972, Shestopalov *et al.*, 1973, Feld, 1975).

Scattering theory deals with the evolution of fields in the time domain. Application of the Laplace transform ties together stationary and non-stationary problems (Lubich, 1966) and enables one to effectively use the methods of modern diffraction theory for solving the problems of wave scattering on diffraction gratings. Let us indicate in this way the classical results of Lax and Phillips (1967) and the works of Muravej (1978a, 1978b) and Weinberg (1982) in which the asymptotic expansions of the solutions to mixed external non-stationary problems were constructed. Calculation of the inverse Laplace transform is connected with taking into account the type and location of the singularities of the analytical continuation of $R(\kappa)$ (from the real axis) considered as a function of the complex (frequency) parameter κ where $R(\kappa)$ is the resolvent of the non-self-adjoint (external) diffraction problem (Sirenko *et al.*, 1985). It has been proved by Sirenko (1986) that for a wide family of diffraction gratings these singularities are isolated poles. Determination of singularities (or spectral points) are reduced on the basis of the methods of the spectral theory of operator-valued functions (Ilyinsky and Shestopalov, 1989) to the generalised dispersion equations and this approach begins to dominate nowadays in many areas of mathematical physics [see, e.g. the work by Sukhinin (1984)].

General methods of the theory of non-self-adjoint operators applied for the solution to diffraction problems (Povzner, 1953) are sometimes unsubstantial to consider in detail some electrodynamic effects treated by the theory of gratings (Koshparenok *et al.*, 1983, 1984). Solutions of external diffraction problem do not belong to the space L_2 (Dolf, 1963). From the viewpoint of the nonstationary approach, there exists an exponentially attenuating time factor corresponding to each unbounded solution (satisfying a generalised condition at infinity) which "overlaps" its growth as a function of co-ordinates. In other words, every solution describing a free oscillation is a finite function and its support grows in time.

The knowledge of spectra distributions are important for correct solutions to the scattering and diffraction problems. In fact, the uniqueness (and existence) theorems in the theory of waves diffraction by gratings (Sirenko and Shestopalov, 1985) are proved under the assumptions of small attenuation, i. e. these theorems are in some sense equivalent to the limiting absorption principle (Sveshnikov, 1951, Eidus, 1962). But Sirenko (1986) has shown that there may exist real values of the spectral parameter (κ) where the corresponding homogeneous problems have nontrivial solutions. Every small perturbation of a grating close to these critical frequencies causes various anomalies strongly violating the scattered field. The researchers have stored a big volume of theoretical and calculation data concerning such anomalous phenomena obtained on the basis of the analysis of generalised dispersion equations. In this way we can assert that now it is not sufficient to point to a "resonance" or "quasi-eigenregime" and it is necessary to attract a method [a good example is given by Levin (1958)] which would enable one to qualitatively and quantitavely describe every such effect from the viewpoint of the spectral theory. Let us note that within the frame of this theory every parameter of the problem may be considered as a spectral one (Voitovich *et al.*, 1977).

Chapter 7

Diffraction by gratings in the complex frequency domain

It is known (Whitmer, 1948) that wave diffraction by gratings and by obstacles in waveguides have much in common. For example, the solution to the problem of an aperture in a plane waveguide (at real frequencies) enables one to solve the diffraction problem for a strip grating with the help of the mirror reflection principle. We will show that a similar approach may be also applied at complex frequencies considering waves diffraction by gratings in the complex frequency domain using the methods of the spectral theory of operator-valued functions developed in Chapters 4-6.

7.1 Statements of the problems. Existence and uniqueness theorems

We will state the problem of excitation of an arbitrary grating (periodical in one dimension) by a mono-chromatic quasi-periodic source (with respect to y) assuming at first that the frequency ω is a real value. Let us denote by S the piece-wise smooth contours coinciding with the transversal cross-sections of the perfectly conducting parts of the grating (see Figure 7.1). The relative dielectric permittivity $\varepsilon = \varepsilon(\mathbf{r})$ is a complex-valued piece-wise smooth function (piece-wise constant for the H-polarisation) satisfying the conditions $\operatorname{Re}\varepsilon > 0$, $\operatorname{Re}\kappa \operatorname{Im}\varepsilon \geq 0$, $\varepsilon(x_2+2\pi, x_3) = \varepsilon(\mathbf{r})$, $\varepsilon(\mathbf{r}) = 1$, $|x_3| > 2\pi\delta$, $\mathbf{r} = (x_2, x_3) \in R = \{\mathbf{r} : \ 0 \leq x_2 \leq 2\pi, \ |x_3| \leq \infty\} \setminus S$. It is necessary to determine the solution $u(\mathbf{r})$ of the Helmholtz equation

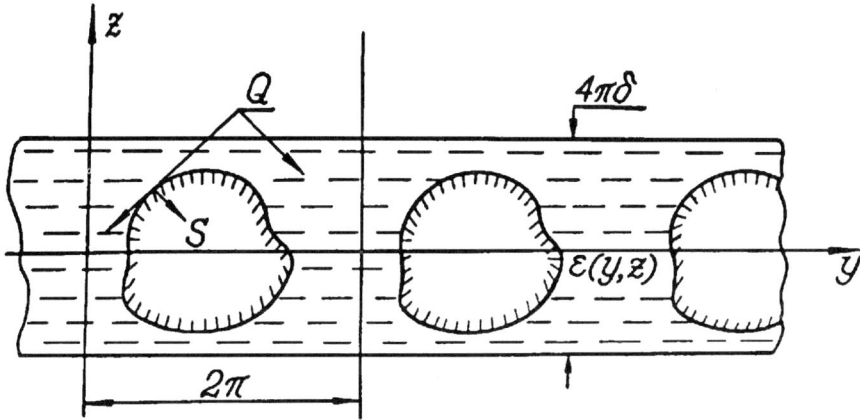

Figure 7.1   *Diffraction grating in a dielectric layer*

$$\left[\frac{\partial^2}{\partial x_2^2} + \frac{\partial^2}{\partial x_3^2} + \kappa^2\varepsilon(\mathbf{r})\right]u(\mathbf{r}) = f(\mathbf{r}), \quad \mathbf{r} \in R \tag{7.1}$$

where $u(\mathbf{r}) = E_{x_1}$ for the E-polarisation and $u(\mathbf{r}) = H_{x_3}$ for the H-polarisation, E_{x_1}, H_{x_1} (E_{tg}, H_{tg}) are the tangential electric and magnetic fields components and $Q = \{\mathbf{r} \in R : \ |x_3| \leq 2\pi\delta\}$ is the the communication domain. The solution of eqn. 7.1 satisfies the radiation condition

$$u(\mathbf{r}) = \sum_{n=-\infty}^{\infty} \begin{Bmatrix} a_n \\ b_n \end{Bmatrix} \exp\{j[\Phi_n x_2 \pm \Gamma_n(x_3 \mp 2\pi\delta)]\} \tag{7.2}$$

$$x_3 \gtrless \pm 2\pi\delta, \quad \operatorname{Im}\Gamma_n \geq 0, \quad \operatorname{Re}\kappa \operatorname{Re}\Gamma_n \geq 0 \quad (n = 0, \pm 1, \pm 2, \dots)$$

the generalised boundary conditions

$$E_{tg} = 0 \quad \text{on} \quad S$$

E_{tg} and H_{tg} are continuous on the gap lines of $\varepsilon(\mathbf{r})$,

$$u\left\{\frac{\partial u}{\partial y}\right\}(2\pi, x_3) = \exp(j2\pi\Phi)u\left\{\frac{\partial u}{\partial x_2}\right\}(0, x_3) \tag{7.3}$$

the edge condition

$$\int_{C\setminus B} \left(|u|^2 + |\operatorname{grad} u|^2\right) d\mathbf{r} < \infty$$

on an arbitrary compactum $C \in \mathbb{R}^2$. Here δ and Φ are the real parameters of the problem, $\Phi_n = n + \Phi$, $\Gamma_n = (\kappa^2 - \Phi_n^2)^{1/2}$, the source function $f(\mathbf{r})$: $f(x_2 + 2\pi, x_3) = \exp(j2\pi\Phi)f(\mathbf{r})$ has a compact support B in R^2 belonging to Q, i.e. $f(\mathbf{r}) \equiv 0$ for $|x_3| > 2\pi$.

We will construct the continuation of the problem given by eqns. 7.1-7.3 into the complex domain κ determining the natural boundaries of the domain $\mathcal{H}$ of the complex frequency parameter κ variation and then re-formulating the radiation conditions given by eqn. 7.2.

Assume that a scattering object with the definite S and $\varepsilon(\mathbf{r})$ is brought into the strip R where the problem is considered. Perturbations caused by this scatterer change the electrodynamic characteristics of the canonical problem given by eqns. 7.1-7.3 with $S = \emptyset$ and $\varepsilon(\mathbf{r}) \equiv 1$. One can follow these changes if the structure of the domain $\mathcal{H}$ does not depend on geometrical and material parameters of the scatterer. This approach (described in Section 4.2) allows us to connect the choice of $\mathcal{H}$ with the determination of the natural boundaries of analytical continuation for the canonical Green function

$$G_0(\mathbf{r}, \mathbf{r}_0, \kappa, \Phi) = \frac{j}{4\pi} \sum_{n=-\infty}^{\infty} \exp\{j[\Phi_n(x_2 - x_{20}) + \Gamma_n|x_3 - x_{30}|]\}\Gamma_n^{-1} \quad (7.4)$$

(of the canonical problem given by eqns. 7.1-7.3). The parameter κ enters (eqn. 7.4) as an argument of an infinite number of the double-valued functions $\Gamma_n(\kappa)$ what enables us to define $\mathcal{H}$ as the infinite-sheet Riemann surface whose first sheet (the pair κ, $\{\Gamma_n(\kappa)\}_{n=-\infty}^{\infty}$) is completely determined by the conditions $\operatorname{Im}\Gamma_n \geq 0$, $\operatorname{Re}\Gamma_n \operatorname{Re}\kappa \geq 0$ and by the cuts

$$(\operatorname{Re}\kappa)^2 - (\operatorname{Im}\kappa)^2 - \Phi_n^2 = 0, \quad \operatorname{Im}\kappa \leq 0 \quad (n = 0, \pm 1, \ldots) \quad (7.5)$$

Other sheets differ from the first by the fact that the signs of $\Gamma_n(\kappa)$ are changed to opposite ones for a finite number of indices n. It completes the definition of the domain of analytical continuation (the surface $\mathcal{H}$) for the diffraction problem given by eqns. 7.1-7.3 considered at complex frequencies.

Theorem 7.1: The canonical Green function $G_0 = G_0(\mathbf{r}, \mathbf{r}_0, \kappa, \Phi)$ admits the natural analytical continuation from the real axis κ, where G_0 satisfies the radiation conditions

$$\text{Im}\,\Gamma_n \geq 0 \quad \text{and} \quad \kappa\,\text{Re}\,\Gamma_n \geq 0 \quad \text{for} \quad \text{Im}\,\kappa = 0 \quad (n = 0, \pm 1, \pm 2, \ldots) \quad (7.6)$$

to the infinite-sheet Riemann surface $\mathcal{H}$. The only singularities of the canonical Green function are the simple poles in the points κ_n: $\Gamma_n(\kappa_n) = 0$ $(n = 0, \pm 1, \pm 2, \ldots)$.

The properties of the canonical Green function are given by the following lemmas.

Lemma 7.1: On the first sheet of $\mathcal{H}$

$$G_0(\mathbf{r}, \mathbf{r}_0, \kappa, \Phi) = \frac{j}{4} \sum_{n=-\infty}^{\infty} H_0^{(1)}(\kappa R_n) \exp(j2\pi\Phi_n) \qquad (7.7)$$

if $\text{Im}\,\kappa > 0$, $\text{Re}\,\kappa = 0$ and where $R_n = \sqrt{(x_2 - x_{20} - 2\pi n)^2 + (x_3 - x_{30})^2}$.

The proof is based on simple calculations

$$\sum_{n=-\infty}^{\infty} H_0^{(1)}\left(\kappa\sqrt{(x_2 - 2\pi n)^2 + x_3^2}\right) \exp(j2\pi\Phi_n)$$

$$= \sum_{m=-\infty}^{\infty} H_0^{(1)}\left(\kappa\sqrt{(x_2 + 2\pi m)^2 + x_3^2}\right) \exp[-j(2\pi m + x_2)\Phi] \exp(j\Phi x_2) \quad (7.8)$$

$$= \exp(j\Phi x_2) \sum_{m=-\infty}^{\infty} f(x_2 + 2\pi m)$$

and, due to the Poisson summation formula, this is equal to

$$\frac{1}{2\pi} \sum_{n=-\infty}^{\infty} \exp(j\Phi_n x_2) \int_{-\infty}^{\infty} H_0^{(1)}\left(\kappa\sqrt{t^2 + x_3^2}\right) \exp(-j\Phi_n t)\,dt \qquad (7.9)$$

because $\int_{-\infty}^{\infty} H_0^{(1)}(\kappa\sqrt{t^2 + x_3^2}) \exp[-j(n + \Phi)t]\,dt < \infty$ and the series in eqn. 7.8 converge uniformly with respect to $0 \leq x_2 \leq 2\pi$ for $\text{Im}\,\kappa > 0$. The required results may be easily obtained if to calculate eqn. 7.9 with the help of eqn. 7.5:

$$\sum_{n=-\infty}^{\infty} H_0^{(1)}\left(\kappa\sqrt{(x_2 - 2\pi n)^2 + x_3^2}\right) \exp(j2\pi n\Phi)$$

$$= \frac{1}{\pi} \sum_{n=-\infty}^{\infty} \exp[j(\Phi_n y + \Gamma_n |x_3|)]\Gamma_n^{-1}$$

The result of lemma 7.1 may be generalised for an arbitrary sheet of the Riemann surface $\mathcal{H}$.

Lemma 7.2: In the regular points $\kappa \neq 0$ the canonical Green function admits the representation

$$
G_0 = \begin{cases}
\dfrac{j}{4} \displaystyle\sum_{n=-\infty}^{\infty} H_0^{(1)}(\bar{\kappa}R_n)\exp(j2\pi n\Phi) \\[2mm]
\quad + \dfrac{j}{2\pi} \displaystyle\sum_{n(\kappa)} \exp[j\Phi_n(x_2 - x_{20})]\cos[\Gamma_n|x_3 - x_{30}|]\Gamma_n^{-1}, & 0 \leq \arg\kappa \leq \pi \\[4mm]
\dfrac{j}{4} \displaystyle\sum_{n=-\infty}^{\infty} H_0^{(1)}(-\bar{\kappa}R_n)\exp(j2\pi n\Phi) \\[2mm]
\quad + \dfrac{j}{2\pi} \displaystyle\sum_{m(\kappa)} \exp[j\Phi_m(x_2 - x_{20})]\cos[\Gamma_m|x_3 - x_{30}|]\Gamma_m^{-1}, & \pi < \arg\kappa < 2\pi
\end{cases}
$$

$$(7.10)$$

where $\bar{\kappa}$ is the projection of κ on the first sheet of $\mathcal{H}$ and the sums with the symbols $n(\kappa)$ and $m(\kappa)$ contain finite number of terms; $n(\kappa)$ is determined only by the number N of the sheet and $m(\kappa)$ both by N and the position of κ on the Nth sheet.

The continuation of G_0 keeps all properties of the canonical Green function.

Theorem 7.2: The function $G_0 = G_0(\mathbf{r}, \mathbf{r}_0, \kappa, \Phi)$ satisfies the following conditions in the regular points $\kappa \in \mathcal{H}$ for all $\mathbf{r} \neq \mathbf{r}_0$:

(i) $\left(\dfrac{\partial^2}{\partial x_2^2} + \dfrac{\partial^2}{\partial x_3^2} + \kappa^2 \right) G_0 = 0$

(ii) $G_0(\mathbf{r}, \mathbf{r}_0, \kappa, \Phi) = G_0(\mathbf{r}_0, \mathbf{r}, \kappa, \Phi) = G_0(\mathbf{r}_0, \mathbf{r}, \kappa, -\Phi) = G^*(\mathbf{r}, \mathbf{r}_0, -\kappa^*, -\Phi)$

(iii) $G_0(\mathbf{r}, \mathbf{r}_0, \kappa, \Phi)$ is once continuously differentiable with respect to the co-ordinates of $\mathbf{r}, \mathbf{r}_0$ and

$$G_0 = -\frac{1}{2\pi}\ln\rho + O(1), \quad \frac{\partial G}{\partial \rho} = -\frac{1}{2\pi\rho} + O(1), \quad \rho = R_0 = |\mathbf{r} - \mathbf{r}_0| \to 0$$

(iv) if a function $f(\mathbf{r}_0)$ has continuous derivatives $\frac{\partial f}{\partial x_{20}}$ and $\frac{\partial f}{\partial x_{20}}$, the Green potential

$$u(\mathbf{r}, \kappa, \Phi) = -\int_Q G_0(\mathbf{r}, \mathbf{r}_0, \kappa, \Phi)f(\mathbf{r}_0)\,d\mathbf{r}_0$$

is a (twice continuously differentiable) solution of the problem given by eqns. 7.1-7.3.

The proofs of the first three points are based on the estimates of $H_0^{(1)}(\kappa R_0)$ and $\frac{\partial H_0^{(1)}(\kappa R_0)}{\partial R_0}$ for $R_0 \to 0$ and on the possibility of term-wise differentiation of the series given by eqn. 7.10. In order to prove the symmetry of $G_0(\mathbf{r}, \mathbf{r}_0, \kappa, \Phi)$ it is sufficient to check that $\Gamma_n(\kappa, \Phi) = \Gamma_{-n}(\kappa, -\Phi)$, $\Gamma_n(\kappa, \Phi) = -\Gamma_n(-\kappa^*, -\Phi)$, $\Phi_n(\Phi) = -\Phi_n(-\Phi)$ and to calculate all combinations of the series given in condition (ii) using eqn. 7.4.

Let us define the resolvent set R of the problem given by eqns. 7.1-7.3 as the points $\kappa \in \mathcal{H}$ where there exists the unique solution to this problem. The complementary set Ω is called the spectral set (the spectrum) of this diffraction problem. The values $\kappa \in \Omega_P \subseteq \Omega$ for which the homogeneous Helmholtz equation

$$\left[\frac{\partial^2}{\partial x_2^2} + \frac{\partial^2}{\partial x_3^2} + \kappa^2 \varepsilon(\mathbf{r}) \right] v(\mathbf{r}) = 0, \quad \mathbf{r} \in R \tag{7.12}$$

has non-trivial solutions (eigenfunctions) satisfying the conditions given by eqns. 7.2 and 7.3 form the point spectrum. It will be proved below that $\Omega = \Omega_P$ and, consequently, eqns. 7.12, 7.2, 7.3 define the spectral problem. Its eigenfunctions, which determine the free quasi-stationary field's oscillations, will be called eigenoscillations corresponding to the fundamental frequencies $\kappa \in \Omega$. Let us note that the radiation conditions stated originally on the real axis and excluding the waves coming to a grating from infinity (at real frequencies) was the only criterion for the constructed "continuation" of the diffraction problem into the complex domain κ and for the correct statement of the corresponding spectral problem.

The following qualitative results (which are similar to the assertions proved in Section 4.2 assuming that the same notation is used) enable one to determine the domains in $\mathcal{H}$ free of (complex) fundamental frequencies for semi-transparent gratings [the field is not equal to zero only in the reflection $(x_3 < 2\pi\delta)$ and transmission $(x_3 > 2\pi\delta)$ zones].

Theorem 7.3: Assume that $w \neq 0$, $V \neq \emptyset$ and $\text{Im}\,\epsilon = 0$ (these conditions define a volume grating filled with an ideal dielectric). Then the following situations are possible for both polarisations:

(i) For $\text{Re}\,w \neq 0$, $\text{Im}\,w \neq 0$ there are no non-trivial solutions of the spectral problem given by eqns. 7.12, 7.2 and 7.3 such that

$$\text{Re}\,w \, \text{Im}\,w \sum_n (|a_n|^2 + |b_n|^2) \, \text{Re}\,\Gamma_n > 0$$

In particular, there are no "outgoing" solutions for $\operatorname{Im}\omega > 0$ and "incoming" solutions for $\operatorname{Im}\omega < 0$.

(ii) For $\operatorname{Re}\omega \neq 0$, $\operatorname{Im}\omega = 0$ there are no non-trivial solutions such that $\sum_n (|a_n|^2 + |b_n|^2)\operatorname{Re}\Gamma_n = 0$. It means, in particular, that the "non-outgoing" and "non-incoming" solutions may have the non-zero amplitudes a_n and b_n b_{ml} only if $\operatorname{Re}\Gamma_n = 0$.

(iii) For $\operatorname{Re}\omega = 0$, $\operatorname{Im}\omega \neq 0$ there are no non-trivial solutions such that $\operatorname{Im}\Gamma_n > 0$ for all $n = 0, \pm 1, \ldots$.

(iv) For $|\operatorname{Im}\omega| \geq |\operatorname{Re}\omega|$ there are no non-trivial solutions such that

$$\operatorname{Re}\omega\,\operatorname{Im}\omega \sum_n (|a_n|^2 + |b_n|^2)\operatorname{Im}\Gamma_n \geq 0$$

In particular, there are no solutions for which $\operatorname{Im}\Gamma_n \geq 0$ for all n.

An account of losses in the dielectric media filling the grating leads to the following specifications.

Theorem 7.4: Assume that $\omega \neq 0$, $V \neq \emptyset$ and $\operatorname{Im}\epsilon > 0$. Then the following situations are possible:

(i) In the case of the E-polarisation there are no non-trivial solutions of the spectral problem given by eqns. 7.12, 7.2 and 7.3 such that

$$\{\operatorname{Im}\epsilon[(\operatorname{Re}\omega)^2 - (\operatorname{Im}\omega)^2 + 2\operatorname{Re}\omega\,\operatorname{Im}\omega\,\operatorname{Re}\epsilon]\} \sum_{n=-\infty}^{\infty} (|a_n|^2 + |b_n|^2)\operatorname{Re}\Gamma_n > 0$$

$$(7.13)$$

everywhere in the domain Q (here it is assumed that $\epsilon = \epsilon(\mathbf{r})$ is a function of the points of the domain Q). In particular, if $\operatorname{Re}\omega \neq 0$, $\operatorname{Im}\omega \neq 0$ and $|\operatorname{Re}\omega| \geq |\operatorname{Im}\omega|$, there are no "incoming" solutions for $\operatorname{Re}\omega < 0$, $\operatorname{Im}\omega < 0$ and "outgoing" solutions for $\operatorname{Re}\omega > 0$, $\operatorname{Im}\omega > 0$. If $|\operatorname{Re}\omega| \leq |\operatorname{Im}\omega|$, there are no "incoming" solutions for $\operatorname{Re}\omega > 0$, $\operatorname{Im}\omega < 0$ and "outgoing" solutions for $\operatorname{Re}\omega < 0$, $\operatorname{Im}\omega > 0$. There are no non-trivial solutions on the imaginary axis and on the real axis such that $\operatorname{Re}\Gamma_n \geq 0$ for all $n = 0, \pm 1, \ldots$.

(ii) In the case of the H-polarisation there are no non-trivial "outgoing" solutions for $\operatorname{Re}\omega > 0$, $\operatorname{Im}\omega > 0$ and $\operatorname{Re}\omega < 0$, $\operatorname{Im}\omega < 0$. There are no non-trivial solutions on the imaginary axis $\operatorname{Re}\omega = 0$ and on the real axis $\operatorname{Im}\omega = 0$ such that $\operatorname{Re}\Gamma_n \geq 0$ for all $n = 0, \pm 1, \ldots$.

Analysis of plane gratings enables us to formulate a number of more precise results concerning the spectrum localisation.

Theorem 7.5: In the case of plane gratings there are no non-trivial solutions of the spectral problem given by eqns. 7.12, 7.2 and 7.3 such that

$$\sum_n (|a_n|^2 + |b_n|^2)\,\text{Re}\,\Gamma_n \neq 0 \quad \text{or (and)} \quad \sum_n (|a_n|^2 + |b_n|^2)\,\text{Im}\,\Gamma_n \neq 0 \quad (7.14)$$

In addition to this, one can essentially narrow the domain of the existence of spectral points (compare with theorem 4.5 from Section 4.2).

Theorem 7.6: In the case of plane gratings the subset of the point spectrum of the spectral problem given by eqns. 7.12, 7.2 and 7.3 situated on the first sheet of H may contain only the threshold points $\kappa : \Gamma_n(\kappa) = 0$.

The conditions determining the absence of spectral points on different sheets of $\mathcal{H}$ formulated as theorems 7.3-7.6 may be naturally considered as the uniqueness theorems for the stationary diffraction problems. In particular, it turns out that the uniqueness is provided by the presence of lossy grating's material without assuming that the surrounding media satisfies this requirement.

All these results remain valid for reflection gratings (when the field is assumed to exist only in the reflection zone) if $b_n = 0$ $(n = 0, \pm 1, \ldots)$.

Further spectra qualitative characteristics are similar to the assertions proved in Section 4.3 for waveguide resonators. Here the limitations concerning the smoothness of $\epsilon(\mathbf{r})$ may be also weakened and it is sufficient to assume that $\epsilon(\mathbf{r})$ is a bounded piece-wise continuous and piece-wise differentiable function in Q.

Lemma 7.3: There exists a domain $\mathcal{K} \subset \mathcal{H}$ such that the sequence defined by eqn. 4.12 uniformly (with respect to $\kappa \in \mathcal{K}$ and $\mathbf{r} \in Q$) converges to the function

$$u(\mathbf{r}, \kappa) = -\int_Q G_0(\mathbf{r}, \mathbf{r}_0, \kappa, \Phi)\{\kappa^2[1 - \epsilon(\mathbf{r}_0)]u(\mathbf{r}_0, \kappa) + f(\mathbf{r}_0)\}\,d\mathbf{r}_0, \quad \mathbf{r} \in Q$$
$$(7.15)$$

which may be continued from the domain $\mathcal{K}$ to the whole surface $\mathcal{H}$ as a meromorphic function of κ where $u(\mathbf{r}, \kappa)$ keeps all its properties as a solution to the problem given by eqns. 7.1-7.3.

Theorem 7.7: The spectrum Ω of the problem given by eqns. 7.1-7.3 of an arbitrary grating made of metal bars placed in a dielectric layer with the piece-wise continuous permittivity consists of not more than a countable number of isolated points situated on the Riemann surface $\mathcal{H}$ and the resolvent of the problem has finite-order poles in these points.

The proof of this spectral theorem which is based on the consideration of Green's potentials in the complex frequency domain (with the use of the results of theorem 7.1 and lemmas 7.2 and 7.3) is similar to the one of theorem 4.7 from Section 4.3. Here it is also assumed that the piece-wise smooth contour S of the perfectly conducting parts of grating's cross-section does not contain "infinitely" sharp edges. The gratings with such edges may be considered by other approaches, e.g. by the semi-inversion method.

The spectral theorem may be generalised for non-ideal metal gratings if to conditionally replace lossy metal surfaces by dielectric layers with large $|\operatorname{Im}\epsilon|$.

The existence of spectral points may be probed only for a limited number of gratings having a simple structure. For example, the residues

$$\operatorname{res} G_0(\kappa_n) = j(4\kappa_n \pi)^{-1} \exp[j\Phi_n(x_2 - x_{20})]$$

of the canonical Green function G_0 in the branch points $\kappa = \kappa_n : \Gamma_n(\kappa_n) = 0$ determine the possible non-trivial solutions $v_n = \exp(j\Phi_n x_2)$ of the canonical spectral problem given by eqns. 7.1-7.3. The fields of corresponding free oscillations do not vary in the direction perpendicular to grating's plane. In the case of knife gratings Levin (1958) has shown that there exist two waves $v^{\pm} = \exp(\pm 0.5jpx_2)$ for $\kappa = \Phi = 0.5p$ corresponding to the partial components in eqn. 7.2 with the indices 0 and p $(p = 0, \pm 1, \ldots)$ whose sum (for the H-polarisation) and difference (for the E-polarisation) form the free oscillations which satisfy all the conditions of the spectral problem given by eqns. 7.12, 7.2 and 7.3. Similar results have been obtained by Sirenko (1968) for certain combinations of knife and strip gratings.

Let us consider the diffraction problem given by eqns. 7.1-7.3 for a grating made of metal bars with $\epsilon(\mathbf{r}) \equiv 1$ denoting by θ the characteristic size of the cross-section of perfectly conducting inclusions formed by the smooth contour $S(\theta)$. The problem with $\theta \neq 0$ will be called the perturbed one (with respect to the canonical problem).

Lemma 7.4: Let γ be a simple closed curve in $\mathcal{H}$ surrounding the branch point κ_n such that there are no other spectral points of the canonical problem inside γ. Then the perturbed problem has at least one fundamental frequency inside the double-sheet circle D bounded by γ for sufficiently small θ.

This lemma is proved in a similar way to lemma 4.5 from Section 4.4.

Hence, as well as in the case of open waveguide resonators, the existence of fundamental frequencies may be shown for arbitrary gratings which are "weakly" perturbed with respect to certain canonical structures.

The following assertion repeats the continuity principle stated by lemma 4.6 from Section 4.4.

Lemma 7.5: Let D and $T(\kappa, \mu)$ be a simple-connected domain of the Riemann surface $\mathcal{H}$ and a family of compact operators acting in a Banach space X defined for $\kappa \in D$, $\mu \in [\bar{\mu}_1, \bar{\mu}_2]$, respectively, where μ is a real parameter. Assume that

(i) $T(\kappa, \mu)$ is a holomorphic function of κ for every $\mu \in [\bar{\mu}_1, \bar{\mu}_2]$;

(ii) $T(\kappa, \mu)$ is a continuous function on $D \times [\bar{\mu}_1, \bar{\mu}_2]$ with respect to the operator norm;

(iii) a bounded operator $R_T(\kappa, \mu) = [I - T(\kappa, \mu)]^{-1}$ (the resolvent of T) exists at least in one point $\kappa \in D$ for all $\mu \in [\bar{\mu}_1, \bar{\mu}_2]$.

Then if κ_0 is a regular point of R_T, the latter is continuous in the vicinity of $\kappa_0, \mu_0 \in [\bar{\mu}_1, \bar{\mu}_2]$ and the poles of R_T continuously depend on μ, i. e. they may emerge (or disappear) only on the boundary of D.

The last lemmas give sufficient conditions for the existence of fundamental frequencies and enable proof for a rather general family $\mathcal{G}$ of arbitrary gratings with piece-wise continuous dielectric permittivity and perfectly conducting inclusions formed in the cross-section by smooth contours. This family is indirectly defined by the limitations inserted in the formulations of these lemmas.

Theorem 7.8: The spectrum Ω of the problem given by eqns. 7.1-7.3 is not empty for every grating belonging to the family $\mathcal{G}$.

The proof here may be carried out in a similar way to the one discussed in Section 4.4. One can find simple exceptions (for the gratings which do not belong to the family $\mathcal{G}$) when it is not possible to apply this perturbation technique. For example, there are no free field oscillations in the periodic structure having the form of a perfectly conducting plane $z = 0$ since the corresponding spectral problem has only trivial solutions for all $\kappa \in \mathcal{H}$ (the case of the E-polarisation).

Applying the properties of Green's functions formulated in theorem 7.2 one can specify the result of theorem 7.8 for a sub-family $\mathcal{G}_- \subset \mathcal{G}$ such that Green's functions of the problem given by eqns. 7.1-7.3 for the gratings from $\mathcal{G}_-$ depend only on κ^2 and the sequence of values $\{\Gamma_n(\kappa)\}_{n=-\infty}^{\infty}$.

Theorem 7.9: Assume that there exists $\kappa = \kappa_- \in \Omega$ with $\operatorname{Im}\kappa_- < 0$ (a resonance attenuating in time). Then there exists a sheet $\mathcal{H}_- \in \mathcal{H}$ such that $-\kappa_- \in \mathcal{H}_-$ and $-\kappa_- \in \Omega$.

7.2 Model problems and numerical algorithms in the spectral theory of gratings

Existence of free oscillations whose energy grows in time may be justified analysing the radiation conditions given by eqn. 7.2 where the signs of $\operatorname{Im}\Gamma_n$ and $\operatorname{Re}\Gamma_n$ determine the "dynamics" of the nth partial component: for $n \in N_1 : \operatorname{Re}\Gamma_n \operatorname{Re}\kappa > 0$ ($n \in N_2 : \operatorname{Re}\Gamma_n \operatorname{Re}\kappa < 0$) the waves propagate off the grating (to the grating) attenuating (growing) in the direction of growing $|x_3|$ when $\operatorname{Im}\Gamma_n > 0$ ($\operatorname{Im}\Gamma_n < 0$). Free oscillations, which emerge only when a part of the waves given by eqn. 7.2 goes to the grating while the other part tears off it, are "supported" by external sources forming the plane waves with the indices $n \in N_1$. Since the sets N_1 and N_2 do not intersect, the package $v_1(\mathbf{r})$ of waves coming to the grating (given by eqn. 7.2 with the summation index $n \in N_1$) transforms to the package $v_2(\mathbf{r})$ of waves going off the grating ($n \in N_1$) at every fundamental frequency $\kappa \in \Omega$.

This natural classification together with the semi-inversion methods developed in the previous chapters will be used for determining the characteristic numbers of Fredholm matrix operator-valued functions of the complex frequency spectral parameter $\kappa \in \mathcal{H}$ (where $\mathcal{H}$ is the Riemann surface of the analytical continuation of the canonical Green function $G_0 = G_0(\mathbf{r}, \mathbf{r}_0, \kappa, \Phi)$ introduced in Section 7.1) arising in the stationary scattering problems on the waveguide and strip gratings (Figure 7.1). In fact, these very operators define, in accordance with the equivalence and uniqueness theorems, (homogeneous) operator equations describing free oscillations. We will consider here only the E-polarisation assuming that the other polarisation may be treated in a similar way.

The spectral problem given by eqns. 7.12, 7.2 and 7.3 for the waveguide grating shown in Figure 7.1 is reduced by the semi-inversion of matrix operators to the equivalent (spectral) problem on characteristic numbers for the Fredholm matrix operator-valued function produced by the coupled system of homogeneous operator equations

$$\begin{cases} \mathbf{t}^\pm = BA\mathbf{t}^\pm + \sum_{l=1}^{2} D_l^\pm \mathbf{v}_l^\pm \\[2mm] \mathbf{v}^\pm = BA\mathbf{t}^\pm + \sum_{l=1}^{2} T_l^\pm \mathbf{v}_l^\pm \end{cases} \tag{7.16}$$

where

$$\mathbf{a} \pm \mathbf{b} = P\mathbf{t}^\pm \qquad \mathbf{a} = (a_n)_{n=-\infty}^{\infty} \in \tilde{l}_2, \quad \mathbf{b} = (b_n)_{n=-\infty}^{\infty} \in \tilde{l}_2$$
$$\tilde{l}_2 = \{\mathbf{a} = (a_n)_{n=-\infty}^{\infty} : \sum_n |a_n|^2(1 + |n|) < \infty\} \tag{7.17}$$

the signs $+$ or $-$ correspond to the symmetric or anti-symmetric oscillations, respectively, with respect to the plane $z = 0$. The latter coincide with the free oscillations in the reflection grating which is obtained if to replace the plane of symmetry $z = 0$ in a semi-transparent structure by a perfectly conducting plane.

The system given by eqn. 7.16 may be re-written with the help of one matrix operator R:

$$\mathbf{d} = R\mathbf{d} \tag{7.18}$$

where

$$\mathbf{d} = \begin{Vmatrix} \tilde{\mathbf{t}} \\ \tilde{\mathbf{v}}_1 \\ \tilde{\mathbf{v}}_2 \end{Vmatrix} \qquad \tilde{\mathbf{t}} = (\sqrt{n}t_n)_{n=-\infty}^{\infty} \qquad R = \begin{pmatrix} \tilde{B}A & \tilde{D}_1 & \tilde{D}_2 \\ \tilde{B}_1A & \tilde{T}_{11} & \tilde{T}_{12} \\ \tilde{B}_2A & \tilde{T}_{21} & \tilde{T}_{22} \end{pmatrix} \tag{7.19}$$

Let us separate the simplest case of a plane strip grating when the matrix operator constructed by Shestopalov (1971) consists of one block $R(\kappa) = \|a_{mn}(\kappa)\|_{m,n=-\infty}^{\infty}$ where

$$a_{mn}(\kappa) = (V_m^n - R_m R_\sigma^{-1} V_\sigma^n) \frac{|\Phi_n| + j\Gamma_n(\kappa) + 2\delta_{nl}\Phi_n}{\Phi_n}$$

$l = [\Phi]$ denotes the entire part of $\Phi > 0$, $\Phi - l \neq 0$.

The estimates

$$
\begin{aligned}
(\tilde{B}A)_{mn} &\approx (\tilde{B}_l A)_{mn} = O(m^{-1}n^{-1}) \\
(\tilde{D}_l)_{mn} &\approx (\tilde{T}_{li})_{mn} = O[n^{-2}(m+n)^{-1}]
\end{aligned}
\qquad (m, n \to \infty) \qquad (7.20)
$$

yield $\sum_{m,n} |(\tilde{B}A)_{mn}|^2 < \infty$. Hence, the blocks of R are the Koch matrices (which form a Banach space with the norm $\|\{a_{mn}\}\| = \sum_m |a_{mm}| + \sqrt{\sum_{m,n} |a_{mn}|^2}$). There exists an infinite determinant $\Delta(\kappa) = \det[I - R(\kappa)]$ and the following assertion holds.

Lemma 7.6: $R(\kappa) : l_2^3 \to l_2^3$, where $l_2^3 = l_1 \times l_2 \times l_2$, is a compact finite-meromorphic operator-valued function of $\kappa \in \mathcal{H}$.

Hence $I - R(\kappa)$ is a canonical Fredholm operator-valued function, $\kappa \in \mathcal{H}$, and due to the meromorhpic Fredholm theorem (Reed and Simon, 1978) $I - R(\kappa)$ may have only isolated characteristic numbers (we take into account that the resolvent set of this operator is not empty in $\mathcal{H}$ due to the equivalence and uniqueness theorems) what yields the spectrum discreteness.

Theorem 7.10: The set Ω of the fundamental frequencies of a waveguide grating consists of not more than a countable number of points in $\mathcal{H}$ which have no finite accumulation points and coincide with characteristic numbers of $R(\kappa)$ having finite multiplicities.

Let us note that theorem 7.10 is not a corollary of theorem 7.7. In fact, one cannot prove the former by the methods of the potential theory since waveguide gratings have infinitely thin ("zero") edges.

Since the operator-valued function in eqn. 7.19 keeps its properties in a wide domain of variation of the (non-spectral) parameters and, in particular, for every finite complex value of ϵ, the results obtained remain valid for Im $\epsilon \gg 1$ which enables one to analyse the bar gratings with lossy bars considering non-ideal metal as a non-ideal dielectric.

Calculation of fundamental frequencies are carried out on the basis of the well-studied truncation procedure: eqn. 7.18 is replaced by the sequence of finite-dimensional problems

$$\mathbf{d}_N = R_N \mathbf{d}_N, \quad R_N = \left\| \begin{matrix} P_N \\ P_N \\ P_N \end{matrix} \right\| R \left\| \begin{matrix} P_N \\ P_N \\ P_N \end{matrix} \right\|$$

$$P_N \mathbf{x} = \mathbf{x}_N \ (\mathbf{x} = (x_m)_{m=-\infty}^{\infty}, \ \mathbf{x}_N = (x_m)_{m=-\infty}^{N}) \tag{7.21}$$

The convergence of this algorithm may be proved in a way similar to that presented in Sections 1.11 and 2.3. Let $\Omega_N = \{\kappa = \kappa_N \in \mathcal{H} : \Delta_N(\kappa_N) \equiv \det[I - R(\kappa_N)] = 0 \ (N = 1, 2 \dots)$ be the set of characteristic numbers of the problem given by eqn. 7.21.

Lemma 7.7: If $\kappa_0 \in \Omega$, then there exists a sequence of complex numbers $(\kappa_N)_{N=0}^{\infty}$ such that $\kappa_N \in \Omega_N$, $\lim_{N \to \infty} |\kappa_N - \kappa_0| \to 0$ and

$$|\kappa_N - \kappa_0| < \text{const} \frac{\Delta(\kappa_N)}{\Delta'(\kappa_N)}, \quad \Delta'(\kappa_N) = \frac{\partial}{\partial \kappa} \Delta(\kappa), \quad \Delta(\kappa_0) = 0, \ \Delta'(\kappa_0) \neq 0 \tag{7.22}$$

On the other hand, if $\kappa_N \in \Omega_N \ (N \geq N_0 > 0)$ and there exists an infinite subsequence $\{\kappa_N'\} \subset (\kappa_N)_{N=0}^{\infty}$ such that $\lim_{N' \to \infty} |\kappa_N' - \kappa_0| \to 0$, then $\kappa_0 \in \Omega$.

The rate of convergence is given by the inequality

$$|\kappa_N - \kappa_0| < \text{const} \frac{1}{\sqrt{N} \Delta'(\kappa)} \quad (N \to \infty) \tag{7.23}$$

which follows from eqn. 7.22 and the known estimates of Sanchez-Palencia (1980)

$$|\Delta(\kappa_N) - \Delta_N(\kappa_N)| \leq \|R(\kappa_N) - R_N(\kappa_N)\| \exp[\|R(\kappa_N)\| + 1]$$

$$|\Delta(\kappa_N)| < \text{const} \|R(\kappa_N) - R_N(\kappa_N)\| \tag{7.24}$$

The method of generalised scattering matrices introduced by Shestopalov and Shcherbak (1968) and applied in Section 5.6 for the analysis of open waveguide resonators enables us to broaden the family of gratings for which one can prove the spectrum discreteness and construct effective justified algorithms for calculation of fundamental frequencies. Let us consider an example of application of this method for solving the problem (given by eqns. 7.1-7.3) of excitation of the grating formed by two "elementary" strip inhomogeneities (see Figure 7.2*d*) by the spatial harmonics

$$u^{inc} = \sum_{n=-\infty}^{\infty} u_n \exp[j(\Phi_n y - \Gamma_n x_3)] \quad (x_3 \geq 0)$$

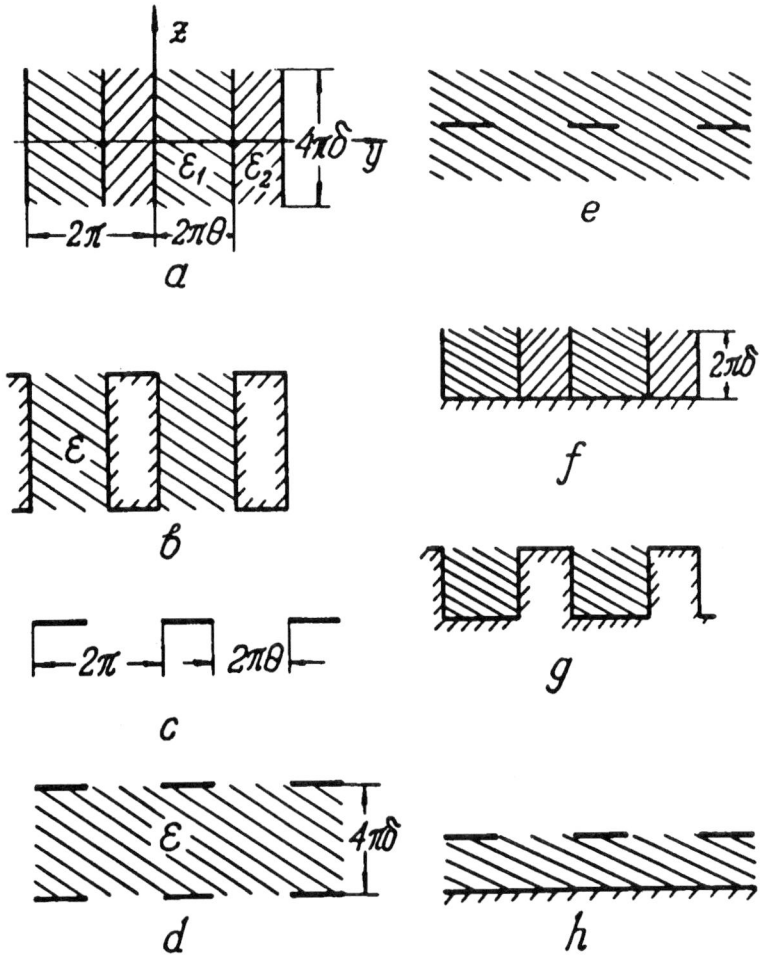

Figure 7.2 *Modifications of diffraction gratings*

The transmission (T) and reflection (R) matrices which determine the amplitude of the incident $[\mathbf{u} = (u_n)]$ and scattered $[\mathbf{a} = (a_n),\ \mathbf{b} = (b_n)]$ fields by the relations

$$\mathbf{a} = R\mathbf{u}, \quad \mathbf{b} = T\mathbf{u} \qquad (7.25)$$

are considered here as bounded finite-meromorphic operator-valued function of $\kappa \in \mathcal{H}$ acting in the space $\tilde{l}_2$. Eqn. 7.25 is reduced to a canonical Fredholm operator equation

$$\mathbf{d} = [R(\kappa)E^{\delta}(\kappa)]^2 \mathbf{d}, \quad \mathbf{d} \in \tilde{l}_2 \qquad (7.26)$$

where $\mathbf{a} = T(\kappa)E^\delta(\kappa)\mathbf{d}$, $\mathbf{b} = T(\kappa)E^\delta(\kappa)R(\kappa)E^\delta(\kappa)\mathbf{d}$ and the diagonal matrix $E^\delta(\kappa) = \|\delta_{mn}\exp(4j\pi\delta\Gamma_n)\|_{m,n=-\infty}^\infty$ defines a compact finite-meromorphic operator-valued function of $\kappa \in \mathcal{H}$ acting in $\tilde{l}_2$. Hence one can generalise theorem 7.10 and establish the spectrum discreteness for the considered types of gratings. The same assertion holds for the possibility to create convergent numerical procedures for calculating the fundamental frequencies.

Let us note in conclusion that all the results obtained in this Section for semi-transparent gratings having the plane of symmetry ($z = 0$) remain valid for reflection gratings when the plane of symmetry is replaced by a perfectly conducting plane.

7.3 Dynamics of the fundamental frequency variation. Classification of free oscillations

We will present in this Section the results of numerical investigations of some model spectral problems for classical waveguide gratings (see Figure 7.2b) in the case of the E-polarisation (corresponding the H-type oscillations when the vector of the electric field has the only non-zero E_{x_1} component). Variation of the fundamental frequencies on the first sheet of $\mathcal{H}$ will be mainly studied since they are situated at the shortest distance from the real axis of the frequency spectral parameter where the monochromatic waves scattering is considered. In particular, the circle $|\kappa| < \kappa_0$ of the first sheet of $\mathcal{H}$ [where $\operatorname{Re}\Gamma_n = 0$, $\operatorname{Im}\Gamma_n > 0$ ($n = 0, \pm1, \pm2, \ldots$) and there is no radiation of energy to the region of large $|x_3|$] is the most perspective domain for the emergence of free oscillations. We will separate in our analysis the oscillations which are symmetric ($\frac{\partial E_{x_1}}{\partial x_3}\big|_{x_3=0} = 0$) or anti-symmetric ($E_{x_1}\big|_{x_3=0} = 0$) with respect to the plane $x_3 = 0$.

The spectral curves 1 and 2 in Figure 7.3 illustrate the behaviour of the real and imaginary parts κ' and κ'' of the complex fundamental frequencies against the relative width of the grating δ (the anti-symmetric and symmetric oscillations are shown in Figures 7.3a and 7.3b, respectively). The bold lines correspond to the first sheet of $\mathcal{H}$, the broken lines with one and two dots correspond to the second and the third sheets, respectively. The numbers of the sheets indicate only the number of the cuts (depicted by bold lines

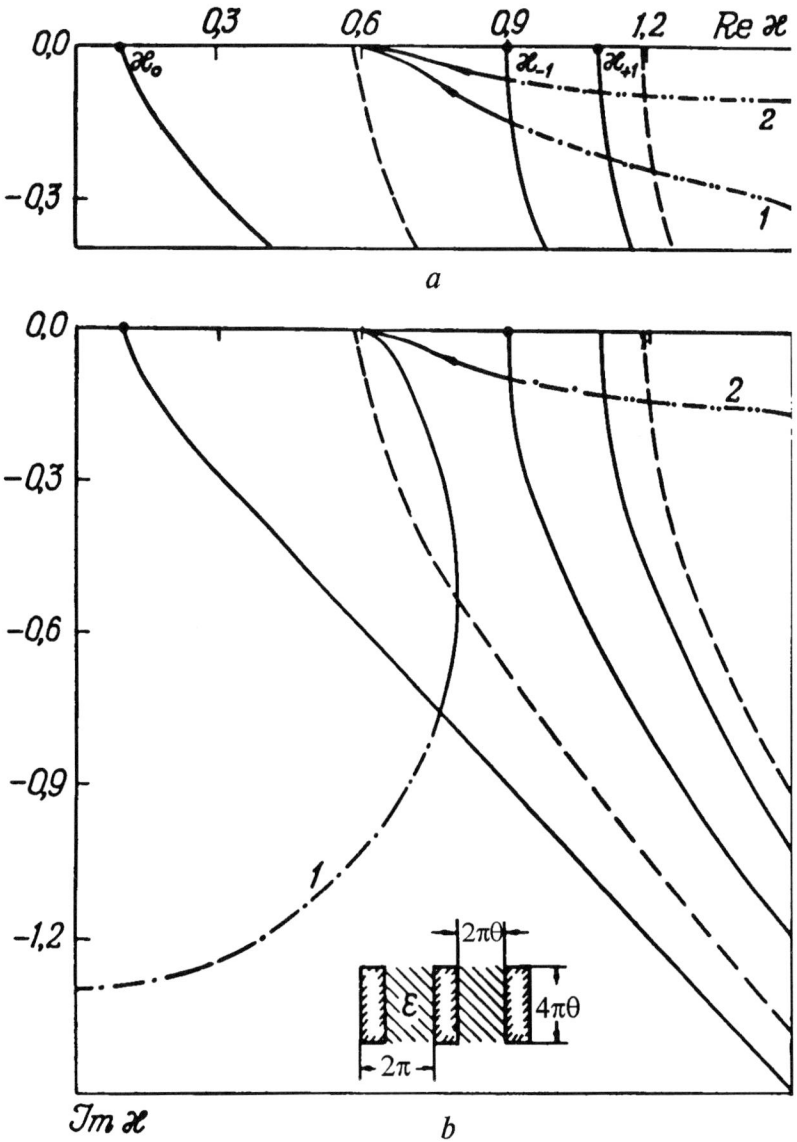

Figure 7.3 *Spectral curves of the first four lowest oscillations of
a grating made of rectangular metal bars against their
height*

starting in the branch points κ_0, $\kappa_\pm$) whose projections are crossed by the curves $\kappa(\delta)$ when they pass from one sheet to another. The curves $\epsilon[(\mathrm{Re}\,\kappa)^2 - (\mathrm{Im}\,\kappa)^2] - m^2(2\theta)^{-2} = 0$, $\mathrm{Re}\,\kappa\,\mathrm{Im}\,\kappa \leq 0$, which start in the points $\mathrm{Re}\,\omega_m = \mathrm{Im}\,\omega_m = 0$ where $\omega_m(\kappa) = \sqrt{\kappa^2\epsilon - m^2(2\theta)^{-2}}$ $(m = 1, 2, \ldots)$ are depicted by dot-dashed lines. The parameters may be chosen in such a way (e.g. $\theta = 0.8, \epsilon = 1.108, \Phi = 0.1$ in Figure 7.3 or $\theta = 0.8, \epsilon = 6.920, \Phi = 0.1$ in Figure 7.6) that exactly m curves $(m = 1, 2, 3)$ starting the m H_{0m} wave threshold points $[\kappa = \kappa_{01} \approx 0.58$ for $m = 1$, $\kappa_{01} \approx 0.58$ and $\kappa_{02} \approx 0.63$ for $m = 2$, (see Figure 7.4 where only the anti-symmetric oscillations are shown) $\kappa_{01} \approx 0.24$, $\kappa_{01} \approx 0.48$ and $\kappa_{02} \approx 0.71$ for $m = 3]$ find themselves in the domain of the first sheet between the cuts connected with κ_0 and $\kappa_\pm$. All the spectral curves situated in this domain do not intersect and tend to the corresponding threshold points when $\delta \to \infty$ and, therefore, for every fixed value of $m = 1, 2, 3$ they may be considered as elements of a single (mth) family of nondegenerate oscillations (in accordance with the index of the threshold point and the number of field's variation along the width of waveguide sections). The oscillations from the second family have much higher diffraction Q-factors (and they essentially grow in the vicinities of intersection points of the spectral curves because in these points the coefficients a_0 and b_0 of the spatial harmonics may vanish) and their spectrum thickens. The growth of Q-factors with decreasing δ is also observed on the second sheet of $\mathcal{H}$ when the spectral curve from the second family crosses the real axis κ and goes to the first quarter of the second sheet where the free oscillation corresponding to this curve do not attenuate in time. Let us note that the passage of spectral curves to the upper half-planes of the sheets of $\mathcal{H}$ does not contradict the uniqueness theorems (see Section 7.1) and the existence of nonattenuating oscillations directly follows from the analysis preceding theorem 7.7.

A superposition of the H_{0m} waves (where $m = 1, 2, 3$ corresponds to the family number) propagating to meet each other in the waveguide communication channels at real frequencies (which are greater than the value of the threshold point) may form the symmetric or anti-symmetric field's distributions (see the isolines $|E_{x_1}(\mathbf{r})| = \mathrm{const}$ in Figure 7.5 where the lines and crosses indicate the direction of amplitude decrease and the maxima of $|E_{x_1}(\mathbf{r})|$, respectively) and at the complex frequencies this mechanism remains approximately the same. The number of field variations in the waveguide apertures

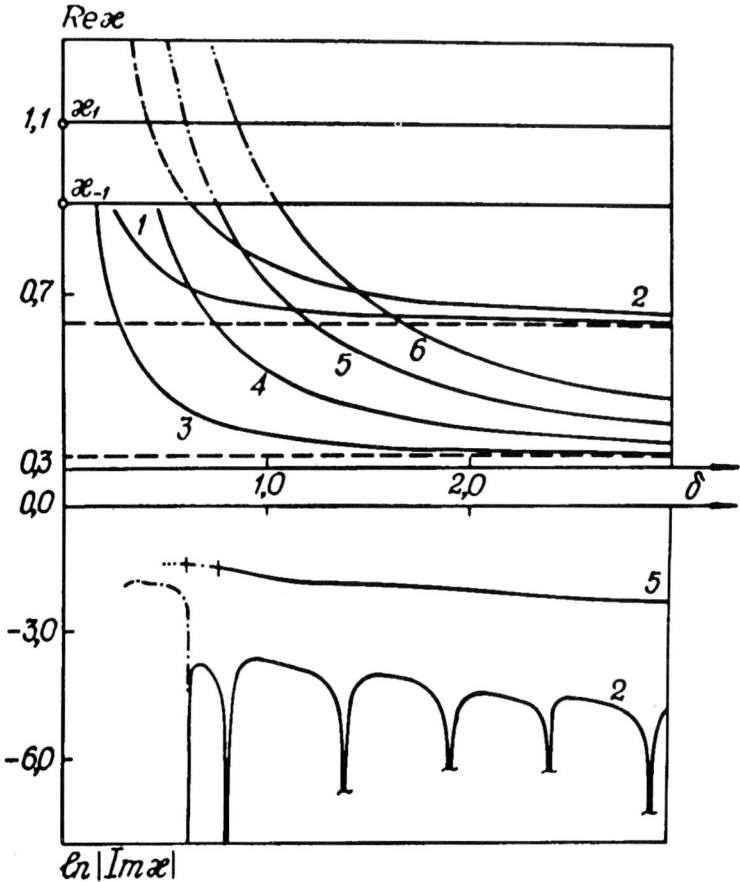

Figure 7.4   *Spectral curves of anti-symmetric oscillations of a bar grating*

is uniquely determined for each of these oscillations by the structure of the H_{01} field. The maximal field concentration is observed close to the plane of symmetry in the communication domain. When δ decreases, the diffraction Q-factor falls ($-\kappa''$ grows), the points of maximal field concentration approach each other, the field's structure becomes less definite and additional maxima of $|E_{x_1}(\mathbf{r})|$ caused by the interference of the zero partial components in the radiation conditions (with the longitudinal wavenumbers $\Gamma_0(x_3)$) propagating in opposite directions and the package of harmonics with $n \neq 0$ appear close to the grating.

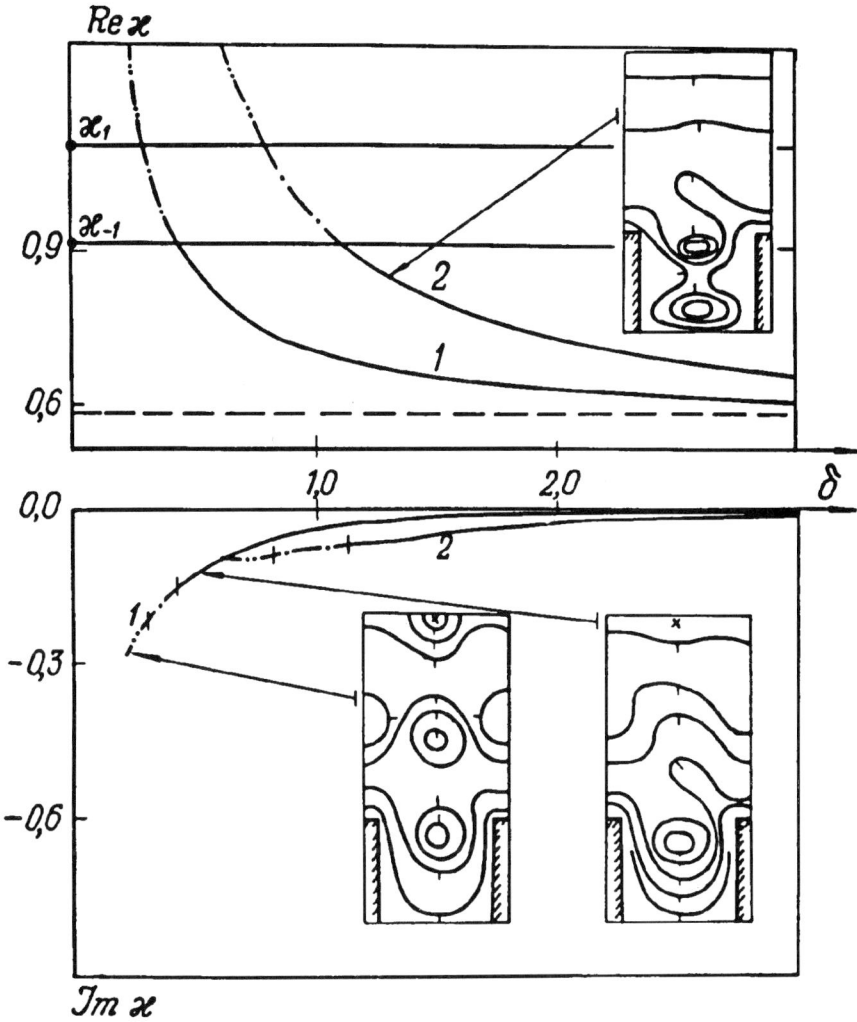

Figure 7.5 *Distribution of E_x-fields over one period for a grating made of rectangular metal bars*

Wave processes in open structures occur in the spatial regions directly adjoining them and admitting only a conditional location in the space. That is why only in a limited number of particular cases one can assert that a resonance takes place in the communication domain and formation of a free oscillation is completely determined by the interaction of the H_{01} waves running

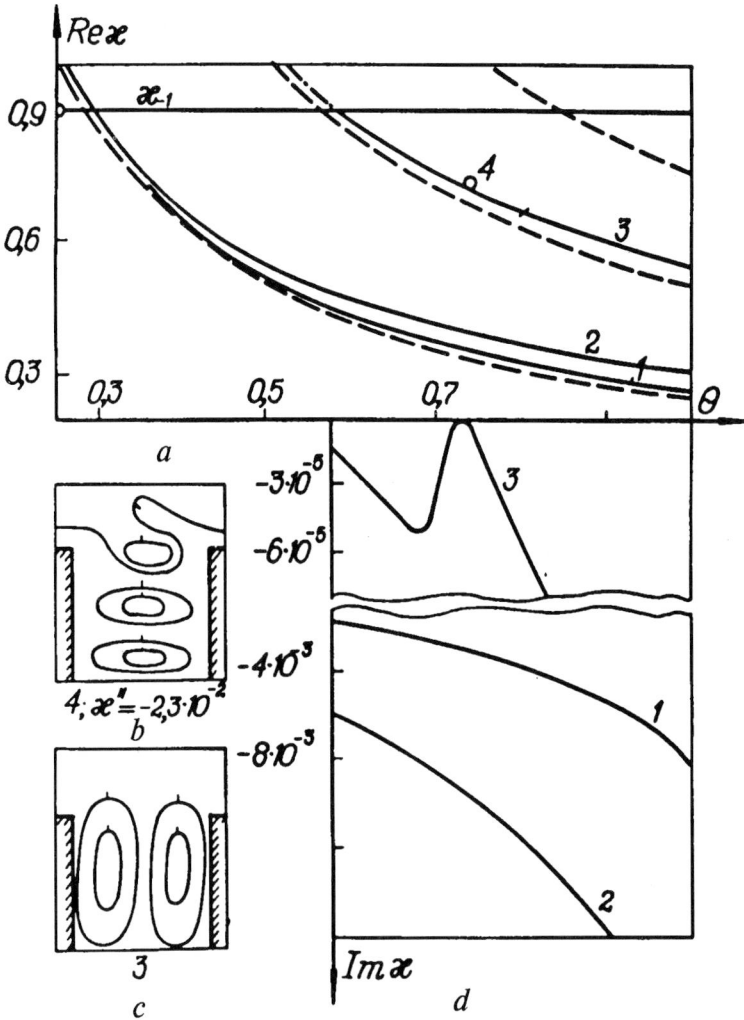

Figure 7.6   *Spectral curves and field distributions for a bar grating*
against the distance between bars

to meet each other without an essential energy exchange with the surrounding space. Such an "ideal" situation (corresponding to the "pure" resonances of the communication domain when the field does not vanish at grating's exit but exponentially falls when moving off the grating) is observed (for example, if $0.278 < \kappa < 0.33$) when δ is sufficiently large and the moduli of the imaginary parts of fundamental frequencies are small (i.e. close to the cut-off frequency). In fact, it follows from the radiation conditions that for large $|x_3|$ the level of $|E_{x_1}(\mathbf{r})|$ is completely determined by $|\operatorname{Im}\Gamma_0(x_3)|)$. The "pure" resonances of the external domain caused by transfomation of the lowest oscillations belonging to different families are also possible.

The fundamental frequencies against the relative width θ of the communication channels are shown in Figure 7.6. The threshold points of the H_{m0} waves are depicted by broken lines and the bold lines 1, 2, 3 correspond to the lowest symmetric and anti-symmetric oscillations belonging to the first family. The $|E_{x_1}(\mathbf{r})|$ distribution is calculated in the point marked by a cross.

The narrower the grating's slots and the less the field variations of a free oscillation along the grating's height, the closer the spectral curves $\operatorname{Re}\kappa$ are to the graphs of threshold points (the former "feel" the directing influence of the latter even for free oscillations with many field variations). A good example of this influence is presented in Figures 7.6a and 7.6b. The isolines $|E_{x_1}(\mathbf{r})| =$ const correspond to the point $\kappa'' = -2.3 \cdot 10^{-2}$ denoted by the number 4. The spectral curves from different families but of equal symmetry intersect close to this point which yields violation of the monotonous behaviour of the curve $\operatorname{Im}\kappa$ from the second family (but not of the curve $\operatorname{Im}\kappa$ from the first family) and a sharp growth of the diffraction Q-factor.

The type of free oscillation combines rather broad characteristics as a "dynamic variable". One can say that a definite type is preserved only in local regions of the parameter's variaton. Its symmetry in particular turns out to be the only stable feature. The correct classification of free oscillations is determined by the position of their fundamental frequencies in the definite region of the definite sheet of $\mathcal{H}$. The uniqueness theorems proved in Section 7.1 enable us to assert that there are no free oscillations whose partial components are "outgoing" ($\operatorname{Re}\Gamma_n \operatorname{Re}\kappa > 0$) or "incoming" ($\operatorname{Re}\Gamma_n \operatorname{Re}\kappa < 0$) plane waves in the domain $|x_3| > 2\pi\delta$. Hence one can expect the existence of wave processes connected with the interaction of waves propagating to meet each other which have been described in this Section. A free oscillation may be also identified by a number of field variations along the height of grating's bars (as done for the families of oscillations described) when the main contribution to the formation of oscillation's field is made by the wave processes (resonances) in the communication domain. An example presented in Figure

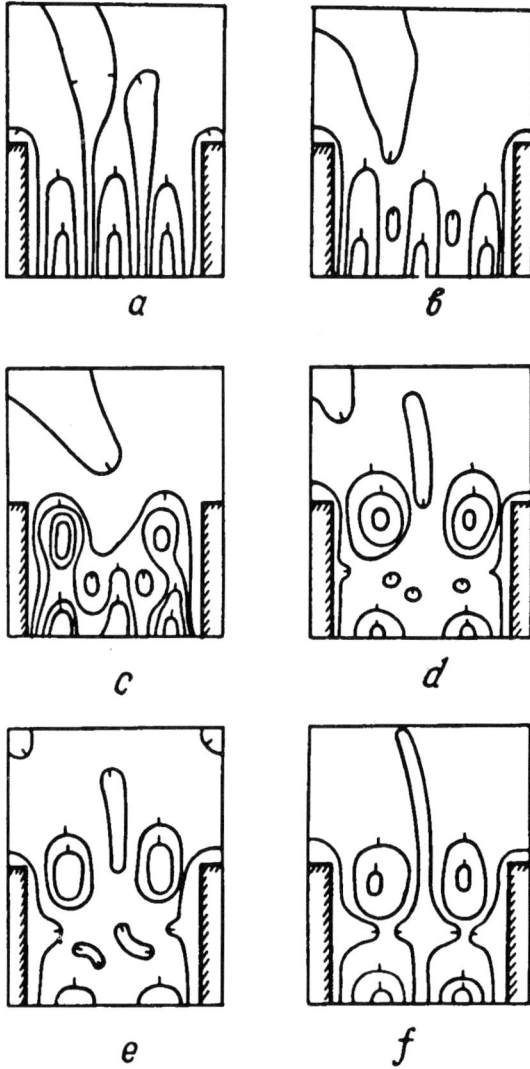

Figure 7.7 *Fields' distributions for the intertype oscillations of a*
bar grating a $\delta = 0.370$ d $\delta = 0.393$
                b $\delta = 0.387$ e $\delta = 0.396$
                c $\delta = 0.390$ f $\delta = 0.410$

7.5 (for $\delta = 0.22$) corresponds to the case when to define the free oscillations
(as the $\hat{H}_{m0n}$, $m = 1$ and $n = 1, 2$) it is sufficient to take into account field dis-

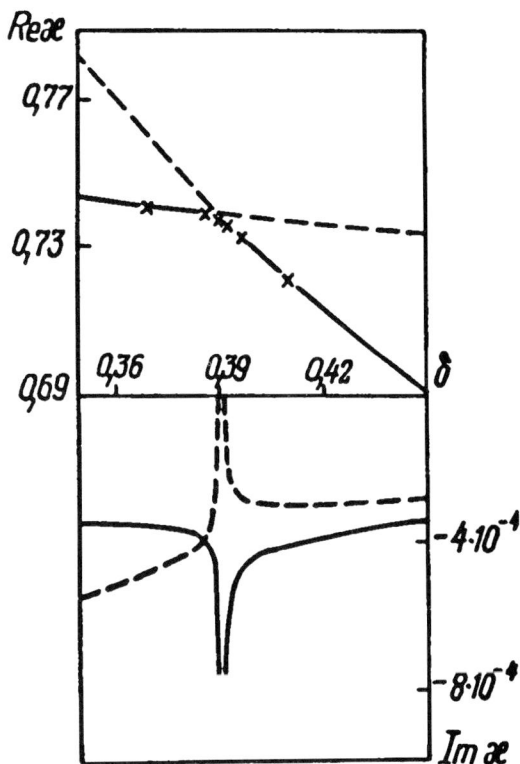

Figure 7.8 *Spectral curves of a bar grating near the point of an intertype oscillation*

tribution (the isolines $|E_{x_1}(\mathbf{r})| = $ const) only in the communication domain.
The $\hat{H}_{m0n}$ oscillations may also have high field concentration under and over
the grating.

The so-called intertype interaction of the different types of free oscillations
may cause the exchange of their characteristics. An example of such an essential exchange which leads to a complete change of the type of a free oscillation
connected with a definite spectral curve ($H_{301} \rightarrow H_{203}$) observed in a small
vicinity of the point $\delta \approx 0.39$ is shown in Figures 7.7 and 7.8 (where the
isolines $|E_{x_1}(\mathbf{r})| = $ const are calculated in the points marked by crosses).

Chapter 8

Stationary resonance wave scattering by gratings

This Chapter is devoted to the study of resonance effects observed in the diffraction frequency domain when various gratings are excited by plane waves. Establishing a reliable qualitative appropriatenesses in this frequency range is possible only on the basis of a big volume of numerical data which have been obtained within the frames of semi-inversion numeric-analytical models and approaches.

8.1 The semi-inversion method in the diffraction problems for a grating with complex period and for an absorbing echelette

The problem of plane wave diffraction on the gratings presented in Figure 8.1 (the statements are given in Section 7.1) is reduced with the help of the procedure described in Section 7.1 to the infinite system of linear algebraic equations

$$A^k \mathbf{a}^k + \sum_{l=1}^{N} B^l \mathbf{a}^l = \mathbf{b}^k \quad (k = 1, 2, \ldots N) \tag{8.1}$$

where $\mathbf{a}^k = (a_n^k)_n \in l_2$ are the unknown expansion coefficients of the sought for function in each of the N regular (partial) domains, A^k and B^k are the matrix operators with the elements $A_{mn}^k = (\Gamma_n^k - \beta_m^k)^{-1}$, $B_{mn}^l = (\Gamma_n^l + \beta_m^k)^{-1} f_{mn}^{k,l}$, β_m^k are the propagation constants in different regular domains, $f_{mn}^{k,l}$ are the

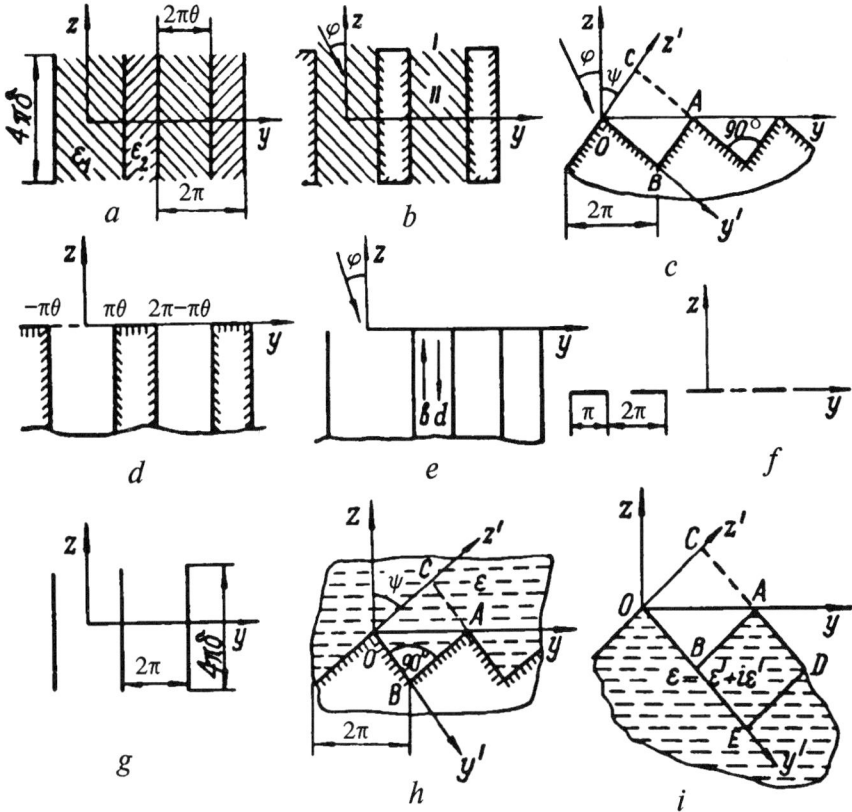

Figure 8.1   *Diffraction gratings for which the numerical-analytic methods for solving the diffraction problems have been elaborated*

given functions.

Before carrying out the semi-inversion of eqn. 8.1 in the general case let us consider the basic example of a grating made of perfectly conducting half-planes $x_2 = 0, \pm 2\pi, \ldots, x_3 \leq 0$. Then eqn. 8.1 takes the form obtained by Mittra and Lee (1971)

$$\sum_{n=-\infty}^{\infty} \frac{a_n}{\Gamma_n - \beta_m} = \frac{1}{\Gamma_0 + \beta_m} \quad (m = 1, 2, \ldots) \tag{8.2}$$

which may be solved explicitly by the complex analysis methods. Here $\kappa = 2\pi\lambda^{-1}$, λ is the wavelength, ϕ is the angle of incidence and

$$\Gamma_n = \sqrt{\kappa^2 - (n + \kappa \sin \phi)^2}, \quad \beta_m = \sqrt{\kappa^2 - 0.25m^2} \quad (\text{Re}\,\sqrt{\cdot} \geq 0, \text{Im}\,\sqrt{\cdot} \geq 0)$$

The operator A of eqn. 8.2 is bounded in l_2 and its right-hand side belongs to the same space. The left $(P = \{P_{mn}\})$ and right $(T = \{T_{mn}\})$ inverse operators to A coincide, $P_{mn} = T_{mn}$ $(n = 1, 2, \ldots, m = 0, \pm 1, \pm 2, \ldots)$ but they are unbounded in l_2 due to the estimate

$$P_{mn} = O\left[\frac{\sqrt{n}}{(\Gamma_m - \beta_n)\sqrt{|m|}}\right] \quad (n, |m| \to \infty) \tag{8.3}$$

obtained by Shestopalov *et al.* (1984). The same features are characteristic for the operator of eqn. 8.1. In order to regularise eqn. 8.1 the system defined by this equation may be rewritten in the equivalent form

$$\mathbf{a}^k + \sum_{l=1}^{N} c^l \mathbf{a}^l = \mathbf{d}^k \quad (k = 1, 2, \ldots N) \tag{8.4}$$

which is reduced to a chain of successively solved equations

$$\mathbf{a} + (R_1 D_1 + P R_2 D_2 R_3 D_3)\mathbf{a} = \mathbf{b} \tag{8.5}$$

where R_l are the scattering matrices which are known explicitly for elementary (canonical) inhomogeneities, D_l $(l = 1, 2, 3)$ are either completely continuous operators with exponentially decreasing matrix elements (if there exists a finite communication domain between elementary inhomogeneities) or unit operators (otherwise), P and $\mathbf{b}$ are the operator and right-hand side obtained as a result of the solution to a preceding equation in the chain.

If D_l are completely continuous operators, eqn. 8.5 is a canonical Fredholm operator equation in l_2 and it is easy to apply and justify a truncation procedure described in previous Sections. If a grating has absorbing inclusions, these operators are even invertible for the real values of κ (since the real axis of the spectral parameter does not contain spectral points).

The main difficulties arise when at least one of D_l is a unit operator.

Let us consider the case when $D_1 = I$ is a unit operator, i.e. the complete operator of eqn. 8.5 has the form $I + R_1 + V$ where V is a completely continuous operator. Then in order to reduce eqn. 8.5 to a canonical Fredholm equation in l_2 it is sufficient to prove that there exists a bounded operator $(I + R_1)^{-1}$. Sometimes it is convenient to use the fact that R_1 is a scattering matrix of an elementary inhomogeneity. As an example, we will treat the problem of the plane E-polarised $[\exp(-i\kappa x_3)]$ wave diffraction on a grating made of "thick" perfectly conducting half-planes $\pi\theta + 2\pi n \leq x_2 \leq -\pi\theta + 2\pi(n+1)$, $0 < \theta < 1$, $x_3 \leq 0$ ($n = 0, \pm 1, \pm 2, \ldots$, see Figure 7.9d). The complex diffraction amplitudes $\mathbf{a} = (a_n)_n \in \tilde{l}_2 : \sum_n |a_n|^2(1+n) < \infty$ in the reflection zone $x_3 > 0$ are determined from the coupled infinite system of linear algebraic equations

$$\sum_{n=-\infty}^{\infty} a_n \cos \pi n\theta \left(\frac{1}{\Gamma_n - \beta_{ml}} - \frac{\beta_l^2}{\Gamma_n + \beta_{ml}} \right) = \frac{1}{\Gamma_0 + \beta_{ml}} - \frac{\beta_l^2}{\Gamma_0 - \beta_{ml}} \quad (8.6)$$

$$(m = 1, 3, 5, \ldots, \; l = 1, 2)$$

$$\Gamma_n = \sqrt{\kappa^2 - n^2}, \quad \beta_{ml} = \sqrt{\kappa^2 - 0.25m^2(\theta + 1 - j)^2}$$

Eqn. 8.6 is reduced by the semi-inversion method to the infinite system

$$c_m = \sum_r f_r(-\beta_{m2})c_r - f(-\beta_{m2}) \quad (m, r = 1, 3, 5, \ldots) \quad (8.7)$$

with respect to the new unknowns $\mathbf{c} = (c_m)_m \in \tilde{l}_2$ such that

$$a_n \cos \pi n\theta = \sum_m \text{res} \, f_m(\Gamma_n)c_m + \text{res} \, f_m(\Gamma_n) \quad (8.8)$$

The meromorphic functions

$$f(w) = \frac{\exp[j(w + \Gamma_0)\psi(\theta)]}{w + \Gamma_0} \prod_{s=0}^{\infty} \left(\frac{\Gamma_s + \Gamma_0}{\Gamma_s - w} \prod_{l=1}^{2} \frac{\beta_{2s+1,l} - w}{\beta_{2s+1,l} + \Gamma_0} \right)$$

$$f_m(w) = -\exp[j(w - \beta_{m2})\psi(\theta)] \prod_{s=0}^{\infty(m)} \left(\frac{\Gamma_s - \beta_{m2}}{\Gamma_s - w} \prod_{l=1}^{2} \frac{\beta_{2s+1,l} - w}{\beta_{2s+1,l} - \beta_{m2}} \right)$$

$$(8.9)$$

are uniquely determined by the given sets of zeros and poles, normalising constants and behaviour at a regular set of contours C_R. Here $\psi(\theta) = \theta \ln \theta - (1-\theta)\ln(1-\theta)$ and the index (m) indicates that the factor $(\beta_{m2} - w)(\beta_{2m} - \beta_{mr})$ in the product is omitted.

The direct solution to the corresponding diffraction problem (see Shestopalov *et al.*, 1984) enables one to conclude that the matrix operator A of eqn. 8.7 with the elements $A_{mr} = f_r(-\beta_{mr})$ ($m, r = 1, 3, 5, \ldots$) coincides with the reflection matrix R arising in the problem of periodical excitation of non-ideal half-planes $x_2 = \pm\pi\theta + 2\pi n$ $x_3 \le 0$ $(n = 0, \pm1, \pm2, \ldots,$ see Figure 7.9e) by the incident waves

$$E_{x_1} \sim \sum_m b_m \exp(jx_3\beta_{2m}) \sin \frac{m(x_2 - \pi\theta + 2\pi n)}{2(1 - \theta)}$$

$$x_3 \le 0, \quad \pi\theta \le x_2 + 2\pi n \le -\pi\theta + 2\pi \quad (n = 0, \pm1, \pm2, \ldots m = 1, 3, 5, \ldots)$$

Then it is easy to present the incident field in each of the exciting waveguides as

$$E_{x_1} = \sum_m [b_m \exp(jx_3\beta_{2m}) - d_m \exp(-jx_3\beta_{2m})] \sin \frac{m(x_2 - \pi\theta + 2\pi n)}{2(1 - \theta)}$$

what yields an operator relation $\mathbf{d} = R\mathbf{b}$ connecting the amplitudes $\mathbf{d} = (d_n)_n$ and $\mathbf{b} = (b_n)_n$. Assuming that $\mathbf{b}$ is an eigenvector of R corresponding to the eigenvalue λ $(\mathbf{b} = \lambda R\mathbf{b})$ and applying the complex power theorem to the total fields in the plane $z = 0$, we finally get the relation

$$\sum_n |t_n| \Gamma_n^* + \theta \sum_m |p_m|^2 \beta_{m1}^* = (1 - \theta) \sum_m |b_m|^2 (1 - |\lambda|^2 + \lambda - \lambda^*) \beta_{m2}^* \quad (8.10)$$

It follows from eqn. 8.10 that $1 - |\lambda|^2 > 0$ and the spectrum of R is situated inside the unit disk (Sanchez-Palencia, 1980). Hence there exists the inverse operator (resolvent) $(I - A)^{-1}$ of eqn. 8.7, its solution may be obtained by the method of successive approximations and the truncation procedures applied for the infinite matrix equation $(I - A)\mathbf{a} = \mathbf{f}$ and based on the use of projectors P_N (see Sections 1.11 and 2.3) will be convergent.

The same approach enables one to estimate the spectrum of a scattering operator for every elementary inhomogeneity and, consequently, to prove resolvent's existence.

The cases when D_2 or all D_l $(l = 1, 2, 3)$ are unit operators require the spectra estimates for the products of scattering operators which may be carried out for every particular (elementary) inhomogeneity.

Let us consider now the specific diffraction problems for waveguide gratings and an absorbing echelette. The problem of the plane E-polarised $\exp\{j\kappa[x_2 \sin\phi - (x_3 - 2\pi\delta)\cos\phi]\}$ wave diffraction on a grating with perfectly conducting inclusions (see Figure 7.9a) is reduced in terms of the complex diffraction amplitudes and with the help of the procedure applied above for solving the problem of diffraction by a system of half-planes to the infinite system of linear algebraic equations

$$\sum_{n=-\infty}^{\infty} c_n^{\pm} A_{nm,l} \left(\frac{1 \mp \lambda_{ml} e_{ml}}{\Gamma_n - \beta_{ml}} - \frac{\lambda_{ml} \mp e_{ml}}{\Gamma_n + \beta_{ml}} \right)$$
$$= A_{0m,l} \left(\frac{1 \mp \lambda_{ml} e_{ml}}{\Gamma_0 - \beta_{ml}} - \frac{\lambda_{ml} \mp e_{ml}}{\Gamma_0 + \beta_{ml}} \right) \qquad (m = 1, 2, 3, \ldots, l = 1, 2)$$

$$(8.13)$$

where

$$A_{nm,l} = \exp[j\Phi_n \pi(2\delta_i^2 - \theta)] - (-1)^m \exp(j\Phi_n \pi\theta)$$

$$\lambda_{ml} = \frac{\tau_{ml} - \beta_{ml}}{\tau_{ml} + \beta_{ml}}$$

$$e_{ml} = \exp(4j\pi\tau_{ml}\delta)$$

$$\tau_{ml} = \sqrt{\kappa^2 \epsilon_l - 0.25m^2(1 - j + \theta)^2}$$

and ϵ_l is the permittivity of dielectric inclusions filling the waveguide channels.

Since the coefficients $A_{nm,l}$ depend on the indices m and j, it is not possible to represent the sums

$$\sum_{n=-\infty}^{\infty} c_n^{\pm} A_{nm,l} \frac{1}{\Gamma_n - \beta_{ml}} \qquad (m = 1, 2, 3, \ldots l = 1, 2)$$

as a pole expansion (similar to eqn. 8.9) of a meromorphic function $f(w)$. In order to overcome this obstacle we will reformulate the problem expanding the tangential components of the total field in the planes $z = \pm 2\pi\delta$ in series in terms of functions

$$\{\sin \frac{1}{2} m(x_2 - \pi\theta)\}_{m=1}^{\infty}$$

which form the complete system over grating's period. Carrying out the semi-inversion enables us finally to obtain the following operator systems

$$\mathbf{t}^{\pm} = BA\mathbf{t}^{\pm} + \sum_{l=1}^{2} D_l^{\pm} \mathbf{v}_l^{\pm} + B\mathbf{u}$$

$$\mathbf{v}^{\pm} = B_l^{\pm} A\mathbf{t}^{\pm} + \sum_{i=1}^{2} T_{li}^{\pm} \mathbf{v}_i^{\pm} + B_l^{\pm}\mathbf{u} \qquad (8.14)$$

$$\mathbf{a} \pm \mathbf{b} = P\mathbf{t}^{\pm} + \mathbf{g}$$

where $\mathbf{u}$, $\mathbf{g}$, A, B, $D_l^{\pm}$, P, $T_{li}^{\pm}$ and $B_l^{\pm}$ are the given vector-functions and matrix operators with the elements

$$A_{mn} = \frac{nF_n \operatorname{res} \phi_{n1}(\gamma_m)}{mF_m}, \quad B_{mn} = \frac{\sin \pi\theta \operatorname{res} \phi_{n2}(\gamma_m)}{\sin m\pi\theta}$$

$$(D_l^{\pm})_{mn} = \pm\frac{(e_{nl} \mp \lambda_{nl}) \operatorname{res} \phi_{n3}^l(\gamma_m)}{\sin m\pi\theta}, \quad (B_l^{\pm})_{mn} = \frac{\sin n\pi\theta \phi_{n2}(-\beta_{ml})}{1 \pm e_{ml}\lambda_{ml}}$$

$$(T_{li}^{\pm})_{mn} = \pm\frac{(e_{ni} \mp \lambda_{ni})\phi_{n3}^l(-\beta_{ml})}{1 \pm e_{ml}\lambda_{ml}}, \quad P_{mn} = \frac{mF_m\phi_{m1}(-\Gamma_n)}{A_n}$$

$$u_m = \frac{A_0 \operatorname{res} f(\gamma_m)}{mF_m}, \quad g_n = \frac{A_0 f(-\Gamma_n)}{A_n}$$

$$f(w) = -\exp[-j(w - \Gamma_0)\ln 2]\prod_{s=1}^{\infty}\frac{(\Gamma_s - w)(\Gamma_{-s} - w)(\gamma_s - \Gamma_0)}{(\Gamma_s - \Gamma_0)(\Gamma_{-s} - \Gamma_0)(\gamma_s - w)}$$

$$\phi_{m3}^l(w) = \exp[2j(w - \beta_{ml})\psi(\theta)]\prod_{s=1}^{\infty}\left(\frac{\gamma_s - \beta_{ml}}{\gamma_s - w}\prod_{i=1}^{2}\frac{\beta_{si} - w}{\beta_{si} - \beta_{ml}}\right)$$

where $m = 1, 2, 3, \ldots$, $n = 0, \pm 1, \pm 2, \ldots$.

In accordance with asymptotic estimates

$$\phi_{m1}(w) = O[w^{0.5}\gamma_m^{-0.5}(w + \gamma_m)^{-1}], \quad \phi_{m2}(w) = O[w^{-0.5}\gamma_m^{0.5}(w + \gamma_m)^{-1}]$$

$$\phi_{m3}^l(w) = O[w^{-0.5}m^{0.5}(w - \beta_{ml})^{-1}], \quad \lambda_{ml} = O(m^{-2})$$

the operators $B_l^{\pm}A$, $D_l^{\pm}$ and $T_{li}^{\pm} : \tilde{l}_2 \to \tilde{l}_2$ are completely continuous and, consequently, eqn. 8.14 is a canonical Fredholm operator equation in $\tilde{l}_2$.

Let us consider the diffraction problem with the third kind boundary conditions taking a "non-ideal" echelette (Figure 7.9c) as a basic example. In order to solve this problem, we will expand the function $u(\mathbf{r})$ satisfying the Helmholtz equation in the domain $0 \leq x_2 \leq 2\pi$, $|x_3| < \delta$ and the boundary-value conditions

$$\alpha u + \beta\frac{\partial u}{\partial \mathbf{n}} = 0, \quad x_2 = 0, 2\pi$$

in the form

$$u(\mathbf{r}) = \sum_{n=0}^{\infty}[c_n \exp(-jx_3\sqrt{\kappa^2 - \mu_n^2}) - d_n \exp(jx_3\sqrt{\kappa^2 - \mu_n^2})]P_n(x_2) \quad (8.15)$$

where

$$P_n(x_2) = \alpha\mu_n^{-1}\sin\mu_n x_2 - \beta\cos\mu_n x_2 \quad (n = 0, 1, 2, \ldots)$$

are the eigenfunctions of the boundary-value problem

$$\frac{d^2 P}{d\,x_2^2} + \mu^2 P = 0$$

$$\alpha P + \beta \frac{d\,P}{d\,x_2} = 0 \quad (x_2 = 0,\ 2\pi) \tag{8.16}$$

corresponding to the (nondegenerated for $\mathrm{Re}(\alpha\beta^{-1}) \neq 0$) eigenvalues $\mu_0 = j\alpha\beta^{-1}$, $2\mu_n = n$.

We will now look for the solution $u(\mathbf{r})$ of the $\exp[j\kappa(x_2 \sin\phi - x_3 \cos\phi)]$ wave diffraction problem on an absorbing echelette (see Figure 7.9c) satisfying the boundary conditions

$$\alpha u + \beta \frac{\partial u}{\partial x_3'} = 0 \quad \text{on} \quad OB$$

$$\gamma u + \delta \frac{\partial u}{\partial x_2'} = 0 \quad \text{on} \quad AB \tag{8.17}$$

in the form

$$u(\mathbf{r}) = \exp[j\kappa(x_2 \sin\phi - x_3 \cos\phi)] \sum_{n=-\infty}^{\infty} a_n \exp[j(\Phi_n x_2 + \Gamma_n x_3)] \quad (x_3 > 0)$$

where $\Phi_n = n + \kappa \sin\phi$, $\Gamma_n = \sqrt{\kappa^2 - \Phi_n^2}$ ($\mathrm{Im}\,\Gamma_n \geq 0$, $\mathrm{Re}\,\Gamma_n \geq 0$) and the diffraction amplitudes $\mathbf{a} = (a_n)_{n=-\infty}^{\infty} \in \tilde{l}_2$ are determined by the boundary and conjugation conditions.

In order to match the fields in the triangle and reflection regions, we will apply the second Green formula in the domain bounded by AB, BO and OA for the solution $u(\mathbf{r})$ and the function

$$G_m(x_2', x_3') = G_m^2(x_2')G_m^1(x_3')$$

$$= \left(\frac{\alpha}{\omega_m} \sin\omega_m x_3' - \beta \cos\omega_m x_3'\right)\left(\frac{\alpha}{\mu_m} \sin\mu_m x_2' - \delta \cos\mu_m x_2'\right) \tag{8.18}$$

where $G_m^1(x_3')$ and $G_m^2(x_2')$ are the elements of the complete systems of eigenfunctions of the boundary-value problems given by eqn. 8.16 on the segments OB and AB with the boundary conditions given by eqn. 8.17. Then the integrals over AB and BO vanish and the integral over AO yields the infinite system of linear algebraic equation with respect to $\mathbf{a}$ which is reduced after the semi-inversion to the following system

$$B_m = \sum_{r=0}^{\infty} B_r \eta_r f_r(-w_m) + f(-w_m) \quad (m = 0, 1, 2, \dots)$$

$$A_n = \sum_{m=0}^{\infty} B_m \eta_m \operatorname{res} f_m(\Gamma_n^+) + \operatorname{res} f_n(\Gamma_n^+) \quad (n = 0, \pm 1, \pm 2, \dots) \tag{8.19}$$

where

$$\Gamma_n^{\pm} = \Gamma_n \cos \psi \pm \Phi \sin \psi, \quad \Phi_n^{\pm} = \Gamma_n \sin \psi \mp \Phi_n \cos \psi$$

$$P_m^{\pm} = \exp[2j\pi(\Phi_0^{\pm} \omega_m \sin \psi - \mu_m \cos \psi)] - 1$$

$$\eta_m = (\beta \omega_m + j\alpha) P_m^+ [(\beta \omega_m - i\alpha) P_m^-]^{-1}$$

$$A_n = a_n \frac{\gamma_n - j\delta\Phi_n^+}{\gamma_n + j\delta\Phi_0}$$

$$f(w) = \frac{(\Gamma_0^+ + \Gamma_0^-) \exp[j(w + \Gamma_0^-)t(\psi)]}{(\Gamma_0^+ - w)(\Gamma_0^- + w)(\Gamma_0^- + \omega_0)} \prod_{s=1}^{\infty} \frac{(\Gamma_0^+ + \Gamma_0^-)(\Gamma_{-s}^+ + \Gamma_0^-)(w_s - w)}{(\Gamma_s^+ - w)(\Gamma_{-s}^+ - w)(w_s + \Gamma_0^-)}$$

$$f_m(w) = \frac{(\Gamma_0^+ - \omega_m)(\omega_0 - w) \exp[j(w - \omega_m)t(\psi)]}{(\Gamma_0^+ - w)(\omega_0 - \omega_m)}$$

$$\times \prod_{s=1}^{\infty (m)} \frac{(\Gamma_s^+ - \omega_m)(\Gamma_{-s}^+ - \omega_m)(\omega_s - w)}{(\Gamma_s^+ - w)(\Gamma_{-s}^+ - w)(\omega_s \omega_m)}$$

Here

$$t(\psi) = 2[\cos \psi \ln 2 \cos \psi - \psi \sin \psi]$$

and the index (m) indicates that the factor $(\omega_s - w)(\omega_s - \omega_m)^{-1}$ in the product is omitted.

Due to the asymptotic estimates

$$f_m(w) = O[|w|^{0.5} \omega_m^{-0.5} (\omega_m - w)^{-1} \exp(-2\pi\omega_m \sin \psi)]$$

$$f(w) = O(w^{-0.5})$$

which hold for $m, |w| \to \infty$ the operators of the problem given by eqn. 8.19 are completely continuous in $\tilde{l}_2$ and, consequently, the complete operator of eqn. 8.19 admits a canonical Fredholm representation

$$I + Q = I + (Q + P_1)(I + P_2) = I + Q_1 + P$$

where

$$Q = \{\eta_r f_r(-\omega_m)\}_{m,r=0}^{\infty}, \quad I + P_2 = \{\delta_m^r (\beta \omega_m + j\alpha)(\beta \omega_m - j\alpha)^{-1}\}_{m,r=0}^{\infty}$$

the operator Q_1 coincides with the matrix describing the reflection from a grating made of perfectly conducting half-planes continuing echelette's sides (in the case of quasi-periodical excitation by the complete set of the waveguide TE modes) to the waveguide channel. Such a choice of Q_1 enables us to prove that the operators P_1, P_2 and P are completely continuous in $\tilde{l}_2$ and there exist bounded operators $(I+Q_1)^{-1}$ and $(I+Q)^{-1}$ (the spectrum of Q_1 may be estimated with the help of the complex power theorem applied in the domain AOB because Q_1 is a scattering operator for an elementary inhomogeneity). Hence we have proved solvability of the problem given by eqn. 8.19 in the space $\tilde{l}_2$.

8.2 Analytical solution of two classical problems of the dynamical theory of gratings

Two example diffraction problems which admit analytical solution on the basis of the analysis of the Wiener-Hopf-Fock equation will be treated in this Section. First, we will reduce the problem of the plane E-polarised $\exp[j\kappa(x_2 \sin \phi - x_3 \cos \phi)]$ wave diffraction on a strip grating with perfectly conducting strips (see Figure 7.9f) to the Wiener-Hopf-Fock equation by matching the tangential field's components in the plane $x_3 = 0$. The initial system of functional equations has the form

$$
\begin{cases}
\displaystyle\sum_{n=-\infty}^{\infty} b_n \exp(j\Phi_n x_2) = 0, & |x_2| < \frac{\pi}{2} \\[4mm]
\displaystyle\sum_{n=-\infty}^{\infty} b_n \Gamma_n \exp(j\Phi_n x_2) = \Gamma_0 \exp(j\Phi_0 x_2), & \frac{\pi}{2} < |x_2| < \frac{3\pi}{2}
\end{cases}
\tag{8.20}
$$

where $\mathbf{b} = (b_n)_n$, $b_n \approx O(|n|^{-1.5})$ are the unknown complex diffraction amplitudes in the transmission zone connected with the diffraction amplitudes $\mathbf{a} = (a_n)_n$, $a_n \approx O(|n|^{-1.5})$ in the reflection zone by the relation $b_n = a_n + \delta_{n0}$, the quantities Φ_n and Γ_n are introduced in Section 8.1 (see eqn. 8.17).

Let us transfer eqn. 8.20 to the complex plane w carrying out the $\exp(-iwx_2)$ transformation

$$
\sum_{n=2m} \frac{B_n}{v_n \mp v} + \cot (w - \Phi_0) \sum_{n=2m+1} \frac{B_n}{v_n \pm v} = -\frac{1}{2(v_2 \pm v)}
\tag{8.21}
$$
$$
(m = 0, \pm 1, \pm 2, \dots)
$$

and consider eqn. 8.21 with respect to the new complex variable

$$v = v(w) = \sqrt{\frac{\kappa - w}{\kappa + w}} \tag{8.22}$$

establishing the conformal mapping of the two-sheet Riemann surface H_g of the function $g(w) = \sqrt{\kappa^2 - w^2}$ to the complex plane C_v of the variable v such that $C_v = L_+ \cup L_-$ where $L_\pm$ are the images of the upper $(\operatorname{Im} g(w) > 0)$ and lower $(\operatorname{Im} g(w) < 0)$ sheets of H_g. Here

$$B_n = b_n \left(\cos \frac{\pi n}{2} + \sin \frac{\pi n}{2} \right) + \frac{1}{2} \delta_{n0}, \quad v_n = v(\Phi_n) \tag{8.23}$$

and $v_n \in L_+$ due to the conditions $\operatorname{Im} \Gamma_n \geq 0$, $\operatorname{Re} \Gamma_n \geq 0$ $(n = 0, \pm 1, \pm 2, \ldots)$.

The functions

$$f_-(v) = \sum_{n=2m} \frac{B_n}{v_n - v}, \quad f_+(v) = \sum_{n=2m+1} \frac{B_n}{v_n + v}, \quad \psi_+(v) = -\frac{1}{2(v_2 + v_0)}$$

are holomorphic in $L_\pm$ since the corresponding series uniformly converge in $L_\pm$ and for $\operatorname{Im} \kappa > 0$ they have a common regularity strip in the vicinity of the contour L (separating the domains $L_\pm$). In addition to this, $f_\pm$ tend to zero for $|v| \to \infty$ in $L_\pm$ as $|v|^{-1-\sigma}$ for a certain positive value of σ and f_+ has a simple pole in the point $v = -v_0$.

Hence eqn. 8.21 may be considered as a functional Wiener-Hopf-Fock equation which may be solved explicitly by factorising the function

$$G(v) = \frac{\pi}{2} \cot (w - \Phi_0) = G_+(v) G_-(v)$$

where

$$G_+(v) = G_-(-v) = \sqrt{\frac{\pi}{2} \cot (\kappa^2 - \Phi_0^2)} \prod_{n=2m} \left(1 + \frac{v}{v_n} \right)^{-1} \prod_{n=2m+1} \left(1 + \frac{v}{v_n} \right)$$

The solution enables us to determine the functions

$$f_+(v) = -[2(v_0 + v) G_+(v) G_-(v)]^{-1}, \quad f_-(v) = \frac{1}{2(v_0 + v)} \left[\frac{G_-(v)}{G_+(-v_0)} - 1 \right]$$

and to find the unknowns B_n by calculating the residues

$$B_{2m+1} = \operatorname{res} f_+(-v_{2m+1}), \quad B_{2m} = -\operatorname{res} f_-(v_{2m}) \quad (m = 0, \pm 1, \pm 2, \ldots)$$

To justify the latter it is sufficient to check that the integrals

$$\frac{1}{2\pi i}\int_{C_R}\frac{f_\pm}{v-\alpha}\,d\alpha = \sum_{n=2m(2m+1)}\frac{\mathrm{res}\,f_\pm(\mp v_n)}{\pm v_n - \alpha} + f_\mp(\alpha)$$

vanish when the radius R of a sequence of regular contours C_R tends to infinity.

The second example problem is connected with reconstructing a meromorphic function by a set of zeros which is given implicitly. Let us consider in this way the problem of the plane E-polarised $\exp\{j\kappa[x_2\sin\phi - (x_3 - 2\pi\delta)\cos\phi]\}$ wave diffraction on a grating presented in Figure 7.9h which is reduced by matching the tangential field's components to the system of functional equations

$$\sum_{n=-\infty}^{\infty}(a_n \pm b_n)\cos\pi n\left[\frac{1}{\Gamma_n - \gamma_m} \pm \frac{\exp(4j\pi\delta\gamma_n)}{\Gamma_n + \gamma_m}\right] = \frac{1}{\Gamma_0 + \gamma_m} \pm \frac{\exp(4j\pi\delta\gamma_m)}{\Gamma_0 - \gamma_m}$$

$$(m = 1, 2, 3, \ldots)$$

(8.24)

where $\mathbf{a} = (a_n)_n$ and $\mathbf{b} = (b_n)_n$ are the diffraction amplitudes in the reflection and transmission zones, $\gamma_n = \sqrt{\kappa^2 - 0.25n^2}$ and other quantities have been introduced above.

The change of variables $(a_n \pm b_n)\cos\pi n = A_n^\pm + \mathrm{res}\,f(\Gamma_n)$ where $f(w)$ is uniquely determined by the following conditions: $f(w) = O(|w|^{-1.5})$ $(|w| \to \infty)$, $f(w)$ has simple poles at the points $w = \Gamma_n$ $(n = 0, \pm 1, \pm 2, \ldots)$, $\mathrm{res}\,f(-\Gamma_0) = 1$ and $f(\Gamma_m) = 0$ $(m = 1, 2, 3, \ldots)$ yields the equivalent system

$$\sum_{n=-\infty}^{\infty}A_n^\pm\left[\frac{\exp(-2j\pi\delta\gamma_m)}{\Gamma_n - \gamma_m} \pm \frac{\exp(2j\pi\delta\gamma_m)}{\Gamma_n + \gamma_m}\right] = \mp f(-\gamma_m)\exp(2j\pi\delta\gamma_m)$$

$$(m = 1, 2, 3, \ldots)$$

(8.25)

Assume that one knows the functions $F^\pm(w)$ satisfying the conditions $F^\pm(w) \to 0$ when $|w| \to \infty$ over a sequence of regular contours, $F^\pm(w)$ has simple poles in the points $w = \Gamma_n$ $(n = 0, \pm 1, \pm 2, \ldots)$ and

$$F^\pm(w)\exp(-2j\pi\delta\gamma_m) \pm F^\pm(-w)\exp(2j\pi\delta\gamma_m) = \pm f(-\gamma_m)\exp(2i\pi\delta\gamma_m)$$

Then

$$A_n^\pm = \mathrm{res}\,F^\pm(\Gamma_n) \quad (n = 0, \pm 1, \pm 2, \ldots)$$

(8.26)

To construct the family $F^{\pm}(w)$ it is sufficient to determine the function $\psi^{\pm}(w)$ such that

$$F^{\pm}(w)\exp(-2j\pi\delta w) \pm F^{\pm}(-w)\exp(2j\pi\delta w) = \psi^{\pm}(w) \qquad (8.27)$$

satisfying the relations

$$\psi^{\pm}(\gamma_n) = \pm f(-\gamma_m)\exp(2j\pi\delta\gamma_m), \quad \psi^{\pm}(-\gamma_n) = f(-\gamma_m)\exp(2j\pi\delta\gamma_m)$$
$$(m = 1, 2, 3, \ldots)$$

and having simple poles in the points $w = \pm\Gamma_n$ $(n = 0, \pm1, \pm2, \ldots)$. The required functions are represented explicitly:

$$\psi^{\pm}(w) = \sum_{s=1}^{m} [\pm(\gamma_s + w)\exp(-2j\pi\delta w) + (\beta_s - w)\exp(2j\pi\delta w)]\,\phi_s(w)$$

where

$$\phi_s(w) = \frac{\exp(4j\pi\delta\gamma_s)f(-\gamma_s)\sin\left(2\pi\sqrt{\kappa^2-w^2}\right)(\kappa^2-\gamma_s^2)[\cos\left(2\pi\kappa\sin\phi\right)-\cos s\pi]}{2\pi\gamma_s\cos s\pi\sqrt{\kappa^2-w^2}(\gamma_s^2-w^2)[\cos\left(2\pi\kappa\sin\phi\right)-\cos 2\pi\sqrt{\kappa^2-w^2}]}$$

The functions $\psi^{\pm}(w)$, $F^{\pm}(w)\exp(-2j\pi\delta w)$ and $F^{\pm}(-w)\exp(2j\pi\delta w)$ have a common regularity strip for $\operatorname{Im}\kappa > 0$ in a vicinity of the real axis $\operatorname{Im} w = 0$ and the second and the third functions are holomorphic in the lower and upper half-planes, respectively. Therefore, one can consider eqn. 8.27 as a functional Wiener-Hopf-Fock equation whose solution may be obtained for $\operatorname{Im} w < 0$ in the form of a standard expansion

$$F^{\pm}(w)\exp(-2j\pi\delta w) = \pm\sum_{s=1}^{\infty}(\gamma_s + v)\exp(-2j\pi\delta w)\phi_s(w)$$

$$-\sum_{p=-\infty}^{\infty}\exp(2j\pi\delta\Gamma_p)\left(\frac{1}{\Gamma_p - w} \mp \frac{1}{\Gamma_p + w}\right)\sum_{s=1}^{\infty}(\gamma_s - \Gamma_p)\operatorname{res}\phi_s(\Gamma_p)$$
$$(8.28)$$

It follows from eqn. 8.28 that

$$F^{\pm}(-\gamma_m)\exp(2j\pi\delta\gamma_m) = -\sum_{p=-\infty}^{\infty}\exp(2j\pi\delta\Gamma_p)\left(\frac{1}{\Gamma_p - \gamma_m}\right.$$

$$\left.\mp\frac{1}{\Gamma_p + \gamma_m}\right)\sum_{s=1}^{\infty}(\gamma_s - \Gamma_p)\operatorname{res}\phi_s(\Gamma_p)$$

and, consequently,

$$F^{\pm}(\gamma_m) = \pm\exp(4j\pi\delta\gamma_m)[f(\gamma_m) - F^{\pm}(-\gamma_m)] \quad (m = 1, 2, 3, \ldots)$$

what enables finally to obtain the explicit representation

$$F^{\pm}(w) = \sum_{m=1}^{\infty} \frac{\Gamma_0 - \gamma_m}{\Gamma_0 - w}\exp[2j\ln 2(w - \gamma_m)]F^{\pm}(\gamma_m)$$

$$\prod_{s=0}^{\infty(n)} \frac{(\gamma_n - w)(\Gamma_n - \gamma_m)(\Gamma_{-n} - \gamma_m)}{(\Gamma_n - w)(\Gamma_{-n} - w)(\gamma_n - \gamma_m)}$$

where the index (m) indicates that the factor $(\gamma_n - w)(\gamma_n - \gamma_m)^{-1}$ with $n = m$ in the product is omitted.

Determination of the diffraction amplitudes is thus reduced to calculating the residues of $F^{\pm}(w)$ with the use of eqn. 8.26.

8.3 Infinite systems of the mode matching method

In this Section we will determine the rules for optimal choices of the truncation parameters for the methods of numerical solution of the infinite systems arising in the diffraction problems studied in the present Chapter.

Let us consider the problem of the plane E-polarised $\exp[j\kappa(x_2\sin\phi - x_3\cos\phi)]$ wave diffraction on a grating made of perfectly conducting half-planes with complex structure of the period (see Figure 7.9e). The corresponding first-kind infinite system of the mode matching method has the form

$$\sum_{n=-\infty}^{\infty} a_n \frac{A_{nm,l}}{\Gamma_n - \beta_{ml}} = \frac{A_{0m,l}}{\Gamma_0 + \beta_{ml}} \quad (m = 1, 2, 3, \ldots, \; l = 1, 2) \qquad (8.29)$$

where $\mathbf{a} = (a_n)_n \in \tilde{l}_2$ are the unknown complex diffraction amplitudes and all the notations are the same as in Section 8.1.

Assume at first that the angle of incidence satisfies the condition $\kappa\sin\phi = p$ $(p = 0, \pm1, \pm2, \ldots)$. Then it is convenient to represent every amplitude as a sum of symmetric and anti-symmetric components $a_n = a_{n1} + a_{n2}$ (with respect to the index $n = -p$) where $a_{n1} = a_{-n-2p,1}$, $a_{n2} = a_{-n-2p,2}$ and to rewrite eqn. 8.29 in the form

$$\sum_{n=-p}^{\infty} a_{n1} \frac{2 - \delta_{n,-p}}{\Gamma_n - \beta_{ml}} \cos \Phi_n \pi\theta = \frac{\cos \Phi_0 \pi\theta}{\Gamma_0 + \beta_{ml}} \quad (m = 1, 3, \ldots) \qquad (8.30)$$

$$2 \sum_{n=-p+1}^{\infty} a_{n2} \frac{\sin \Phi_n \pi\theta}{\Gamma_n - \beta_{ml}} = \frac{\sin \Phi_0 \pi\theta}{\Gamma_0 + \beta_{ml}} \quad (m = 2, 4, \ldots) \qquad (8.31)$$

The systems given by eqns. 8.30 and 8.31 may be solved explicitly by the complex analysis methods if to consider (as well as in the previous Section) the series as meromorphic functions with given poles and zeros having definite behaviour at infinity (on a system of regular contours):

$$(2 - \delta_{n,-p})a_{n1} \cos \Phi_n \pi\theta = \text{res } f_1(\Gamma_n), \quad 2a_{n2} \sin \Phi_n \pi\theta = \text{res } f_2(\Gamma_n) \qquad (8.32)$$

where

$$f_1(w) = \frac{\Gamma_0 + \Gamma_{-p}}{(\Gamma_0 + w)(\Gamma_{-p} - w)} \exp[j\psi(\theta)(w + \Gamma_0)] \cos \Phi_0 \pi\theta$$

$$\times \prod_{s=1}^{\infty} \frac{(\beta_{2s-1,1} - w)(\beta_{2s-1,2} - w)(\Gamma_{-p+s} + \Gamma_0)}{(\beta_{2s-1,1} + \Gamma_0)(\beta_{2s-1,2} + \Gamma_0)(\Gamma_{-p+s} - w)}$$

$$f_2(w) = \frac{\sin \Phi_0 \pi\theta}{(\Gamma_0 + w)} \exp[j\psi(\theta)(w + \Gamma_0)]$$

$$\times \prod_{s=1}^{\infty} \frac{(\beta_{2s,1} - w)(\beta_{2s,2} - w)(\Gamma_{-p+s} + \Gamma_0)}{(\beta_{2s,1} + \Gamma_0)(\beta_{2s,2} + \Gamma_0)(\Gamma_{-p+s} - \Gamma_0)}$$

Let us "cut" the system given by eqn. 8.31

$$2 \sum_{n=-p+1}^{N-p} a_{n2}^N \frac{\sin \Phi_n \pi\theta}{\Gamma_n - \beta_{ml}} = \frac{\sin \Phi_0 \pi\theta}{\Gamma_0 + \beta_{ml}}$$

$$m = \begin{cases} 2, 4, \ldots, 2M, & l = 1 \\ 2, 4, \ldots, 2L, & l = 2 \end{cases} \quad (M + L = N) \qquad (8.33)$$

and represent the solution in the form

$$2a_{n2}^N \sin \Phi_n \pi\theta = \text{res } f_2^N(\Gamma_n)$$

where

$$f_2^N(w) = \frac{\sin \Phi_0 \pi\theta}{(\Gamma_0 + w)} \prod_{s=1}^{N} \frac{\Gamma_{-p+s} + \Gamma_0}{\Gamma_{-p+s} - w} \prod_{s=1}^{M} \frac{\beta_{2s,1} - w}{\beta_{2s,1} + \Gamma_0} \prod_{s=1}^{L} \frac{\beta_{2s,2} - w}{\beta_{2s,2} + \Gamma_0}$$

and this representation is possible due to the estimate $f_2^N(w) = O(|w|^{-1})$ ($|w| \to \infty$).

If, e.g. $L < M$, then the following asymptotic estimate holds

$$
\begin{aligned}
\frac{a_{n2}^N}{a_{n2}} = \exp\Bigg\{ & j(\Gamma_0 + \Gamma_n)\Bigg[(1 - \theta)\ln\left(1 + \frac{\epsilon_N}{1 - \theta}\right) \\
& + \theta\ln\left(1 - \frac{\epsilon_N}{\theta}\right) + \frac{1}{2N} + \frac{(L - M)\epsilon_N}{4LM}\Bigg]\Bigg\} \\
& \times \exp\left\{\frac{\Gamma_n^2 - \Gamma_0^2}{2}\left[\frac{1}{N(N-1)} + \frac{N\epsilon_N^2}{LM}\right]\right\} + O(N^{-3})
\end{aligned}
\tag{8.34}
$$

where $\epsilon_N = \theta - \frac{M}{N} = \frac{L}{N} - (1 - \theta)$. It is clear that the ratio given by eqn. 8.34 asymptotically (for $N \to \infty$) tends to unit if and only if $\epsilon_N \to 0$ and, consequently, (approximate) solutions of eqn. 8.33 will converge to an exact solution of eqn. 8.31 if the truncation order is chosen in accordance with the limiting condition

$$
\lim_{N \to \infty} \frac{M}{L} = \frac{\theta}{1 - \theta}
\tag{8.35}
$$

Assume that $|\epsilon_N| \ll 1$. Then it follows from eqn. 8.34 (for $\operatorname{Im}\Gamma_n = 0$) that

$$
\begin{aligned}
\arg a_{n2}^N - \arg a_{n2} &= (\Gamma_0 + \Gamma_n)\left[\frac{1}{2N} + \frac{(L - M)\epsilon_N}{4LM} - \frac{\epsilon_N^2}{\theta(1 - \theta)}\right] \\
&\quad + O(N^{-3}) + O(\epsilon_N^3)
\end{aligned}
\tag{8.36}
$$

$$
\frac{|a_{n2}^N| - |a_{n2}|}{|a_{n2}|} = \frac{\Gamma_n^2 - \Gamma_0^2}{2}\left[\frac{1}{N(N-1)} - \frac{N\epsilon_N^2}{LM}\right] + O(N^{-3})
$$

and hence the error of determination of the nth amplitude decreases one order greater than that of the corresponding phase. But for attenuating harmonics ($\operatorname{Im}\Gamma_n > 0$) the rates of error's decrease for phases and amplitudes become equal and the phase error does not depend on the harmonic index (within the limits of asymptotic estimates given by eqn. 8.36).

Assume now that $\epsilon_N = O(1)$ and $N \to \infty$. Then for $\operatorname{Im}\Gamma_n = 0$ $|a_{n2}^N|$ will tend to $|a_{n2}|$ but the rate of convergence will be one order less than that in the previous case while the phase error

$$
(\Gamma_0 + \Gamma_n)\left[(1 - \theta)\ln\left(1 + \frac{\epsilon_N}{1 - \theta}\right) + \theta\ln\left(1 - \frac{\epsilon_N}{\theta}\right)\right] + O(N^{-2})
$$

will be practically the same.

The same conclusions are valid for the system given by eqn. 8.30. In fact, assuming that eqn. 8.35 holds and returning from this system to the initial one given by eqn. 8.29, the corresponding "correctly" truncated system will have the form

$$\sum_{n=-N-p}^{N-p} a_n^N \frac{A_{nm,l}}{\Gamma_n - \beta_{ml}} = \frac{A_{0m,l}}{\Gamma_0 + \beta_{ml}}$$

$$m = \begin{cases} 2, 4, \ldots, 2M, & l = 1 \\ 2, 4, \ldots, 2L, & l = 2 \end{cases} \quad (M + L = 2N + 1)$$

(8.37)

and it is easy to see (repeating the above considerations) that such a way of truncation provides the convergence $|a_n^N - a_n| \to 0$ $(N \to \infty)$ and the fulfillment of asymptotic conditions given by eqns. 8.34 and 8.36.

The present results essentially specify the methods of constructing the truncated systems. In particular, they confirm that the energy scattering characteristics are much more stable with respect to the methods of truncation than the phase properties and the energy balance cannot be a unique criterion for estimating the accuracy of approximate solutions to the diffraction problems.

Let us consider the case when $\kappa \sin \phi$ is an arbitrary real number. Then the problem given by eqn. 8.29 may be reduced by the semi-inversion method to the equivalent second kind system

$$\mathbf{t} = BA\mathbf{t}, \qquad \mathbf{a} = P\mathbf{t} + \mathbf{g} \tag{8.38}$$

The corresponding truncated system has the form

$$\mathbf{t}_N = B_N A_N \mathbf{t}_N + B_N \mathbf{u}_N \tag{8.39}$$

where $\mathbf{t}_N = (t_m^N)_{m=1}^N$, $\mathbf{u}_N = (d_m^N)_{m=1}^N$,

$$u_m^N = -\frac{A_0}{mF_m} \operatorname{res} f^N(\gamma_m)$$

$$A_N = \left(\frac{nF_n}{mF_m} \operatorname{res} \phi_{m1}^N(\gamma_m) \right)_{m,n=1}^N \qquad B_N = \left(\frac{\sin n\pi\theta}{\sin m\pi\theta} \operatorname{res} \phi_{m2}^N(\gamma_m) \right)_{m,n=1}^N$$

where $\phi_{ml}^N(w)$ and $f^N(w)$ are the rational functions of the complex variable w analogous to the "cut" meromorphic functions $f_{1,2}(w)$ and $\phi_{nl}(w)$ introduced in this Section.

The estimate

$$
\begin{aligned}
R = \frac{f^N(w)}{f(w)} &= \exp\left\{ j(w - \Gamma_0)\left[2\ln 2 + \ln\left(1 - \frac{2(\bar{p} - p)}{N - \bar{p} + p}\right)\right.\right. \\
&- 2\ln\left(2 + \frac{2(1 + 2p - 2\bar{p})}{N - 1 - 2p + 2\bar{p}}\right) - \frac{2}{N} \\
&\left.\left. + \frac{N - 1}{N^2 - (p - \bar{p})^2} + O(N^{-2})\right]\right\}
\end{aligned}
\tag{8.40}
$$

(where $\bar{p} = $ entier $\kappa \sin \phi$) yields

$$
\frac{f^N(w)}{f(w)} \to 0 \quad (N \to \infty)
$$

if $\bar{p} - p \ll N$ and the error reaches the smallest value when $\bar{p} = p$, as well as the relation

$$
R_l = \frac{\phi_{nl}^N(w)}{\phi_{nl}(w)}, \quad l = 1
$$

[due to approximation of $f(w)$ by a rational function $f^N(w)$]. In fact,

$$
\begin{aligned}
R &= 1 + \frac{2j(w - \Gamma_0)}{N - 1} O(N^{-2}) \to 1 \\
R_l &= 1 + \frac{2j(w + \gamma_n)}{N - 1} O(N^{-2}) \to 1, \quad (l = 1, 2)
\end{aligned}
\tag{8.41}
$$

if $p = $ entier $\kappa \sin \phi$ and eqn. 8.35 holds. The latter determines the rule for truncating the initial first kind systems (given by eqn. 8.29) which provides the co-ordinate convergence of an approximate solution $\mathbf{a}^N$ ($|a_n - a_n^N| \to 0$, $N \to \infty$). Assume that eqn. 8.39 is solvable for sufficiently big N and consider the difference

$$
\begin{aligned}
\mathbf{t} - \mathbf{t}_N = (I - BA)^{-1}&[(BA - P_N BAP_N)\mathbf{t}_N + BR_N\mathbf{u} + R_N B\mathbf{u}_N \\
&+ (P_N BAP_N - B_N A_N)\mathbf{t}_N + B(P_N\mathbf{u} - \mathbf{u}_N) + (P_N BP_N - B_N)\mathbf{u}_N] \quad (8.42)
\end{aligned}
$$

where $P_N + R_N = I$. It follows from the estimates given by eqn. 8.41 that the vectors $(P_N BAP_N - B_N A_N)\mathbf{t}_N$ and $B(P_N\mathbf{u} - \mathbf{u}_N)$ weakly converge to a zero element when $N \to \infty$ and the vectors $(BA - P_N BAP_N)\mathbf{t}_N$, $BR_N\mathbf{u}$, $R_N B\mathbf{u}_N$ converge to a zero element with respect to the $\tilde{l}_2$ norm (the operator $BA : \tilde{l}_2 \to \tilde{l}_2$ is completely continuous). Therefore, the right-hand side of eqn. 8.42 and, consequently, the vector $\mathbf{t} - \mathbf{t}_N$ also weakly converge to a zero element what implies the required co-ordinate convergence.

We may conclude that the conditions $p - \text{entier } \kappa \sin \phi \ll N$, $\epsilon_N \to 0$ are close to the optimal choice of the truncation parameters. If $p - \text{entier } \kappa \sin \phi = O(N)$, $\epsilon_N = O(1)$, the growth of the number of equations in the system given by eqn. 8.37 does not reduce calculation error (the norm of the difference of the exact and approximate solutions does not decrease when N grows).

8.4 Reduction to co-ordinate boundaries. Echelette in a dielectric layer and a dielectric echelette

This Section is devoted to the study of the second-kind systems obtained within the frame of the approaches which combine the mode-matching technique on nonco-ordinate (in particular, triangle) boundaries, re-expansion and the partial domains methods.

It is convenient to illustrate the method representing the solution to the boundary-value problem

$$(\Delta + \kappa^2 \epsilon)u(\mathbf{r}) = 0, \quad \mathbf{r} \in \Pi = \{\mathbf{r} = (x_2, x_3) : 0 \le x_2 \le 2\pi, \ 0 \le x_3 \le 2\pi\theta\}$$

$$u|_{\delta\Pi} = F(\mathbf{r})$$

($F(\mathbf{r})$ is a given function defined on $\delta\Pi$) in the form $u = u_1 + u_2$ where

$$u_1(x_2, 0) = u_1(x_2, 2\pi\theta) = 0 \qquad (0 \le x_2 \le 2\pi)$$
$$u_2(0, x_3) = u_2(2\pi, x_3) = 0 \qquad (0 \le x_3 \le 2\pi\theta)$$

and $u_{1,2}$ satisfy the initial boundary conditions for u on the complementary parts of the boundary $\delta\Pi$. Then each of the additional functions and the solution u itself may be expanded on the walls of Π with the help of the complete system of eigenfunctions of the corresponding parallel-plane waveguides

$$u(\mathbf{r}) = \sum_{n=1}^{\infty} \{a_{n,1} \exp(j\omega_{n,1} x_2) + b_{n,1} \exp[-j\omega_{n,1}(x_2 - 2\pi)]\} \sin \frac{nx_3}{\theta}$$

$$+ \sum_{n=1}^{\infty} \{a_{n,2} \exp(j\omega_{n,2} x_3) + b_{n,2} \exp[-j\omega_{n,2}(x_2 - 2\pi\theta)]\} \sin \frac{nx_2}{2}$$

$$(8.43)$$

where

$$\omega_{n,1} = \sqrt{\kappa^2\epsilon - \frac{n^2}{4\theta^2}}, \quad \omega_{n,2} = \sqrt{\kappa^2\epsilon - \frac{n^2}{4}}$$

We will apply this approach for the problem of the plane E-polarised $\exp\{j\kappa[x_2\sin\phi - (x_3 - 2\pi\delta)\cos\phi]\}$ wave diffraction on the grating depicted in Figure 7.9h looking for its solution in the form similar to the one used in Section 5.2:

$$v(\mathbf{r}) = \begin{cases} \displaystyle\sum_{n=1}^{\infty} a_n \exp\{j[\Phi_n x_2 + \Gamma_n(x_3 - 2\pi\delta)]\} \\ \quad + \exp\{j\kappa[x_2\sin\phi - (x_3 - 2\pi\delta)\cos\phi]\}, \quad x_3 \geq 2\pi\delta \\ \displaystyle\sum_{n=-\infty}^{\infty} \{b_n e_n \exp(-j\Gamma_{n,\epsilon}x_3) \\ \quad + d_n \exp(j\Gamma_{n,\epsilon}x_3)\}\exp(j\Phi_n x_2), \quad 0 \leq x_3 \leq 2\pi\delta \\ \displaystyle j\sum_{n=1}^{\infty}\left[c_n \sin\frac{nx_2'}{2\theta}\sinh(j\omega_n x_3') + \bar{c}_n \sin\frac{nx_3'}{2\theta}\sinh j\omega_{n,1}(x_2' - 2\pi\theta)\right] \\ \quad \mathbf{r}' = (x_2', x_3') \in OBAC \end{cases}$$

(8.44)

where $v(\mathbf{r}) = E_{x_1}(\mathbf{r})$, the diffraction amplitudes $\mathbf{a} = (a_n)_n$, $\mathbf{b} = (b_n)_n$ and $\mathbf{d} = (d_n)_n$ belong to the space $\tilde{l}_2$,

$$\Phi_n = n + \kappa\sin\phi, \quad \Gamma_n = \sqrt{\kappa^2 - \Phi_n^2}, \quad \Gamma_{n,\epsilon} = \sqrt{\kappa^2\epsilon - \Phi_n^2},$$

$$\omega_n = \sqrt{\kappa^2 - \frac{n^2}{4\theta^2}}, \quad e_n = \exp(2j\pi\delta\Gamma_{n,\epsilon})$$

and the local co-ordinate system

$$\begin{cases} x_2 = x_2'\theta + x_3'\bar{\theta} \\ x_3 = -x_2'\bar{\theta} + x_3'\theta \end{cases} \quad \begin{cases} x_2' = x_2\theta - x_3\bar{\theta} \\ x_3' = x_2\bar{\theta} + x_3\theta \end{cases} \quad (\theta = \cos\psi, \ \bar{\theta} = \sin\psi)$$

(the real and imaginary parts of the square roots are assumed to be positive).

The problem is reduced with the help of mode matching on the lines OC, CA and $x_3 = 2\pi\delta$ to the infinite system of linear algebraic equations

$$\mathbf{d} = (AB + \bar{A}\bar{B})\mathbf{d} + \mathbf{v}$$

(8.45)

where

$$a_n(\Gamma_{n,\epsilon}\Gamma_m) = 2\bar{d}_n e_n \Gamma_{n,\epsilon} + \delta_{n0}(\Gamma_n - \Gamma_{n,\epsilon}), \quad d_n = \sqrt{n}\bar{d}_n$$

(8.46)

$\mathbf{d} = (d_n)_n$ and $\mathbf{v} = (v_n)_n$ are the vector of unknowns and right-hand side, A and B are the matrix operators with the elements

$$A_{mn} = \sqrt{\frac{n}{m}} \frac{\omega_m^-[T_n(\beta_m^+) - T_n(-\beta_m^+)] + \omega_m^+[T_n(\beta_m^-) - T_n(-\beta_m^-)]}{8\pi^2\theta\Gamma_{n,\epsilon}\sin(2\pi\omega_m\bar{\theta})\left[1 + \frac{e_n^2(\Gamma_n - \Gamma_{n,\epsilon})}{\Gamma_n + \Gamma_{n,\epsilon}}\right]}$$

$$B_{mn} = \sqrt{\frac{m}{n}}\left[g_n^+ F_m(\theta, -\bar{\Gamma}_n) - e_n^2 g_n^- F_m(\theta, \bar{\Gamma}_n^+)\frac{\Gamma_n - \Gamma_{n,\epsilon}}{\Gamma_n + \Gamma_{n,\epsilon}}\right] \tag{8.47}$$

where

$$g_n^\pm = \exp(\pm 2i\pi\bar{\theta}\Gamma_n^\pm), \quad \Gamma_n^\pm = \theta(\Gamma_{n,\epsilon} \pm \Phi_n)$$

$$\omega_m^\pm(\omega_m) = \omega_m\theta \pm \frac{m\bar{\theta}}{2\theta}, \quad \beta_m^\pm(\omega_m) = \omega_m\bar{\theta} \pm \frac{m}{2}$$

$$T_n(w) = -j\frac{\exp[2j\pi(w - \Phi_n)] - 1}{w - \Phi_n}, \quad F_m(\alpha, w) = -\frac{m}{2\alpha}\frac{(-1)^m\exp(2i\pi w\alpha) - 1}{\frac{m}{2\alpha} - w^2}$$

and the elements of $\bar{A}$ and $\bar{B}$ have a similar form if $\omega_m^\pm(\omega_m)$ and $\beta_m^\pm(\omega_m)$ are replaced by $\omega_m^\pm(\bar{\omega}_m)$ and $\beta_m^\pm(\bar{\omega}_m)$.

The operators $A : \tilde{m} \to \tilde{m}$ and $B : \tilde{m} \to \tilde{m}$ where $\tilde{m} = \{\mathbf{b} = (b_n)_n : \sup_n |b_n| < \infty\}$ are completely continuous and bounded, respectively (as well as $\bar{A}$ and $\bar{B}$), because

$$\lim_{k \to \infty} \sup_{k \le |n| < \infty} \sum_{m=1}^\infty |A_{mn}| = 0, \quad \sup_{1 \le |m| < \infty} \sum_{p=1}^\infty |B_{mp}| < \infty \tag{8.48}$$

due to the estimates

$$A_{mn} = O\left(\frac{\sqrt{nm}}{m^2 + c_1 n^2}\right), \quad B_{mn} = O\left(\frac{\sqrt{m^3 p^{-1}}}{m^2 + c_2 p^2}\right) \quad (m, |p| \to \infty)$$

In addition to this, it is not difficult to check that $\mathbf{v} \in \tilde{m}$ and every solution of eqn. 8.45 belongs to the space $\tilde{l}_2$. Hence eqn. 8.45 is a (uniquely solvable in $\tilde{l}_2$) canonical Fredholm equation which may be solved numerically by the (convergent) truncation method described in Sections 1.11 and 2.3.

The same approach has been applied for the problem of plane wave diffraction on a dielectric echelette (see Figure 7.9i). Since matrix operators turn out to be very bulky, we will limit ourselves to the description of the general scheme of the method. The fields are expanded in the regions $OCAB$

and $BADE$ in the form given by eqn. 8.43. The mode matching procedure on the segments OC, CA, AB, BE, ED (for the E_{x_1} component) and on the segments OA, AB, BD (for the H_{tg} components) taking into account the quasi-periodicity on OB and AD yields the system of operator equations with respect to ten sets (infinite vectors) of the unknown diffraction amplitudes (Sirenko and Shestopalov, 1982) which is reduced to a canonical Fredholm system of two coupled operator equations. This model enables one also to analyse absorbing echelette gratings with nonideal metal inclusions.

Chapter 9

Local excitation of diffraction gratings

Scattering of surface waves of a periodic structure by an inhomogeneity (bounded in the plane y, z) is a key problem for the study of the local excitation of diffraction gratings. But until recently even a rigorous statement of this problem has not been given. We will present in what follows an approach for solving the problems of local excitation which is based on investigating the dispersion properties of gratings treated as waveguide structures and the use of spectral data for the Fourier analysis of the corresponding diffraction problems.

9.1 Eigenwaves of periodic waveguide structures

We will define eigenwaves of periodic waveguide structures (Figure 9.1) in terms of nontrivial solutions of the homogeneous problem given by eqns. 7.1–7.3 [where it is assumed that $f(\mathbf{r}) \equiv 0$, $\mathbf{r} = (x_2, x_3)$, see Section 7.1] with respect to the spectral parameter Φ varying in the infinite-sheet Riemann surface $\mathcal{F}$ whose structure is similar to that of $\mathcal{H}$ introduced in Section 7.1. $\mathcal{F}$ coincides with the Riemann surface of the analytical continuation from the real axis $\mathrm{Im}\,\Phi = 0$ of the canonical Green function $G_0 = G_0(\mathbf{r}, \mathbf{r}_0, \kappa, \Phi)$ given by eqn. 7.4 (see Section 7.1). $\mathcal{F}$ has the second order real branch points

$\Phi = \Phi_n^b = -n \mp \kappa : \Gamma_n(\Phi_n^b) = 0$ and the cuts are determined by the conditions

$$d_n = \kappa^2 - (n + \operatorname{Re}\Phi)^2 + (\operatorname{Im}\Phi)^2, \quad \operatorname{Im}\Phi \operatorname{Re}(n + \Phi) \geq 0$$

$$|\operatorname{Re}\Gamma_n(\Phi)| = \sqrt{0.5\sqrt{d_n^2 + c_n^2} + d_n}, \quad |\operatorname{Im}\Gamma_n(\Phi)| = \sqrt{0.5\sqrt{d_n^2 + c_n^2} - d_n}$$

$$(n = 0, \pm 1, \pm 2, \ldots)$$

(9.1)

$\Gamma_n(\Phi) = \Gamma_{-n}(-\Phi)$ on the first sheet of $\mathcal{F}$ and other sheets differ from the first by the fact that the signs of $\operatorname{Re}\Gamma_n(\Phi)$ and $\operatorname{Im}\Gamma_n(\Phi)$ are changed to opposite ones for a finite number of indices n (as well as the condition $\operatorname{Im}\Gamma_n \leq 0$).

Eigenwaves of a grating may propagate with or without attenuation. The direction of propagation and phase velocities $v_n = c\kappa(n + \Phi)^{-1}$ are different. The eigenwaves with $|v_n| > c$ ($|v_n| < c$) are called the "fast" ("slow") waves. The eigenwaves with exponentially decaying (growing) fields in the direction off the grating are called the surface ("leaky") waves. The waves whose fields are constant in the direction perpendicular to the grating are called "piston" waves (Sirenko and Shestopalov, 1982).

In accordance with the position of Φ on the Riemann surface, eigenwaves of a grating may form the following groups:

(i) "nonincoming" eigenwaves, nonattenuating with respect to y and non-growing with respect to z [$\operatorname{Re}\Gamma_n \geq 0$, $\operatorname{Im}\Gamma_n \geq 0$, $\operatorname{Re}(n + \Phi)\operatorname{Im}\Phi \leq 0$];

(ii) "nonincoming" eigenwaves, nonattenuating with respect to z and non-growing with respect to y [$\operatorname{Re}\Gamma_n \geq 0$, $\operatorname{Re}\Gamma_n \operatorname{Im}\Gamma_n \geq 0$, $\operatorname{Re}(n + \Phi)\operatorname{Im}\Phi \geq 0$];

(iii) "nonoutgoing" eigenwaves, nonattenuating with respect to z and non-growing with respect to y [$\operatorname{Re}\Gamma_n \leq 0$, $\operatorname{Re}\Gamma_n \operatorname{Im}\Gamma_n \leq 0$, $\operatorname{Re}(n + \Phi)\operatorname{Im}\Phi \geq 0$];

(iv) "nonoutgoing" eigenwaves, nonattenuating with respect to y and non-growing with respect to z [$\operatorname{Re}\Gamma_n \leq 0$, $\operatorname{Re}\Gamma_n \operatorname{Im}\Gamma_n \geq 0$, $\operatorname{Re}(n + \Phi)\operatorname{Im}\Phi \leq 0$].

General classification of eigenwaves may be carried out with the help of

Lemma 9.1: All information on eigenwaves of a given grating considered as nontrivial solutions $v = v(\mathbf{r})$ of the homogeneous problem given by eqns. 7.1–7.3 with respect to the spectral parameter $\Phi \in \mathcal{F}$ may be obtained limiting oneself to analysis of the spectral problem in an arbitrary unit strip (parallel

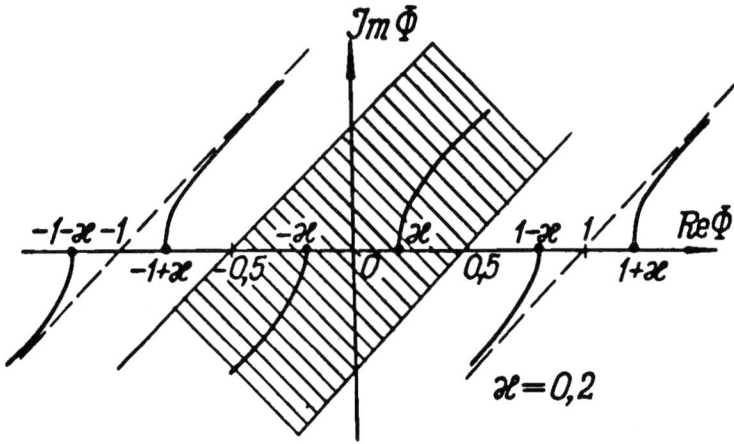

a

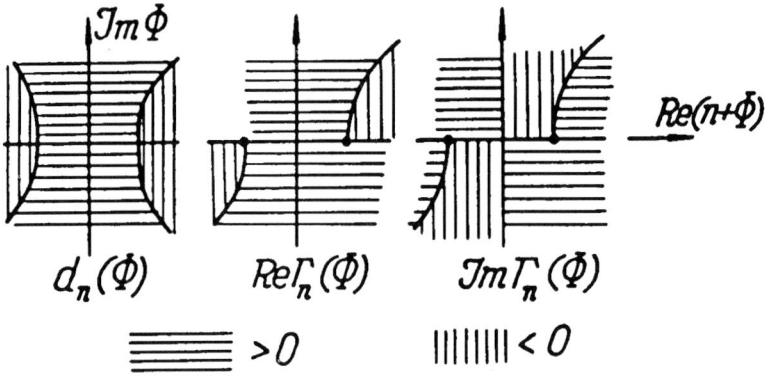

b

Figure 9.1 *The Riemann surface and the curves of cuts*

to the imaginary axis). For symmetric gratings it is sufficient to consider the half-strips $\operatorname{Im} \Phi \geq 0$ $(\operatorname{Im} \Phi \leq 0)$.

The use of the complex power theorem enables one to prove the following

Corollary 9.1: Assume that $\operatorname{Im}\Phi = 0$, then in the E-case there are no non-trivial solutions $v(\mathbf{r})$ of the spectral problem given by eqns. 7.1–7.3 such that

$$2\pi \sum_n \operatorname{Re}\Gamma_n(|a_n|^2 + |b_n|^2) \neq -\kappa^2 \int_V |\mathbf{E}|^2 \operatorname{Im}\epsilon\, dV$$

and (or)

$$2\pi \sum_n \operatorname{Im}\Gamma_n(|a_n|^2 + |b_n|^2) \neq \kappa^2 \int_V \left(-\frac{\mu_0}{\epsilon_0}|\mathbf{H}|^2 + \operatorname{Re}\epsilon|\mathbf{E}|^2\right) dV$$

In the H-case there are no non-trivial solutions such that

$$2\pi \sum_n \operatorname{Im}\Gamma_n(|a_n|^2 + |b_n|^2) \neq \kappa^2 \int_V \left(-\frac{\mu_0}{\epsilon_0}\operatorname{Im}\epsilon|\mathbf{E}|^2\right) dV$$

and (or)

$$2\pi \sum_n \operatorname{Im}\Gamma_n(|a_n|^2 + |b_n|^2) \neq \kappa^2 \int_V \left(-\frac{\epsilon_0}{\mu_0}|\mathbf{E}|^2 + |\mathbf{H}|^2\right) dV$$

Corollary 9.2: Assume that $\operatorname{Im}\Gamma_n(\Phi) > 0$ $(n = 0, \pm1, \pm2, \ldots)$, then both in the E and H cases there are no non-trivial solutions $v(\mathbf{r})$ of the spectral problem given by eqns. 7.1–7.3 such that

$$\operatorname{Re}P[\exp(-4\pi\operatorname{Im}\Phi)-1] \neq -\kappa \int_{V_\infty} \sqrt{\frac{\epsilon_0}{\mu_0}}|\mathbf{E}|^2 \operatorname{Im}\epsilon\, dV, \quad V_\infty = Q_\infty \times \{|\kappa| \leq 0.5\}$$

and (or)

$$\operatorname{Re}P[\exp(-4\pi\operatorname{Im}\Phi) - 1] \neq \kappa \int_{V_\infty} \left(\sqrt{\frac{\mu_0}{\epsilon_0}}|\mathbf{H}|^2 - \sqrt{\frac{\epsilon_0}{\mu_0}}|\mathbf{E}|^2 \operatorname{Re}\epsilon\right) dV$$

where P is the flow of the Umov-Pointing vector through the x_1, x_3 cross-section in the x_2 direction calculated for $|\kappa| \leq 0.5$.

An account of losses in the dielectric media leads to the following specifications:

Corollary 9.3: If $\operatorname{Im}\epsilon \neq 0$, then there are no real eigenvalues Φ on the first sheet of $\mathcal{F}$. If $V = 0$ (flat gratings) the (real) eigenvalues Φ (if they exist) coincide with the branch points $\Phi = \Phi_n^b = -n \mp \kappa$. If $\mathrm{Im}\,\epsilon = 0$ and $V \neq 0$ (ideal volume gratings), then all the eigenwave's partial amplitudes a_n and b_n with the indices n such that $\mathrm{Re}\,\Gamma_n > 0$ are equal to zero and in the E (H) case the eigenwave's electric field energy $\int_V \mathrm{Re}\,\epsilon |\mathbf{E}|^2 \, dV$ in the communication domain is always greater (less) than the eigenwave's magnetic field energy $\int_V \sqrt{\frac{\mu_0}{\epsilon_0}} |\mathbf{H}|^2 \, dV$.

Corollary 9.4: If $\mathrm{Im}\,\epsilon \neq 0$, there are no "correct" eigenwaves carrying the energy in the positive (for $\mathrm{Im}\,\Phi < 0$) or negative (for $\mathrm{Im}\,\Phi > 0$) directions of the axis x_2. If $\mathrm{Im}\,\epsilon = 0$, there are no "correct" complex eigenwaves carrying the energy along the grating.

Corollary 9.5: All eigenwaves of the types of gratings considered whose eigenvalues Φ lie on the real axis of the first sheet of $\mathcal{F}$ and do not coincide with the branch points are the "correct" slow waves, nonattenuating along the grating.

Consider an arbitrary grating made of metal bars placed in a dielectric layer with the piece-wise continuous permittivity assuming that the piece-wise smooth contour S of the perfectly conducting parts of its cross-section does not contain "infinitely" sharp edges. In accordance with the notation introduced in Section 7.1 such a periodic structure belongs to the family $\mathcal{G}$.

Then one can naturally generalise the spectral theorems proved in Sections 4.3 and 7.1.

Theorem 9.1: The spectrum Ω_Φ of the spectral problem given by eqns. 7.1–7.3 for every grating from the family $\mathcal{G}$ consists of not more than a countable number of isolated points situated on the Riemann surface $\mathcal{F}$ and the resolvent of the problem has the finite-order poles in these points.

The proof of this spectral theorem is based on the consideration of Green's potentials in the complex frequency domain with the use of the results of theorem 7.1 and lemmas 7.2 and 7.3. and it is similar to the proofs of theorems 4.7 and 7.7 (see Sections 4.3 and 7.1) since the potentials have the same kernels in the form of the canonical Green function given by eqn. 7.4.

Corollary 9.6: (radiation principle for one-dimensional periodic structures)

The radiation conditions of the excitation problem given by eqns. 7.1–7.3 separates its unique solution [describing a monochromatic quasi-periodic excita-
tion by a source function $f(\mathbf{r}) \neq 0$, $f(x_2 + 2\pi, x_3) = \exp(2j\pi\Phi)f(\mathbf{r})$, supp $f \in$
Q] for all $\Phi \in \mathcal{F}$ except for not more than a countable set Ω_Φ which has no
finite accumulation points.

Since the canonical Green function $G_0 = G_0(\mathbf{r}, \mathbf{r}_0, \kappa, \Phi)$ given by eqn. 7.4
from Section 7.4 is meromorphic with respect to $\Phi \in \mathcal{F}$ it is easy to show that
every characteristic $\psi = \psi(\Phi)$ of a one-dimensional periodic structure linearly
dependent on G_0 may be represented in the local co-ordinates

$$\tau = \begin{cases} \sqrt{\Phi - \bar{\Phi}}, & \bar{\Phi} = -n \pm \kappa \\ \Phi - \bar{\Phi}, & \bar{\Phi} \neq -n \pm \kappa \end{cases} \quad (n = 0, \pm 1, \pm 2, \dots)$$

in a vicinity of $\bar{\Phi}$ as

$$\psi(\Phi) = \sum_{m=-N}^{N} \psi_m \tau^m \tag{9.2}$$

where N is the pole's order in the point $\Phi = \bar{\Phi}$. Eqn. 9.2 enables one to
approximately estimate the contribution made to excitation characteristics
by the "eigenregime" corresponding to the singularity $\bar{\Phi}$.

Existence of eigenwaves may be proved by repeating the considerations
of Section 7.1 (lemmas 7.4 and 7.5) based on the principle of continuous
dependence.

Theorem 9.2: The spectrum Ω_Φ of the spectral problem given by eqns.
7.1–7.3 is not empty for every grating belonging to the family $\mathcal{G}$.

In fact, the residues

$$\text{res } G_0(-m \pm \kappa) = \frac{1}{4\pi\kappa} \exp[\pm j\kappa(x_2 - x_{20})]$$

define the eigenfunctions $v_m(-m \pm \kappa) = \exp(\pm j\kappa x_2)$ of the canonical spectral
problem (standing or "piston" waves) and the proof of theorem 9.2 is based
on constructing a suitable "deformation" of the poles of the resolvent of the
canonical problem which is described in Section 7.1.

9.2 The point source Green function and local excitation of diffraction gratings

Let us denote by $G = G(\mathbf{r}, \mathbf{r}_0, \Phi)$ the point source Green function of the problem given by eqns. 7.1–7.3 describing a monochromatic quasi-periodic excitation by a source function

$$f(\mathbf{r}, \Phi) = - \sum_{n=-\infty}^{\infty} \delta(x_2 - x_{20} - 2\pi n) \exp(2j\pi n\Phi)\delta(x_3 - x_{30}) \quad (0 \le x_2 \le 2\pi)$$

of a one-dimensional periodic grating depicted in Figure 9.2a. Since

$$\bar{f}(\mathbf{r}) = -\delta(x_2 - x_{20})\delta(x_3 - x_{30}) = \int_{-0.5}^{0.5} f(\mathbf{r}, \Phi)\, \mathrm{d}\,\Phi$$

Green's function of this problem with a local source in the point $\mathbf{r}_0 = (x_{20}, x_{30})$ may be represented in the form

$$\bar{G}(\mathbf{r}, \mathbf{r}_0) = \int_{-0.5}^{0.5} G(\mathbf{r}, \mathbf{r}_0, \Phi)\, \mathrm{d}\,\Phi \quad (\kappa > 0) \tag{9.3}$$

where the segment of integration Γ (partially) situated on the real axis of the first sheet of the Riemann surface $\mathcal{F}$ contains the set

$$M = \{\bar{\Phi}_k\} = \cup_{l=1}^{M_P} M_l \cup \bar{\Phi}^- \cup \bar{\Phi}^+$$

of all the singularities of G in Γ: two branch points $\Phi^{\pm}$ and a finite number (M_P) of poles M_l.

Before constructing Green's function let us indicate some important characteristics of grating's modes. First, the notion of the direction of propagation determined by the sign of $\operatorname{Re}\Phi$ cannot be used for periodic structures since a mode corresponding to an eigenvalue $\bar{\Phi}$ is simultaneously an eigenwave corresponding to $\bar{\Phi} + p$ for all $p = 0, \pm 1, \pm 2, \ldots$. Let us denote by $P = P(\tilde{x}_2, \mathbf{q})$ the flow of the Umov-Pointing vector through the complete grating's cross-section in the plane $x_2 = \tilde{x}_2$ calculated for $|\kappa| \approx 0.5$ in the $\mathbf{q}$ direction. Then the magnitude $\operatorname{Re} P(\tilde{x}_2, \mathbf{q})$ (and not the direction of mode's propagation) which determines the value and direction of energy carried by an eigenwave through grating's cross-section has a direct physical meaning.

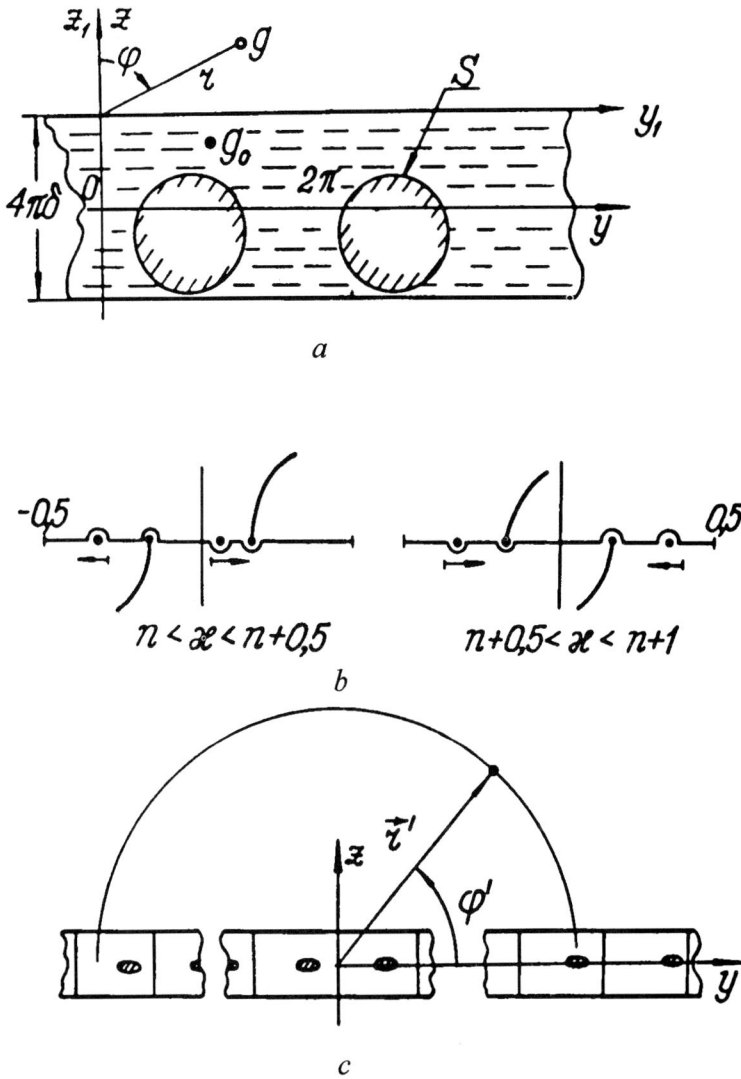

a

$n < \varkappa < n + 0.5$ $n + 0.5 < \varkappa < n + 1$

b

c

Figure 9.2   *Excitation of a grating by a point source and integration contours*

Lemma 9.2: Let $v(\mathbf{r})$ be a correct wave of a grating corresponding to an eigenvalue Φ situated on the real axis of the first sheet of the Riemann surface $\mathcal{F}$. Then $\operatorname{Re} P(\tilde{x}_2, \mathbf{q})$ does not depend on x_2, i.e. the value and direction of the energy transfer is constant in every grating's cross-section.

It is sufficient to prove this lemma for $\tilde{x}_2 \in [0, 2\pi]$. Actually, the complex power theorem applied in the volume

$$\tilde{V}_\infty = \{\mathbf{r} : \mathbf{r} \in V_\infty,\ 0 \le \tilde{x}_2 \le 2\pi\}$$

yields $\operatorname{Re} P(0, \mathbf{q}) - \operatorname{Re} P(\tilde{x}_2, \mathbf{q}) = 0$ since eigenwave's field exponentially decreases in the direction perpendicular to the grating.

It is natural to assume now that the arcs of the integration contour Γ in eqn. 9.3 surrounding the poles of Green's function are determined by the direction of the energy transfer: if $\operatorname{Re} P(\bar{\Phi}_k, \mathbf{q}) > 0$ ($\operatorname{Re} P(\bar{\Phi}_k, \mathbf{q}) < 0$), the arcs lie in the lower (upper) half-plane. We will consider in what follows the presence of simple poles in Γ excluding from consideration the eigenwaves which do not carry the energy, i.e. the eigenwaves with $v(\mathbf{r}) \ne 0$ and $\operatorname{Re} P(\bar{\Phi}, \mathbf{r}) = 0$ ($\operatorname{Im} \Phi = 0$) whose eigenvalues $\bar{\Phi}$ are the nth order poles of G ($n \ge 2$) connected with the existence of associated waves.

In order to prove the existence of $\bar{G}$ and to obtain its explicit asymptotic representation, we will carry out the integration in eqn. 9.3 for $x_3 > 2\pi\delta$ taking into account the choice of Γ and limiting ourselves to the case $M_P = 2$:

$$\bar{G}(\mathbf{r}, \mathbf{r}_0) = -\pi j \sum_{\Phi_k \in M_1} \delta_k \operatorname{res} G(\mathbf{r}, \mathbf{r}_0, \bar{\Phi}_k) + \pi j \sum_{\Phi_k \in M_2} \delta_k \operatorname{res} G(\mathbf{r}, \mathbf{r}_0, \bar{\Phi}_k)$$

$$+ \sum_{n=-\infty}^{\infty} \text{v.p.} \int_{-0.5}^{0.5} a_n(\mathbf{r}_0, \Phi) \exp\{jr[(n - \Phi)\sin\phi \qquad (9.4)$$

$$+ \sqrt{\kappa^2 - (n + \Phi)^2}\cos\phi\}\,d\,\Phi$$

where $x_3 - 2\pi\delta = r\cos\phi$, $x_2 = r\sin\phi$, $\delta_k = 0$ or $\delta_k = 1$ depending on the position of singularities (inside the interval or in its endpoints). Calculating the last integral of eqn. 9.4 by the stationary phase method we finally get (for $r \to \infty$, $|\phi| < \frac{\pi}{2}$, $x_2 \to \infty$, $|\phi| = \frac{\pi}{2}$)

$$\bar{G}(\mathbf{r}, \mathbf{r}_0) = \exp\left(jr\kappa - j\frac{\pi}{2}\right) \sqrt{\frac{2\pi\kappa}{r}}\cos\phi$$

$$\times \sum_{|\kappa \sin\phi - n| \le 0.5} \delta_n a_n(\mathbf{r}_0, \kappa\sin\phi - n) + O(r^{-1})$$

$$+ \begin{cases} -2\pi j \sum_{\bar{\Phi}_k \in M_1} \delta_k \operatorname{res} G(\mathbf{r}, \mathbf{r}_0, \bar{\Phi}_k), & x_2 < 0 \\ 2\pi j \sum_{\bar{\Phi}_k \in M_2} \delta_k \operatorname{res} G(\mathbf{r}, \mathbf{r}_0, \bar{\Phi}_k), & x_2 > 0 \end{cases} \qquad (9.5)$$

and it is easy to see that only the real eigenwaves exponentially decaying at infinity and having the order $O(1)$ for finite values of x_3 and $x_2 \to \infty$ contribute to the formation of the field of $\bar{G}$. This fact enables us to formulate the radiation conditions for Green's function: waves propagating off the grating carry away the energy, i.e. the energy does not come from without.

For $r \to \infty$, $|\phi| < \frac{\pi}{2}$ the first term of eqn. 9.5 is an inhomogeneous cylindrical wave propagating off the grating with the amplitude (far-field pattern)

$$F(\phi) = \cos \phi \sum_{|\kappa \sin \phi - n| \leq 0.5} \delta_n a_n (\mathbf{r}_0, \kappa \sin \phi - n) \tag{9.6}$$

and the far-field pattern is thus completely determined by the amplitudes of propagating harmonics.

The complete solution of the excitation problem considered is connected with obtaining additional information concerning the distribution of real eigenvalues.

Lemma 9.3: If $\mathrm{Im}\, \epsilon \equiv 0$, the points of the set Ω_Φ form the fours $\bar{\Phi}, -\bar{\Phi}, \bar{\Phi}^*, -\bar{\Phi}^*$ symmetric with respect to the real and imaginary axes of Φ.

Let us give the scheme of the proof of this lemma. Assume that $\bar{\Phi} \in \Omega_\Phi$ lies on the first sheet of $\mathcal{F}$ and $v(\mathbf{r}, \bar{\Phi})$ is the corresponding eigenfunction. Then the function $v^*(\mathbf{r}, \bar{\Phi})$ satisfies the generalised boundary conditions and the radiation conditions given by eqn. 7.2 with $\bar{\Phi} = -\bar{\Phi}^*$, $\Gamma_n = -\Gamma_n^*$, $a_n = a_n^*$, $b_n = b_n^*$. If $d_n(\bar{\Phi}) < 0$ for all $n = 0, \pm 1, \pm 2, \ldots$, then $\Gamma_{-n}(-\bar{\Phi}^*) = -\Gamma_n^*(\bar{\Phi})$ for the point $-\bar{\Phi}^*$ from the same sheet (see Figure 9.2b) and the radiation conditions for $v^*(\mathbf{r}, \bar{\Phi})$ coincides with the radiation conditions for $v(\mathbf{r}, -\bar{\Phi}^*)$. If $d_n(\bar{\Phi}) > 0$ for a finite number of $n = n_1, n_2, \ldots, n_k$, the same result is valid for $v(\mathbf{r}, -\bar{\Phi}^*)$ if $-\bar{\Phi}^*$ belongs to that sheet of $\mathcal{F}$ where $\Gamma_{n_l}(\Phi)$ $(l = 1, 2, \ldots, k)$ have the opposite signs. It follows from the second Green's formula applied in the volume Q that

$$G(\mathbf{r}, \mathbf{r}_0, \Phi) = G(\mathbf{r}, \mathbf{r}_0, -\Phi) \tag{9.7}$$

if $\Gamma_{-n}(-\Phi) = \Gamma_n(\Phi)$ (which holds, in particular, for all pairs $\{\Phi, -\Phi\}$ from the first sheet of $\mathcal{F}$) and the proof is completed.

Corollary 9.7: The eigenwaves $v(\mathbf{r}, \bar{\Phi})$ and $v(\mathbf{r}, -\bar{\Phi})$ carry the energy in opposite directions.

We have

$$2\operatorname{Re} P[v(\Phi), \mathbf{r}] = \frac{1}{i\omega\mu} \int_{-\infty}^{\infty} W[v^*, v] \, d x_3 = -2\operatorname{Re} P[v^*(\Phi), \mathbf{r}]$$

$$W[v^*, v] = v^* \frac{\partial v}{\partial x_2} - v \frac{\partial v^*}{\partial x_2} \tag{9.8}$$

but $v(\mathbf{r}, -\bar{\Phi}) = v^*(\mathbf{r}, \bar{\Phi})$ due to lemma 9.3 and hence eqn. 9.8 yields $\operatorname{Re} P[v(\bar{\Phi}), \mathbf{r}] = \operatorname{Re} P[v(-\bar{\Phi}), \mathbf{r}]$.

Lemma 9.4: Let $v(\mathbf{r}) = \sum_{i=1}^{k} v_i(\mathbf{r}, \bar{\Phi}_i)$ where $v_i(\mathbf{r}, \bar{\Phi}_i)$ are the correct real eigenwaves corresponding to different eigenvalues $\bar{\Phi}_i$. Then

$$\operatorname{Re} P(u, \mathbf{r}) = \sum_{i=1}^{k} \operatorname{Re} P[v_i(\bar{\Phi}_i), \mathbf{r}]$$

i.e. eigenwaves are orthogonal in the energy space.

It is sufficient to prove lemma 9.4 for $k = 2$. Let $u = v_1 + v_2$. Then

$$\operatorname{Re} P(u, \mathbf{r}) = \operatorname{Re} P(v_1, \mathbf{r}) + \operatorname{Re} P(v_2, \mathbf{r}) + \frac{1}{j\omega\mu} \int_{-\infty}^{\infty} (W[v_1^*, v_2] + W[v_2^*, v_1]) \, d x_3$$

and the required result follows from the equality

$$[1 - \exp(-2j\pi\Phi_1)\exp(2j\pi\Phi_2)] \int_{-\infty}^{\infty} (W[v_1^*, v_2] + W[v_2^*, v_1]) \, d x_3$$

$$= \int_{Q_\infty} (v_1^* \Delta v_2 - v_2 \Delta v_1^*) \, d\mathbf{r} = 0$$

which in turn is a consequence of the second Green's formula applied in the domain Q_∞.

Hence the correct real and complex eigenwaves of a one-dimensional periodic grating are similar to the normal waves of periodic waveguides. In particular, it follows from lemma 9.3 that when an eigenvalue leaves the real axis, another (symmetric) eigenvalue simultaneously goes to the opposite direction and such a "bifurcation" may cause a splitting of spectral curves or a "collision" of eigenwaves: two real eigenwaves "meet" each other in one point $\bar{\Phi}$ and become complex waves with conjugated eigenvalues.

Integration in eqn. 9.3 may be now repeated for $x_3 < -2\pi\delta$ and $|x_3| < 2\pi\delta$ to give a representation of Green's function in the polar co-ordinates r', ϕ' corresponding to the point $\mathbf{r} = (x_2, x_3)$ as a superposition of a radiating cylindrical wave and correct real eigenwaves $(v_k^{\pm})$ carrying away the energy to the right $(x_2 \to \infty)$ or to the left $(x_2 \to -\infty)$ from the source

$$G(\mathbf{r}, \mathbf{r}_0) = \frac{1}{\sqrt{r'}} \exp(jr'\kappa) A(\kappa, \mathbf{r}_0, \phi') \frac{1}{r''}$$
$$+ \begin{cases} \sum_k v_k^-(\mathbf{r}) B_k^-(\mathbf{r}_0), & x_2 < 0 \\ \sum_k v_k^+(\mathbf{r}) B_k^+(\mathbf{r}_0), & x_2 > 0 \end{cases} \tag{9.9}$$

It is clear that due to the relation

$$\bar{u}(\mathbf{r}) = -\int \bar{G}(\mathbf{r}, \mathbf{r}_0) \bar{f}(\mathbf{r}_0) \, d\mathbf{r}_0 \tag{9.10}$$

the solution $\bar{u}(\mathbf{r})$ of the problem of grating's excitation by an arbitrary source described by a compactly supported function $\bar{f}(\mathbf{r}) : \operatorname{supp} \bar{f} \in Q \subset \{\mathbf{r} : |x_3| < 2\pi\delta\}$ has the same physically correct behaviour at infinity determined by eqn. 9.9. If $\operatorname{supp}\bar{f}$ occupies several periods, the formula for calculating Green's function should be modified:

$$\bar{G}(\mathbf{r}, \mathbf{r}_0) = \int_{-0.5}^{0.5} G(\mathbf{r}, \mathbf{r}_0, \Phi) \exp(-2j\pi m\Phi) \, d\Phi \quad [2\pi(m-1) < x_2 \leq 2\pi m] \tag{9.11}$$

Lemma 9.5: The solution $\bar{u}(\mathbf{r})$ to the problem of diffraction of the field of a local source on an arbitrary one-dimensional periodic grating satisfying the radiation conditions given by eqn. 9.9 is unique.

Assume that the homogeneous excitation problem $(\bar{f}(\mathbf{r}) \equiv 0)$ has a non-trivial solution. Then, due to the complex power theorem and lemma 9.4

$$\operatorname{Re} P(u, \mathbf{r}) + \sum_k |B_k^-|^2 \operatorname{Re} P(v_k^-, -\mathbf{r})$$
$$+ \sum_k |B_k^-|^2 \operatorname{Re} P(v_k^+, \mathbf{r}) + \operatorname{Re} \tilde{P}(\Sigma, \mathbf{r})$$
$$= -\int_V \sqrt{\frac{\epsilon_0}{\mu_0}} \operatorname{Im} \epsilon |\mathbf{E}|^2 \, dV \tag{9.12}$$

where V is the volume bounded by the planes $|x_1| = 0.5$, cylinder's surface $r' = R$ and the perfectly conducting surfaces S (see Figure 9.2c). The values $\operatorname{Re} P(u, \mathbf{r})$, $\operatorname{Re} P(v_k^-, -\mathbf{r})$ and $\operatorname{Re} P(v_k^+, \mathbf{r})$ are strictly positive and they describe the flows of energy carried away by the radiating cylindrical wave and correct real grating's eigenwaves. $\operatorname{Re} \tilde{P}(\Sigma, \mathbf{q})$ which is much less than the first three terms of eqn. 9.12 corresponds to the total energy of the cylindrical wave interacting with grating's eigenwaves. Hence the left-hand and right-hand sides of eqn. 9.12 have different signs and this contradiction completes the proof of lemma 9.5.

Now we can formulate the main assertion of the present Section which has been proved in the process of the above considerations.

Theorem 9.3: Assume that Green's function $G(\mathbf{r}, \mathbf{r}_0)$ of the problem given by eqns. 7.1–7.3 describing a monochromatic quasi-periodic excitation of an arbitrary one-dimensional periodic grating have only simple poles and the frequency κ does not take critical values (i.e. among the real eigenwaves there are no waves carrying the energy along the grating). Then the radiation conditions given by eqn. 9.9 separates a unique solution of the problem of grating's excitation by a local source (in the plane x_2, x_3).

In fact, Green's function $G(\mathbf{r}, \mathbf{r}_0) = G(\mathbf{r}, \mathbf{r}_0, \kappa)$ may be considered at a wider complex manifold of κ.

Lemma 9.6: Green's function $G(\mathbf{r}, \mathbf{r}_0, \kappa)$ of the problem given by eqns. 7.1–7.3 describing a monochromatic quasi-periodic excitation of an arbitrary one-dimensional periodic grating may be continued to the Riemann surface $\mathcal{H}$ of the Hankel function $H_0^{(1)}(\kappa)$ as a meromorphic function of κ.

The complete proof would duplicate the results of Chapter 7 and so we will only make some important remarks: the canonical Green's function $\bar{G}(\mathbf{r}, \mathbf{r}_0, \kappa)$ (of the whole plane without a scattering object) coincides with $\frac{i}{4} H_0^{(1)}[\kappa \sqrt{(x_2 - x_{20})^2 + (x_3 - x_{30})^2}]$ $(0 < x_{20} < 2\pi)$ and for $\operatorname{Im} \epsilon \neq 0$ it has no positive real poles on the first sheet of $\mathcal{H}$ (together with $G(\mathbf{r}, \mathbf{r}_0, \kappa)$); if $\operatorname{Im} \epsilon \equiv 0$, the critical values of κ (when there exist real eigenwaves carrying the energy along the grating) are the poles of $G(\mathbf{r}, \mathbf{r}_0, \kappa)$ and they continuously depend on nonspectral parameters.

Numerical investigations of the spectral problem for various types of gratings which essentially complement theoretical analysis, enables us to assert that the κ-eigenvalues come to the real axis for certain real values of Φ corresponding to the regions where there exist the free field harmonics propagating in the direction $|x_3| \to \infty$. For real Φ and κ the fields of free oscillations and correct grating's eigenwaves are identical. Hence the latter may appear in the "prohibited" zone $\kappa > \Phi$. In fact, when it was proved that grating's surface waves could not exist in the domain $\kappa > \Phi$, it was not taken into account the possibility of vanishing the amplitudes of propagating harmonics.

Chapter 10

Intertype oscillations and waves of gratings and the Morse critical points

Smooth variations of spectral characteristics as functions of nonspectral parameters may be violated in the regions where interaction of oscillations (and waves) of gratings corresponding to different eigenvalues (fundamental frequencies κ or longitudinal wavenumbers Φ) is observed. This phenomenon takes place when spectral points approach each other in the complex space and the conditions providing existence of eigenoscillations or eigenwaves belonging to one symmetry class but different families become close and they exchange their spectral characteristics.

Analytical theory describing the intertype interaction of oscillations which has been developed in Chapters 1 and 3 for open resonators is completely transferable to the case of gratings. We will analyse in this Chapter the interaction of oscillations and waves in gratings using both the numerical data and the methods of the theory of singularities of smooth mappings.

10.1 Intertype oscillations of gratings

Let us consider intertype oscillations of gratings depicted in Figures 10.1 and 10.2.

The oscillations whose fundamental frequencies are situated on the parts

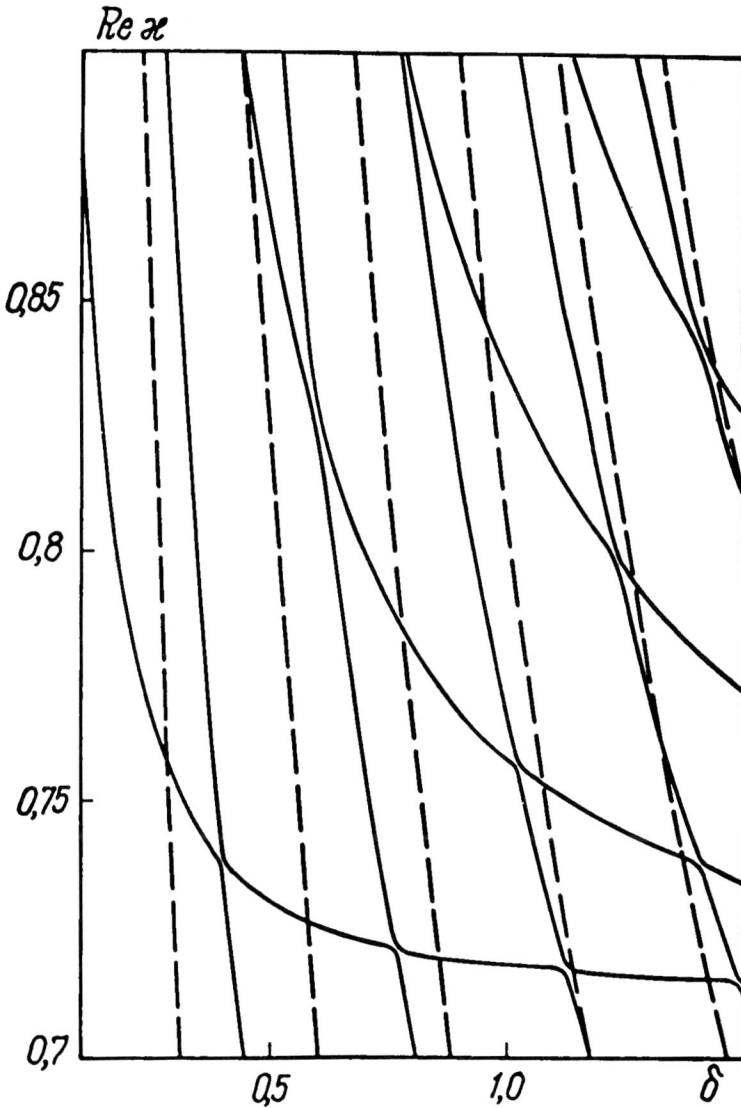

Figure 10.1   *Fundamental frequencies of a bar grating against the height of bars*

of spectral curves belonging to a definite family are identified by the cut-off points where the spectral curves start (these cut-off points correspond to the H_{0m} oscillations where the index m determines the number of field's

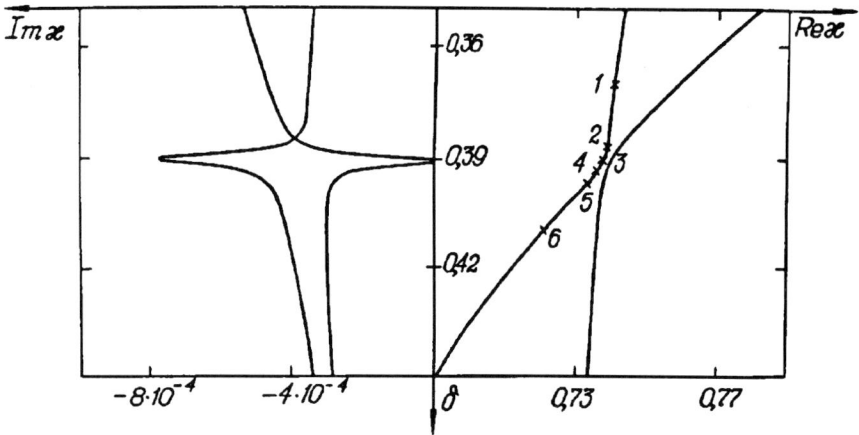

Figure 10.2   *The same data as in Figure 10.1 on enlarged scale*

variations along the width of communication channels) and by curve's number in the family (determining the number of field's variations along the height of communication channels).

One can see that close to the point $\delta \approx 0.39$ the direction of one of the spectral curves gradually changes, the imaginary parts of fundamental frequencies $\mathrm{Im}\,\kappa(\delta)$ are equalised and the spectral curve acquires the features of the curves from the third family (accumulating in the vicinity of the H_{03} cut-off frequency $\kappa \approx 0.71$), i.e. after the interaction point the oscillations from different families have exchanged their characteristics connected with the dynamics of variation (with respect to the parameter δ) and oscillation's type. Interaction of oscillations with equal Q-factors leads to the growth of the Q-factor of one of them and the fall of the other (especially in the point of their maximal rapprochement). Out of the interaction zone the field of one of the interacting oscillations is determined by the H_{03} wave structure. Even after the intersection of the imaginary values of fundamental frequencies oscillation's type is mainly preserved. For $\delta \approx 0.393$ the isolines $|E_{x_1}| = \mathrm{const}$ are similar to those of the H_{023} oscillation and for greater values of δ the observed eigenoscillation may be classified as H_{023} (beginning from $\delta \approx 0.41$), i.e. as the oscillation has the same type (H_{023}) as before the interaction corresponding to the H_{023} spectral curve.

The spectral surfaces $\kappa(\Phi, \delta)$ with $\mathrm{Re}\,\kappa > 0$, $\mathrm{Im}\,\kappa \leq 0$, $\mathrm{Re}\,\Gamma_0 > 0$, $\mathrm{Re}\,\Gamma_n \leq 0$ ($n \neq 0$) situated on the first sheet of the Riemann surface $\mathcal{H}$ between the first

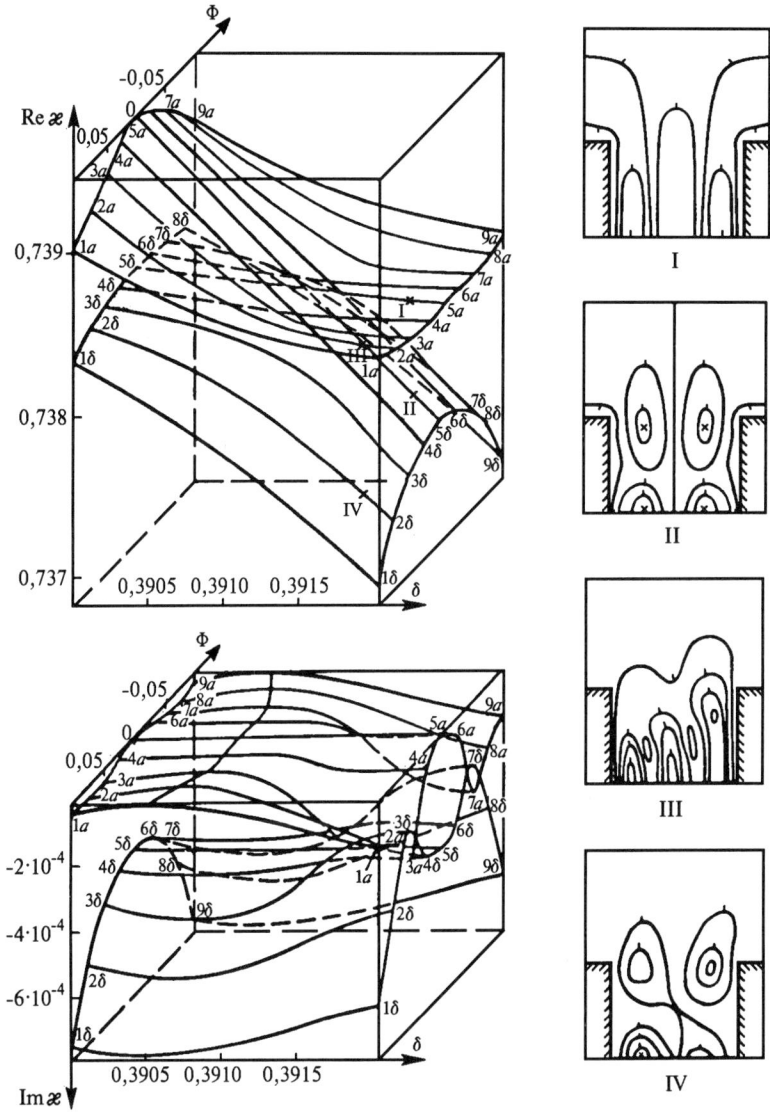

Figure 10.3 *Spectral surfaces of eigenoscillations and fields' distri-butions on a grating*

two cuts and corresponding to the eigenoscillations symmetric with respect to the plane $x_3 = 0$ (and having only one radiation channel open to the free space, i.e. only one harmonic of the radiation field carries away the energy to

the region $|x_3| > 2\pi\delta$) are depicted in Figure 10.3 (for $\epsilon = 6.92$ and $\theta = 0.8$).
Crosses indicate the points ($\Phi = 0.075, \delta = 0.3918, \kappa = 0.73833 - j1.722 \cdot 10^{-4}$
for the third fragment and $\kappa = 0.73741 - j4.218 \cdot 10^{-4}$ for the fourth fragment)
where the isolines $|E_{x_1}| = $ const are calculated. Let us note that the results
presented in Figure 10.2 may be obtained by constructing the cross-section of
the spectral surface $\kappa(\Phi, \delta)$ by the plane $\Phi = 0.1$. Interaction of oscillations
with practically equal Q-factors is observed in Figure 10.3 for $|\Phi| > 0.05$.

Two classes of symmetry appear when $\Phi \to 0$: the H_{031} and H_{023} oscillations are symmetric with respect to the axis $x_2 = 2\pi n$ ($n = 0, \pm 1, \pm 2, \ldots$)
for $\Phi = 0$. And what is more, the free oscillations for $\Phi \neq 0$ (or $\Phi = 0$) are
symmetric only with respect to the plane $x_3 = 0$ [or to the axis $x_2 = 2\pi n$
($n = 0, \pm 1, \pm 2, \ldots$)]. Analysis of the cross-section of the spectral surface
$\kappa(\Phi, \delta)$ by the plane $\Phi = 0$ enables us to assert that although the oscillations
from the different classes of symmetry do not interact the super-high Q-factor
oscillations may appear (the line on the surface $\kappa(\Phi, \delta)$ joins the points with
$\text{Im}\,\kappa(\Phi, \delta) = 0$).

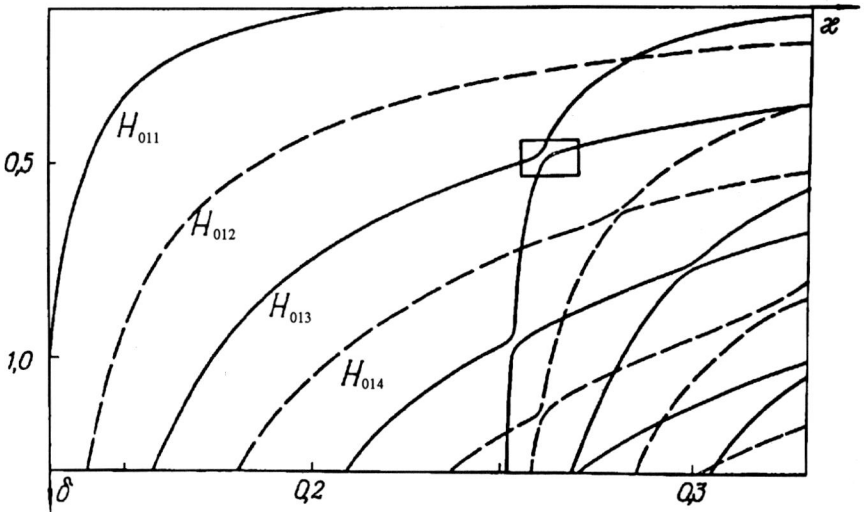

Figure 10.4   *Frequency characteristics of the symmetric and anti-*
symmetric oscillations of a grating

Oscillations from the same class of symmetry may interact if their diffraction Q-factors are equalised. Hence for complex fundamental frequencies only

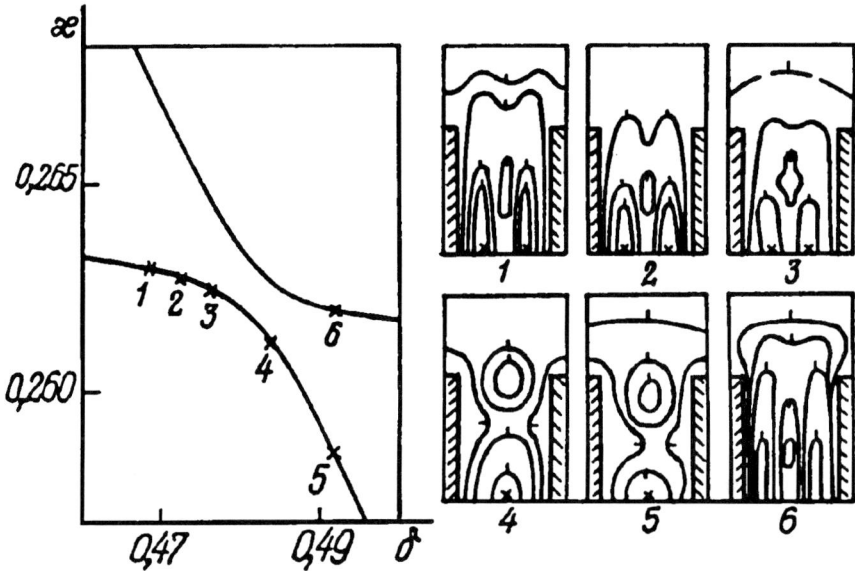

Figure 10.5 *Spectral curves of the intertype oscillations and corresponding field distributions of a grating*

those oscillations interact which are formed by the higher order counter-H_{0m} waves ($m > 1$) propagating in the waveguide channels, i.e. belonging to the mth family. The results presented in Figures 10.4 and 10.5 confirm the possibility of appearance of the interacting high Q-factor oscillations at real fundamental frequencies. The spectral lines of the different symmetry classes intersect and form typical interaction pictures (in Figure 10.5 for the cases of H_{021}-H_{013} and H_{013}-H_{021} interactions) while the lines of the same class have no intersection points The bold (dashed) lines in Figure 10.4 correspond to the fundamental frequencies of the oscillations which are symmetric (antisymmetric) with respect to the plane $x_3 = 0$. In fact, for the real values of κ and Φ the fields of free oscillations and correct (surface) eigenwaves of the gratings (considered as open waveguides) are identical and the results discussed above may be thus understood both in terms of eigenoscillations and eigenwaves.

Interaction of oscillations and waves may be connected with their degeneration (an example is shown in Figure 10.4) but the latter is not always realised in the "physical" domain of the variation of nonspectral parameters.

Degeneration is violated by small deviations of the parameters from the "ideal" values (for example, by variation of the period) which may result in the intertype interaction of oscillations (belonging to one symmetry class).

Interaction of oscillations is a characteristic which is stable with respect to certain (independent) parameters, e.g. with respect to $\text{Im}\,\epsilon$ varying in the interval $0, \hat{\epsilon}$ where $\hat{\epsilon} \ll 1$. Numerical investigation of the H_{031}-H_{023} interaction when $\text{Im}\,\epsilon \in (0, 0.05)$ confirmed this conclusion.

10.2 Super-high Q-factor oscillations of gratings with open radiation channels

Interaction of oscillations corresponding to the values of complex spectral parameter κ varying in the domain $\text{Re}\,\kappa > 0$, $\text{Im}\,\kappa \leq 0$, $\text{Re}\,\Gamma_0 > 0$, $\text{Re}\,\Gamma_n \leq 0$ ($n \neq 0$) situated on the first sheet of the Riemann surface $\mathcal{H}$ between the first two cuts (when only one radiation channel is open to the free space, i.e. only one harmonic of the radiation field carries away the energy to the region $|x_3| > 2\pi\delta$) is presented in Figure 10.6 (for $\epsilon = 6.92$, $\theta = 0.8$ and $\Phi = 0.10$) Let us consider the (symmetric) oscillations as functions of the increasing δ. In the first case the H_{035} (dot-dashed line) and H_{023} (dashed line) oscillations interact. They belong to the third and second families, respectively. As a result of interaction, the Q-factor of the oscillation from the second family sharply grows in the points marked by short lines parallel to the axis $\text{Im}\,\kappa$, the imaginary part of the fundamental frequency vanishes and the super-high Q-factor oscillation appears. The subsequent super-high Q-factor oscillations emerge after the interaction of the oscillations from the first family (bold line) with the oscillations from the third, the second and again the third families (see Figure 10.6). Some quantitative differences of the interaction processes are caused here by the different Q-factors of interacting oscillations.

The uniqueness theorems do not exclude the generation of the super-high Q-factor oscillations for $\text{Im}\,\epsilon = 0$ in the cases when the formally open radiation channels are actually closed. One can assert that the formation of super-high Q-factor oscillations takes place when the radiation field [in open waveguide structures with noncompact (periodic) boundaries] contains a finite number of harmonics carrying away the energy. Let us note that the interaction of

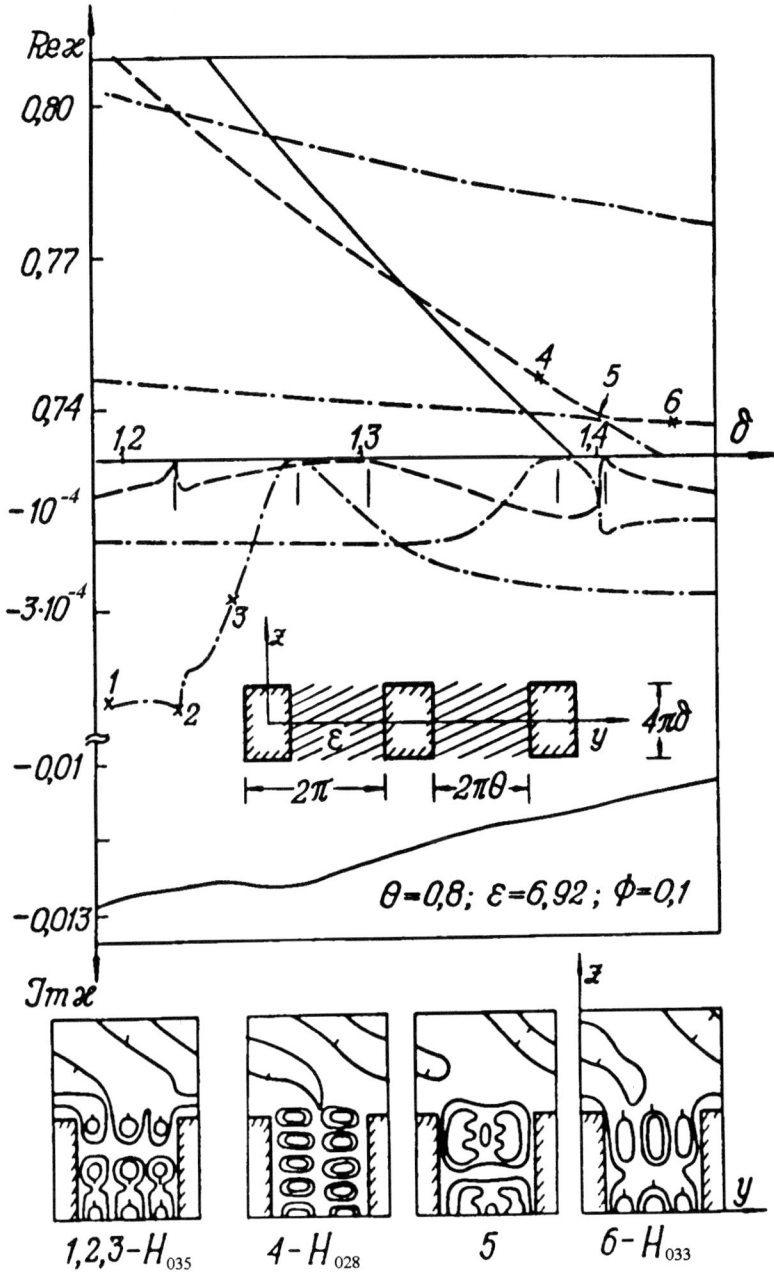

Figure 10.6 *Different types of interaction leading to generation of the super-high-Q-factor oscillations of a grating*

oscillations when only one radiation channel is open to free space is the
best perspective for realisation of various "super-high Q-factor" effects. If
a greater number of channels are open, the probability of generating super-
high Q-factor oscillations falls because it is connected with carrying out the
parameter optimisation of the problem.

Let us analyse the emergence of eigenoscillations with real fundamental frequencies in terms of the functional properties of dispersion equations. Assume
that two waves (H_{0m} and H_{0p}) propagate in the slots without attenuation for
$Im\,\kappa = 0$. Then the dispersion equation will have the form

$$[R_{mn} \exp(4j\pi\omega_n\delta \mp 1][R_{mn} \exp(4j\pi\omega_n\delta \mp 1]$$
$$= R_{pm}R_{mp} \exp[4j\pi(\omega_n + \omega_p)] \tag{10.1}$$

where R_{mn} is the amplitude $H_{0m} \to H_{0n}$ transformation coefficient in the
planes $|x_3| = 2\pi\delta$. In order to prove the solvability of eqn. 10.1, we will
assume that the following phase relations are valid:

$$\arg R_{mm} + \omega_m 4\pi\delta \mp 1 = \pi n, \quad \arg R_{pp} + \omega_p 4\pi\delta \mp 1 = \pi k \tag{10.2}$$

$$\arg R_{mm} + \arg R_{pp} = 2\arg R_{mp} \tag{10.3}$$

where $n = 2a, k = 2b$ ($n = 2a+1, k = 2b+1$) for the eigenoscillations symmetric (anti-symmetric) with respect to the plane $x_3 = 0$ ($a, b = 0, \pm 1, \pm 2, \dots$).
Eqns. 10.2 and 10.3 enable one to determine the "centre" of the H_{0nm}-H_{0pk}
interaction zone, the necessary conditions for the H_{0m}-H_{0p} longitudinal resonance and to reduce eqn. 10.1 to a simple relation

$$(|R_{mm}| - 1)(|R_{pp}| - 1) = |R_{mm}||R_{pp}| \tag{10.4}$$

Due to the known properties of the corresponding scattering matrix eqn. 10.4
holds when one radiation channel is open to free space, i.e. we have proved that
eigenoscillations with real fundamental frequencies may appear in the zones of
interaction of different types of oscillations belonging to one symmetry class.

The spectrum of the super-high Q-factor oscillations of a grating with open
radiation channels is much more rarefied than in the case when $Re\,\Gamma_n = 0$ for
all $n = 0, \pm 1, \pm 2, \dots$. In fact, if radiation channels are closed, the dispersion
curves $\kappa(\delta)$ form a regular net covering the plane κ, δ and the structure of this

net is determined by the number of nonattenuating waves in the communication channel Q (see Figure 10.6). If one radiation channel is open [$\mathrm{Re}\,\Gamma_0 > 0$ and $\mathrm{Re}\,\Gamma_n = 0$ $(n = 0, \pm 1, \pm 2, \dots)$], super-high Q-factor oscillations may appear only in the knots of corresponding nets and the growing number of open radiation channels inevitably leads to the rarefying of the spectrum of super-high Q-factor oscillations.

For the real values of κ and Φ the super-high Q-factor oscillations are simultaneously the correct (surface) eigenwaves of a grating propagating in the direction of its periodicity. The domain $\kappa > |\Phi|$ corresponds for $|\Phi| < 0.5$ to one or more open radiation channels and is classified as a "prohibited" region for the correct real ($\mathrm{Im}\,\Phi = 0$) slow eigenwaves. If only one radiation channel is open to free space ($\mathrm{Re}\,\Gamma_0 > 0$), the amplitudes a_0 and b_0 of the plane waves (the partial components of a super-high Q-factor eigenoscillation or a correct eigenwave) vanishes in the domain $|x_3| > 2\pi\delta$, the region $\kappa > |\Phi|$ does not differ from the domain of slow eigenwaves $\kappa < |\Phi|$ and the existence of a slow eigenwave does not look unusual.

We may conclude that the main qualitative feature of the effects discovered is their "point-wise" character: these effects are observed only when at least two conditions are combined which are determined by the spectral curves of open periodic waveguides and by the number of open (closed) radiation channels.

10.3 Intertype waves in gratings

Intertype interaction of slow waves with equal real propagation constants have been discussed in Section 10.1. Let us consider here the interaction of waves corresponding to the complex longitudinal wavenumbers Φ beginning with the analysis of the grating supporting two slow H_{013} waves with longitudinal wavenumbers Φ and $1 - \Phi$ up to the value $\kappa \approx 0.262$ [see Figure 10.7 where the cross-section of the spectral surface $\Phi(\kappa, \delta)$ by the plane $\delta = 0.489$ is considered and crosses indicate the points of the spectral curves where the isolines $|E_{x_1}| = \mathrm{const}$ have been calculated]. This case is especially interesting because here both interacting eigenwaves change types ($H_{013} \to H_{021}$) while before we have observed either local changes of eigenwaves (pure type

→ hybrid type → reconstruction of the initial pure type) or exchange of types (two different pure types → hybrid types → exchange of pure types). When $\Phi = 0.5$ the real eigenwaves "collide" and become complex eigenwaves. After the collision point $\text{Re}\,\Phi = \text{const} = 0.5$ and the values of $\text{Im}\,\Phi$ have different signs. The fields also transform in opposite directions (see fragments 2-5 in Figure 10.7) and acquire the shape of the H_{021} wave. The next collision occurs in the vicinity of the point $\kappa \approx 0.262$ where the complex eigenwaves become the real slow H_{021} waves (fragment 6 in Figure 10.7).

Inevitability of the appearance of new ways of interaction (and, in particular, the case considered above) is caused by the structure of the spectral surface $\Phi(\kappa, \delta)$ and by the fact that the spectral components Ω_Φ cannot disappear in any finite part of $\mathcal{F}$. For example, the slow waves corresponding to the spectral curves $\Phi(\kappa, 0.487)$ and $\Phi(\kappa, 0.489)$ shown in Figures 10.4, 10.5 and 10.7 have definite constant types. At the same time, different cross-sections of the space $\kappa \times \delta \times \Phi$ produce different versions of overcoming of the non-transmitting zones by slow waves. One such version, when two H_{021} waves collide and pass to the domain of complex Φ as hybrid waves (influenced by the H_{01} wave) has been considered in this Section.

Slow waves with the propagation constants $\Phi = \pm 0.1$ carry the energy in opposite directions. The lines $\pm\Phi(\delta)$ $(\pm\Phi(\kappa))$ lie on the first sheet of $\mathcal{F}$. The series of pictures along the axes $\text{Re}\,\Phi$ and $\text{Im}\,\Phi$ illustrate the formation of slow waves after the collision of two complex waves with $\text{Im}\,\Phi > 0$ and $\text{Im}\,\Phi < 0$ close to the points $\Phi = 0$, $\kappa = 0.6098$ and $\Phi = 0$, $\kappa = 0.7149$; the same phenomenon is presented in Figure 10.3 where the slow H_{023} waves (transforming to the hybrid waves with growing δ and to the H_{031} waves with decreasing δ) have the propagation constants $\Phi = \pm 0.07$ and their spectral curves are symmetric with respect to the imaginary axis.

The points of formation of the slow H_{023} waves (curves $1a$ and $1b$ and fragments 1-3 in Figure 10.8) and the slow H_{041} waves (curves $2a$ and $2b$ in Figure 10.8 and fragments 4,5 in the region where the lines with $\text{Im}\,\kappa = 0$ intersect) may be found in the cross-section of the surface $\kappa(\Phi, \delta)$ by the plane $\kappa = \text{const}$. Crosses in Figure 10.8 indicate the points where the spectral curves $\Phi(\delta)$ $(1a, b$ and $2a, b)$ rest against the cuts of the first sheet of $\mathcal{F}$. The propagation constants of the H_{041} waves acquire a small complex addition with decreasing δ which does not exceed $2.5 \cdot 10^{-4}$, while the imaginary part of the

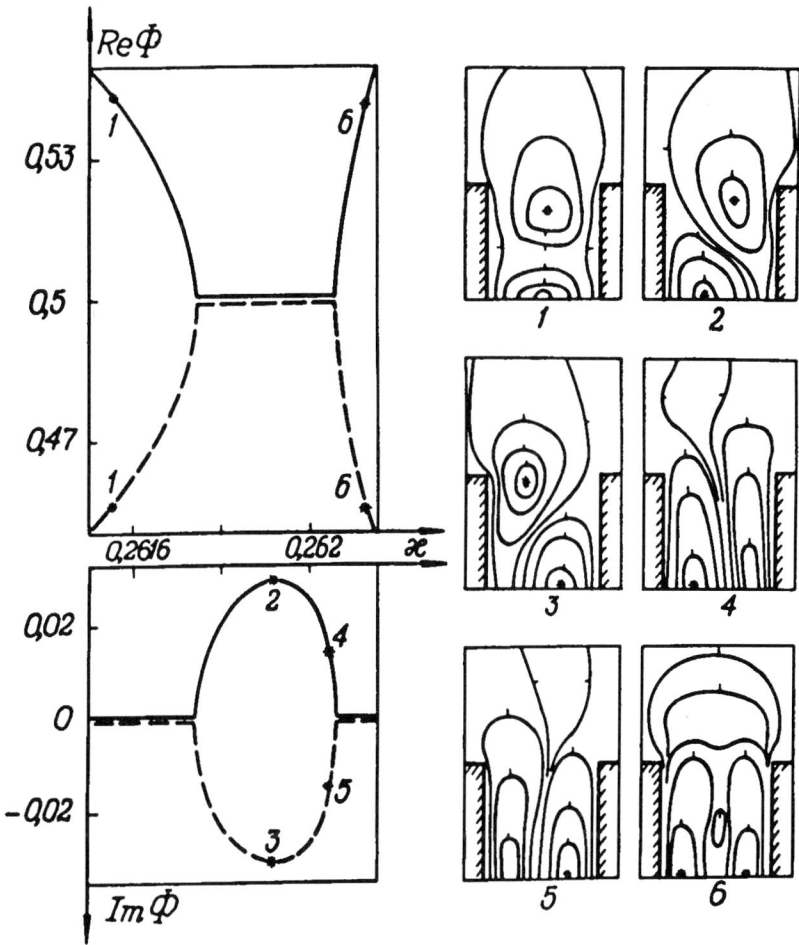

Figure 10.7 *Spectrum of propagation constants and corresponding field distributions of a grating*

propagation constants Φ of the H_{023} waves (transforming to the H_{041} waves) is characterised by an anomalous behaviour with increasing $\delta \in (0.2465, 0.2474)$. Further growth of $\delta > 0.2474$ leads to the reconstruction of the initial type (fragment 3 in Figure 10.8).

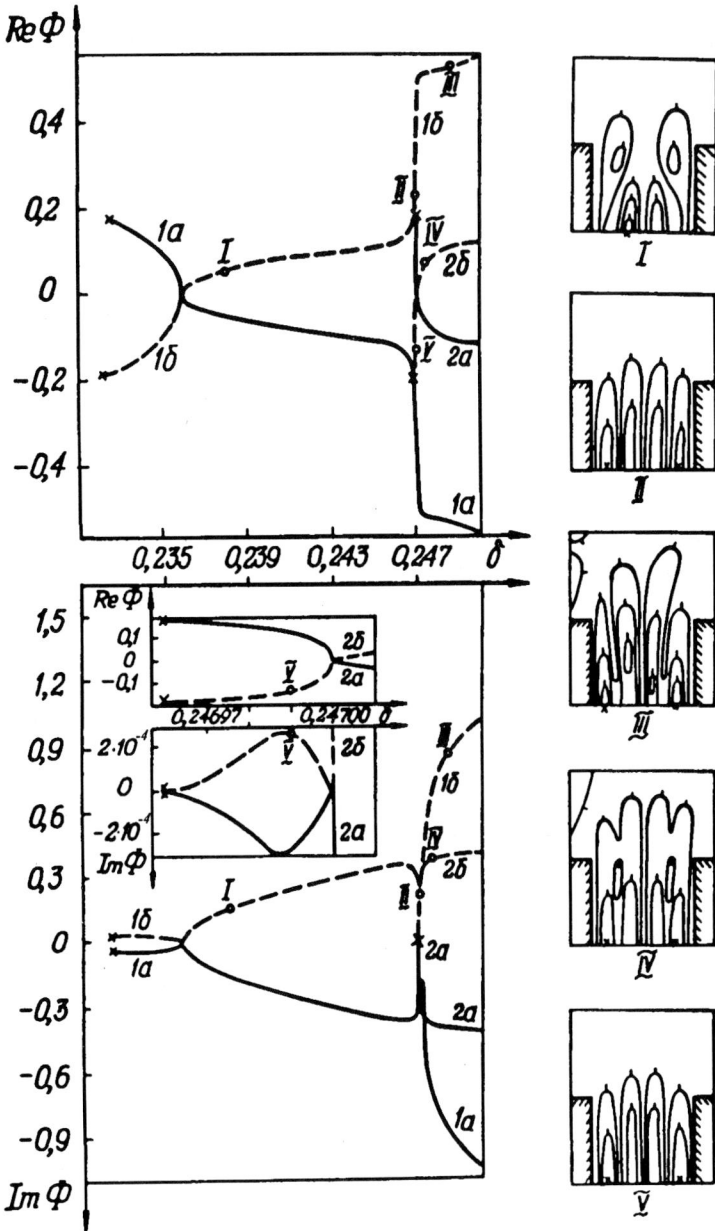

Figure 10.8   *Propagation constants of a bar grating against the height of bar and field distributions*

10.4 The Morse critical points and free oscillations and waves of gratings

In this Section we will study the spectral characteristics of gratings with respect to two spectral parameters κ and Φ formally introducing the vector spectral parameter $\mathbf{k} = (\kappa, \Phi)$. Analytical descriptions of spectral properties will be carried out by means of the analysis of singularities (or the Morse critical points) of complex mappings given by functions $f(\mathbf{k}, \tau)$ of several complex variables where τ denotes a chosen nonspectral parameter.

As has been shown in Chapter 7, the study of spectral characteristics of open periodic waveguides is reduced to the problem on characteristic numbers for a Fredholm matrix operator-valued functions $I + A(\mathbf{k}, \tau)$ which, as well as in Section 1.10, may be rewritten in the form of a dispersion equation

$$F(\mathbf{k}, \tau) \equiv \det\{I + A(\mathbf{k}, \tau)\} = 0 \tag{10.5}$$

where $\mathbf{k} \in \Omega = \Omega(\tau) \subset \mathcal{H} \otimes \mathcal{F}$ is the set of zeros of a meromorphic function $F(\mathbf{k}, \tau)$ considered for $\tau \in R$ (Im $\tau = 0$) in the Cartesian product of the Riemann surfaces $\mathcal{H}$ and $\mathcal{F}$ where the spectral parameters κ and Φ vary. Due to the analytical properties of $I + A(\mathbf{k}, \tau)$ established in Sections 7.1 and 7.2 Ω coincides with the set of its characteristic numbers and, consequently, $\Omega = \Omega(\kappa, \Phi)$ may be called the spectral surface of the operator-valued function $I + A(\mathbf{k}, \tau)$ (or of a given open waveguide resonator). The set Ω may be naturally considered as a hypersurface

$$\Omega(\kappa, \Phi, \tau) = \{(\kappa, \Phi, \tau) : \ F(\kappa, \Phi, \tau) = 0\}$$

which coincides with the spectral surfaces $\Omega_\kappa(\Phi, \tau)$ and $\Omega_\Phi(\kappa, \tau)$ in the "physical" domain of the spectral parameters (where Im $\Phi = 0$ and Im $\kappa = 0$).

Violation of the regular behaviour of spectral curves $\kappa(\tau) \subset \Omega_\kappa$ and $\Phi(\tau) \subset \Omega_\Phi$ may be connected with the presence of the Morse critical point $\mathbf{k}_0, \tau_0$ of the complex mapping given by the function $F(\mathbf{k}, \tau) : \ C \times C \to C$. In a vicinity of this point

$$\frac{\partial F}{\partial \mathbf{k}} = 0, \quad \frac{\partial F}{\partial \tau} = 0, \quad (\mathbf{k}, \tau) = (\mathbf{k}_0, \tau_0) \tag{10.6}$$

$$\frac{\partial^2 F}{\partial \mathbf{k}^2} \frac{\partial^2 F}{\partial \tau^2} - \frac{\partial^2 F}{\partial \mathbf{k} \partial \tau} \frac{\partial^2 F}{\partial \tau \partial \mathbf{k}} \neq 0, \quad (\mathbf{k}, \tau) = (\mathbf{k}_0, \tau_0) \tag{10.7}$$

Hence the Hesse matrix $J = \|J_{il}\|_{i,l=1}^3$, $J_{il} = F_{z_i z_l}$ [where all the derivatives are calculated in the Morse critical point $(\mathbf{k}_0, \tau_0)$, $z_1 = \kappa$, $z_2 = \Phi$, $z_3 = \tau$] is nondegenerated, the spectral curves may be reconstructed in a vicinity of the Morse critical point from the quadratic equation

$$\frac{\partial^2 F}{\partial \mathbf{k}^2}(\mathbf{k}-\mathbf{k}_0)^2 + 2\frac{\partial^2 F}{\partial \mathbf{k} \partial \tau}(\mathbf{k}-\mathbf{k}_0)(\tau-\tau_0) + \frac{\partial^2 F}{\partial \tau^2}(\tau-\tau_0)^2 + 2F(\mathbf{k}_0, \tau_0) = 0 \quad (10.8)$$

up to the accuracy of small cubic terms and the hypersurface $\Omega(\kappa, \Phi, \tau)$ may be represented in the form

$$\frac{\partial^2 F}{\partial \kappa^2}(\kappa - \kappa_0)^2 + \frac{\partial^2 F}{\partial \Phi^2}(\Phi - \Phi_0)^2 + \frac{\partial^2 F}{\partial \tau^2}(\tau - \tau_0)^2 + 2\frac{\partial^2 F}{\partial \kappa \partial \Phi}(\kappa - \kappa_0)(\Phi - \Phi_0)$$
$$+ \frac{\partial^2 F}{\partial \kappa \partial \tau}(\kappa - \kappa_0)(\tau - \tau_0) + \frac{\partial^2 F}{\partial \Phi \partial \tau}(\tau - \tau_0)(\Phi - \Phi_0) + 2F(\mathbf{k}_0, \tau_0) = 0$$
$$(10.9)$$

Eqn. 10.9 enables one to determine the dynamics of variation of the spectral characteristics of both the eigenwaves and eigenoscillations in a vicinity of one and the same Morse critical point.

Numerical investigations of the behaviour of spectral characteristics in the (isolated and nondegenerated) Morse critical points have been carried out on the basis of approximate solution to eqn. 10.9 (with verifying the non-degenerate conditions given by eqns. 10.6 and 10.7) by Newton's method (the average computation accuracy is 10^{-6}). In order to illustrate the application of this technique, we will compare the spectral characteristics of gratings made of perfectly conducting bars filled with dielectric which have been obtained by the direct solution of the dispersion equation 10.5 and the data which have been reconstructed in the vicinity of the Morse critical point with the help of eqn. 10.8. Typical calculations are presented in Figures 10.9 and 10.10 where the nonspectral parameter τ equals to grating's height. The spectral curves $\kappa(\tau)$ (Figure 10.9) and $\Phi(\tau)$ (dashed lines in Figure 10.10) calculated close to the Morse critical points $\kappa_0 = 0.26212$, $\tau_0 = 0.48754$ and $\Phi_0 = 0.49862$, $\tau_0 = 0.49156$, respectively, graphically coincide with the results of numerical solution to eqn. 10.5.

Considering gratings with open radiation channels, the singular points of the hypersurfaces $\Omega(\kappa, \Phi, \tau)$ (determined by eqn. 10.9) and the Morse critical points where the spectral curves are reconstructed analytically (by means of eqn. 10.8) may "pass" to the "nonphysical" (complex) domain and the

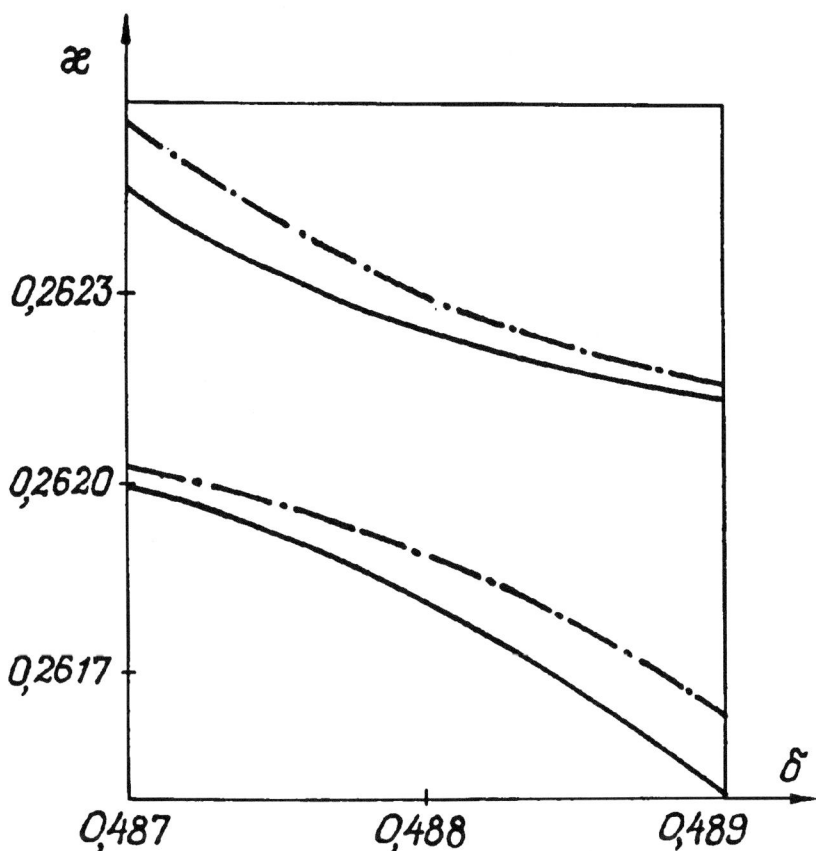

Figure 10.9 *Spectral curves of the eigenoscillations of a grating near the Morse critical point*

distance from a singular point to a given spectral characteristic may become greater than the accuracy of the method. As an example, one may consider the fragment of the spectral surface $\Omega_\kappa(\Phi, \tau)$ calculated by the dispersion equation 10.5 whose points determine the conditions of the existence of the interacting H_{031} and H_{023} oscillations (see Figure 10.3). The dynamics of variation of the eigenoscillations on this hypersurface may be reconstructed in the vicinity the singular point $\kappa_0 = 0.73839 - j3.9712 \cdot 10^{-4}$, $\Phi_0 = 0, \tau_0 = 0.39110 + j4.2320 \cdot 10^{-4}$, while calculation of the spectral characteristics $\Omega_\Phi(\kappa, \tau)$ carried out by eqn. 10.9 does not give satisfactory results. In such a situation the spectral surface $\Omega_\Phi(\kappa, \tau)$ may be recreated by fragments of the spectral curves

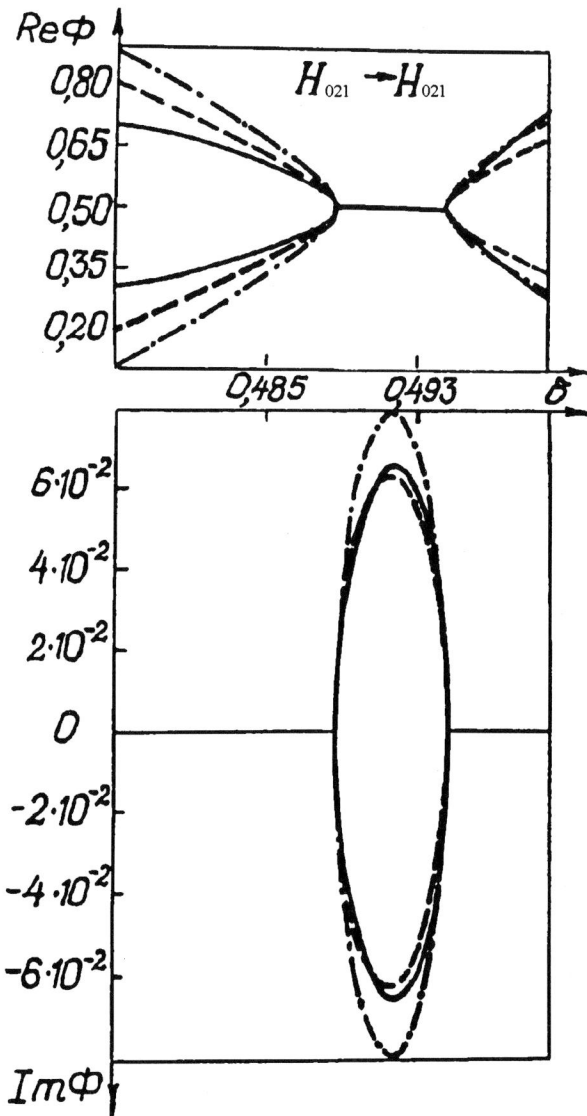

Figure 10.10   *Spectral curves of propagation constants of a grating*
near the Morse critical point

determined (numerically by Newton's method) with the help of eqn. 10.8 in
the vicinity of the Morse critical point (Φ_0, τ_0) [and a singular point of the
hypersurface $\Omega(\kappa, \Phi, \tau)$ is taken as the initial value in Newton's method].

References

ADAMS, R. (1975): 'Sobolev spaces' (Academic Press)

AGRANOVICH, Z.S., MARCHENKO, V.A., and SHESTOPALOV, V.P. (1962): 'Diffraction of electromagnetic waves by plane metal gratings', *Zhurnal Tekhnicheskoi Fiziki.* **32**, pp. 381-394

AKHIEZER, N.I. (1970): 'Foundations of the theory of elliptic functions' (Nauka, Moscow)

ANDREWARSHA, J.R., FOX, J.R., and WILSON, I.J. (1979): 'Resonance anomalies in the lamelian gratings', *Opt. Acta.* **26**, pp. 69-89

ANDROSOV, V.P., VELIEV, E.I. and VERTIJ, A.A (1983): 'Polarization and spectral characteristics of open resonators with internal inhomogeneities', *Izvestiya VUzaved.–Radiofizika.* **26**, pp. 318-328

ARNOLD, V.I. (1972): 'Modes and quasimodes', *Funkcionalny Analis i Ego Prilozheniya.* **6**, pp. 12-20

ARNOLD, V.I., VARCHENKO, A.N., and GUSEIN-ZADE, S.M. (1973): 'Singularities of differentiable mappings' (Nauka, Moscow)

ARSENJEV, A.A. (1976): 'On existence of resonant poles and resonances for the scattering in the case of the second and third kind boundary-value conditions', *Zhurnal Vychiclitel'noi Matematiki i Matematicheskoi Fiziki.* **16**, pp. 718-724

ASTAPENKO, A.A. (1971): 'Asymptotics of the solution to the problem of a long plane electromagnetic wave diffraction on a grating of arbitrary bars', *Radiotekhnika i Elektronika.* **16**, pp. 1589-1596

ASTAPENKO, A.A., and MALUZHINEC, G.D. (1970): 'Diffraction of plane sound wave by a grating with small period', *Akustichesky Zhurnal.* **16**, pp. 354-362

AVDEEV, E.V., and VOSKRESENSKY, G.V. (1969): 'Electromagnetic excitation of a periodic lattice structure made of conducting strips', *Radiotekhnika i Elektronika.* **14**, pp. 839-860

AVERBAKH, V.S., VLASOV, S.N. and TALANOV, V.M. (1965): 'On the influence of the first and second order aberrations on the characteristics of open resonators', *Radiotekhnika i Elektronika.* **10**, pp. 1150-1159

BABICH, V.M., and BULDIREV, V.S. (1972): 'Asymptotic methods for the problems of short waves diffraction' (Nauka, Moscow)

BAGBAYA, I.D. (1972): 'On a history of diffraction grating', *Uspekhi Fizicheskikh Nauk.* **108**, pp. 335-337

BALDWIN, G.L. (1954): 'On the diffraction of a plane wave by an infinite plane grating', *Mathematica Scandinavica.* **2**, pp. 103-119

BASS, F.G., and GUREVICH, YU.G. (1975): 'Hot electrons and strong electromagnetic waves in the plasm of a gas discharge and semi-conductors' (Nauka, Moscow)

BASS, F.G, POCHANINA, I.E., and SHESTOPALOV, V.P. (1993): 'Nonlinear evolution equations in a vicinity of the Morse critical points', *Doklady RAN.* **334**, pp. 32-33

BASS, F.G, VERBITSKY, V.I., and SHESTOPALOV, V.P. (1994): 'On automodel solutions of evolution equations close to the Morse critical points', *Doklady RAN.* **335**, pp. 232-235

BELOSTOTSKY, V.V, VASILJEV, E.N., and SEBELNIKOVA, E.V. (1983): 'Open resonator with a dielectric disc'. In: *Sbornik Nauchno-Metodicheskikh Statei po Prikladnoi Elektrodinamike. Vipusk 5*, pp. 225-251 (Nauka).

BIKOV, V.P. (1965): 'Geometric optics of three-dimesional oscillations in open resonators'. In: *Elektronika Bolshikh Moshchnostei. Vipusk 4*, pp. 105-108 (Nauka).

BIKOV, V.P. (1966): 'Beam theory of open resonators and open waveguides', *Radiotekhnika i Elektronika.* **11**, pp. 477-487

BIKOV, V.P. (1968): 'Aberrations in geometric optics of open resonators'. In: *Electronika Bolshikh Moshchnostei. Vipusk 5*, pp. 117-135 (Nauka).

BOITZOV, V.F. (1968): 'Optical resonator with a lens co-ordinated with mirrors', *Optica i Spectroscopia.* **24**, pp. 122-130

BOITZOV, V.F., and FRADKIN, E.E. (1969): 'Diffraction effects in confocal and semi-confocal optical resonators', *Optica i Spectroscopia.* **25**, pp. 765-778

BOLOTOVSKY, Z.S., and LEBEDEV, A.N. (1967): 'On threshold phenomena in classical electrodynamics', *Zhurnal Eksperimental'noi i Teoreticheskoi Fiziki.* **53**, pp. 1349-1359

BORN, N.,and WOLF, E. (1973): 'Foundations of optics' (Nauka, Moscow)

BOROVIKOV, V.A., and KINBER, Y.E. (1978): 'Geometric diffraction theory' (Svyaz, Moscow)

BOROVSKY, I.V., and KHIZHNYAK, N.A. (1983): 'Diffraction of electromagnetic waves by a dielectric comb and grating made of rectangular bars', *Izvestiya VUzaved.-Radiofizika.* **26**, pp. 231-239

BORZENKOV, A.V., and SOLOGUB, V.G. (1975): 'On a scattering of plane wave by two strip resonators', *Radiotekhnika i Elektronika.* **20**, pp. 925-935

BOYD, G.D., and GORDON, I.P. (1961): 'Confocal multimode resonator for millimeter through optical wavelength masers', *Bell Syst. Tech. J.* **40**, pp. 489-507

BOYD, G.D., and KOGELNIK, H. (1962): 'Generalized confocal resonator theory', *Bell Syst. Tech. J.* **41**, pp. 1347-1358

BREKHOVSKIKH, L.M. (1952): 'Waves diffraction by rough surface', *Zhurnal Eksperimental'noi i Teoreticheskoi Fiziki.* **23**, pp. 275-304

BROVENKO, A.V. (1988): 'Numerical algorithm for calculation of spectral characteristics of two-dimensional open resonator with inhomogeneous dielectric inclusions', *Doklady AN USSR. Ser. A.* (1), pp. 51-54

BROVENKO, A.V., MELEZHIK, P.N., POYEDINCHUK, A.E., and SHESTOPALOV, V.P. (1993): 'Electrodynamic characteristics of open resonators with dielectric layers', *Ukrainskii Fizicheskii Zhurnal.* **10**, pp. 812-820

BULDIREV, V.S., and FRADKIN, E.E. (1964): 'Integral equations of resonators', *Optica i Spectroscopia.* **24**, pp. 583-596

BURSHTEIN, E.L., and VOSKRESENSKY, G.V. (1970): 'Linear accelerators of electrons with intensive pencils' (Atomizdat, Moscow)

CARLSON, J.F., and HEINS, A.E. (1947): 'The reflection of an electromagnetic plane wave by an infinite set of plates', *Quart. Appl. Math.* **4**, pp. 313-329

CHERNOKOZHIN, E.V., and SHESTOPALOV, Y.V. (1982): 'Fredholm property of an integral operator with kernel having a weak singularity', *Vestnik MGU, Seriya 15.* **1**, pp. 23-28

CHERNOKOZHIN, E.V., and SHESTOPALOV, Y.V. (1985): 'A method of calculating the characteristic numbers of a meromorphic operator-function', *Zhurnal Vychiclitel'noi Matematiki i Matematicheskoi Fiziki.* **25**, pp. 1413-1416

CORNMAN, V.I. (1973): 'Nonlinear waves in disperse media' (Nauka, Moscow)

COSTABEL, M. (1988): 'Boundary integral operators of Lipschitz domains: elementary results', *SIAM J. Math. Anal.* **19**, pp. 613-626

CULLEN, A.L., NAGENTHIRAN, P., AND WILLIAMS, D. (1972): 'A variational approach to the theory of open resonators', *Proc. Royal Soc. A.* **329**, pp. 153-169

DERYUGIN, L.N. (1952): 'An equation for the coefficients of the reflection of waves from a periodic rough surface', *Doklady AN SSSR.* **87**, pp. 913-916

DERYUGIN, L.N. (1954): 'On a surface resonance in a reflection grating', *Doklady AN SSSR.* **94**, pp. 203-206

DERYUGIN, L.N. (1956): 'On the theory of diffraction by a reflection grating', *Doklady AN SSSR.* **109**, pp. 1003-1006

DERYUGIN, L.N., and FRIDMAN, G.H. (1956): 'Resonance curves of a double resonance in a reflection grating', *Doklady AN SSSR.* **111**, pp. 1209-1211

DOLF, L. (1963): 'Modern development of some non-self-adjoint problems of mathematical physics'. In: *Sbornik Perevodov. Matematika. Vipusk 7*, pp. 79-136 (Nauka).

EIDUS, D.M. (1969): 'Limiting amplitude principle', *Uspekhi Matematicheskikh Nauk.* **24**, pp. 91-156

ENA, M.L., and ILYINSKY, A.S. (1991): 'Computation of the propagation constants and field components of slot transmission lines with regard to the thickness of conductors', *Radiotekhnika i Elektronika.* **36**, pp. 290-296

EPISHIN, V.A. (1978): 'Open resonator with holes in reflectors', *Kvantovaya Elektronika.* **5**, pp. 1263-1271

ERICKSON, C.W. (1977): 'Perturbation theory generalized to arbitrary (p,1) modes of a Fabry-Perot resonator', *IEEE Trans.* **MTT-25**, pp. 958-963

FELD, Y.N. (1975): 'Diffraction of electromagnetic waves by unclosed screens', *Radiotekhnika i Elektronika.* **20**, pp. 28-38

FELD, Y.N., and SUKHAREVSKY, I.V. (1969): 'Application of the non-resonant Green functions to the construction of the integral equations for diffraction problems on unclosed screens', *Radiotekhnika i Elektronika.* **14**, pp. 1362-1368

FOCK, V.A. 1973: 'The problems of diffraction and propagation of electromagnetic waves' (Sovetskoye Radio, Moscow)

FONG, T.T., and SHUNG WU LEE (1971): 'Theory of Fabry-Perot resonators with dielectric medium', *IEEE J. Quant. Electron.* **7**, pp. 1-11

FORSYTHE, G.E., MALCOLM, M.A., and MOLER, C.B. (1976): 'Computer methods for mathematical computations' (Prentice-Hall)

FOX, G.D., and LEE, T. (1961): 'Resonant modes in maser interferometer', *Bell Syst. Tech. J.* **40**, pp. 453-488

GALISHNIKOVA, T.N., and ILYINSKY, A.S. (1987): 'Numerical methods in diffraction problems' (MGU, Moscow)

GAPONOV, A.V., PETELIN, M.I. and YUMATOV, V.K. (1967): 'Induced radiation of excited classical oscillators and its application in classical electrodynamics', *Izvestiya VUzaved.-Radiofizika.* **10**, pp. 1414-1454

GILMORE, R. (1984): 'Applied theory of catastrophes' (Mir, Moscow)

GLOGE, D (1966): 'A general method for calculating optical resonators and periodic lens systems'. In: Quasioptica, pp. 280-314 (Mir).

GOHBERG, I.T. (1951): 'On linear operators analytical with respect to a parameter', *Doklady AN SSSR.* **78**, pp. 629-633

GOHBERG, I.T., and KREIN, M.G. (1965): 'Introduction to the theory of linear non-self-adjoint operators' (Nauka, Moscow)

GOHBERG, I.T., and SIGAL, E.I. (1971): 'Operator generalization of logarithmic residue theorem and Roushet's theorem', *Matematicheskii Sbornik.* **84**, pp. 607-629

GOLDSTEIN, C.I. (1969): 'Eigenfunctions expansions associated with the laplacian for certain domains with infinite boundaries', *Trans. Amer. Math. Soc.* **135**, pp. 1-50

GRIBANOV, Y.I. (1963): 'Co-ordinate spaces and infinite systems of linear equations', *Izvestiya VUzaved.–Matematika.* **3**, pp. 27-39

HAIKONEN, T. (1975): 'On multiple scattering by small particles in open resonators', *Acta Polytechnical Scandinavica.* **35**, pp. 217-223

HARDY, G.G., LITTLEWOOD, D.E., and POLYA, G. (1948): 'Inequalities' (Izdatelstvo Inostrannoi Literatury, Moscow)

HARRINGTON, R.F., and MAUTZ, I. (1971): 'Theory of characteristic modes for conducting bodies', *IEEE Trans.* **AP-19**, pp. 622-639

HESSEL, A., and OLINER, A.A (1965): 'A new theory of Wood anomalies on optical gratings', *Appl. Optics.* **4**, pp. 1275-1297

HIGHMAN, U. (1966): 'Meromorphic functions' (Mir, Moscow)

HILLE, L., and PHILLIPS, R. (1962): 'Functional analysis and semi-groups' (Izdatelstvo Inostrannoi Literatury, Moscow)

HURWITZ, A., and COURANT, R. (1964): 'Vorlesungen Über Allgemeine Funktionentheorie und Elliptische Funktionen' (Springer)

HÖNL, H., MAUE, A.W., and WESTPFAHL, K. (1961): 'Theorie der Beugung' (Springer)

HÖRMANDER, L. (1986): 'Foundations of the theory of linear partial differential operators' (Mir, Moscow)

IGNATOVSKY, V.S. (1938): 'On the theory of grating', *Doklady AN SSSR.* **20**, pp. 105-108

ILYIN, V.A. (1956): 'Kernels of fractional order', *Matematicheskii Sbornik.* **69**, pp. 319-351

ILYINSKY, A.S. (1973): 'Direct method for calculation of periodic structures', *Zhurnal Vychiclitel'noi Matematiki i Matematicheskoi Fiziki.* **13**, pp. 119-126

ILYINSKY, A.S. (1978): 'Electrodynamic theory of antenna gratings', *Izvestiya VUzaved.–Radioelektronika.* **21**, pp. 9-21

ILYINSKY, A.S., and REPIN, V.M. (1975): 'On a method of integral equations in the problems of diffraction by periodic structures'. In: *Vychislitel'nye Metody i Programmirovanie. Vipusk 24.*, pp. 249-262 (MGU).

ILYINSKY, A.S., and SHESTOPALOV, Y.V. (1989): 'Applications of the methods of spectral theory in the problems of wave propagation' (MGU, Moscow)

ILYINSKY, A.S., SLEPYAN, G.YA., and SLEPYAN, A.YA. (1993): 'Propagation, scattering and dissipation of electromagnetic waves' (Peter Peregrinus, London)

ILYINSKY, A.S., and SMIRNOV, YU.G. (1987): 'Mathematical modelling of the propagation of electromagnetic waves in a slotted transmission line', *Zhurnal Vychiclitel'noi Matematiki i Matematicheskoi Fiziki.* **27**, pp. 163-170

ILYINSKY, A.S., and SVESHNIKOV, A.G. (1977): 'Numerical methods in the problems of diffraction by inhomogeneous periodical structures'. In: Prikladnaya Elektrodinamika. Vipusk 1., pp. 51-93 (Nauka).

IVANOV, V.P. (1968): 'Diffraction of electromagnetic waves by two bodies' (Nauka i Tekhnika, Minsk)

IVANOV, V.P. (1970): 'Solution to the problem of plane wave diffraction by a periodic grating', *Zhurnal Vychiclitel'noi Matematiki i Matematicheskoi Fiziki.* **10**, pp. 673-684

JULL, E.V., and EBBESON, G.R. (1977): 'The reduction of interference from large reflecting surfaces', *IEEE Trans.* **AP-25**, pp. 565-570

KANTOROVICH, L.V., and AKILOV, G.P. (1977): 'Functional analysis' (Nauka, Moscow)

KAPITSA, P.L. (1965): 'Transformation of H-waves to E-waves'. In: *Elektronika Bolshikh Moshchnostei. Vipusk 4*, pp. 7-52 (Nauka).

KATO, T. (1966): 'Perturbation theory for linear operators' (Springer)

KATZENELENBAUM, B.Z. (1955): 'Perturbation of electromagnetic field by small deformations of metal surfaces', *Zhurnal Tekhnicheskoi Fiziki.* **25**, pp. 546-549

KATZENELENBAUM, B.Z. (1961): 'Theory of waveguides with slowly varying parameters' (AN SSSR, Moscow)

KELDYSH, M.V. (1971): 'On the completeness of eigenfunctions of some classes of non-self-adjoint linear operators', *Uspekhi Matematicheskikh Nauk.* **26**, pp. 15-41

KEVANISHVILI, G.S. (1978): 'On a theory of the waveguide T-junction', *Izvestiya VUzaved.–Radiofizika.* **21**, pp. 1669-1674

KEVANISHVILI, G.S., and TZAGAREYSHVILI, O.P. (1977): 'The orthogonalization method in the theory of diffraction by gratings made of cylinders', *Radiotekhnika i Elektronika.* **21**, pp. 498-506

KHRUSLOV, E.Y. (1968): 'On resonance phenomena in one diffraction problem'. In: *Teoriya Funkcii, Funkcionalny Analis i Ego Prilozheniya. Vipusk 6*, pp. 111-129 (AN SSSR).

KIRILENKO, A.A. (1978): 'On the main characteristics and physical nature of the resonances at "locked" modes', *Doklady AN USSR. Ser. A.* (12), pp. 1121-1125

KIRILENKO, A.A. (1980): 'Theory and applications of semi-inversion method for internal problems of electrodynamics'. *Abstract of doctoral thesis* (KhGU, Khar'kov)

KIRILENKO, A.A., LITVINOV, V.R., and RUD', L.A. (1979): 'A truncated break of a rectangular waveguide in the H-plane', *Radiotekhnika i Elektronika.* **24**, pp. 1043-1050

KIRILENKO, A.A., and MASALOV, S.A. (1972): 'The H-polarized waves diffraction by a jalousie-type strip grating', *Izvestiya VUzaved.–Radiofizika.* **15**, pp. 83-97

KIRILENKO, A.A., and MASALOV, S.A. (1982): 'On calculation of some types of infinite products', *Radiotekhnika.* **32**, pp. 155-161

KIRILENKO, A.A., MASALOV, S.A., and SERGIENKO, YU.I. (1980): 'Analysis and opimization of waveguide-type polarization gratings'. In: *Prikladnaya Elektrodinamika. Vipusk 4.*, pp. 29-68 (Nauka).

KIRILENKO, A.A., and RUD', L.A. (1978): 'A symmetric break of a rectangular waveguide as a wave transformator', *Izvestiya VUzaved.– Radiofizika.* **21**, pp. 1379-1380

KIRILENKO, A.A., and RUD', L.A. (1980): 'A waveguide turn with coordinating cuts', *Radiotekhnika i Elektronika.* **25**, pp. 1359-1369

KIRILENKO, A.A., RUD', L.A., and YASHINA, N.P. (1978): 'On boundaries of application of heuristic methods in waveguide problems'. In: Volni i Difrakciya. Tesisi Dokladov 8 Vsesoyuznogo Simposiuma (Moskva, 1978). Tom 3, pp. 243-246 (AN SSSR).

KIRILENKO, A.A., and SENKEVICH, S.L. (1986): 'New type of the high-Q-factor oscillations in open waveguides', *Pis'ma v Zhurnal Tekhnicheskoi Fiziki.* **12**, pp. 876-879

KIRILENKO, A.A., and SENKEVICH, S.L. (1985): 'Resonances at "locked" modes in gratings covered by dielectric', *Doklady AN USSR. Ser. A.* (12), pp. 51-54

KIRILENKO, A.A., and YASHINA, N.P. (1980): 'On the communication of the resonance phenomena at locked modes with excitation of quasi-eigenregimes of unclosed volumes', *Pis'ma v Zhurnal Tekhnicheskoi Fiziki.* **6**, pp. 1512-1516

KIRILENKO, A.A., and YASHINA, N.P. (1982): 'Resonance phenomena in a sharp widening of a circular waveguide', *Radiotekhnika i Elektronika.* **27**, pp. 2140-2147

KOROBKIN, V.A., and OSINCEV, V.V. (1985): 'Eigenoscillations of a dielectric resonator in a waveguide junction', *Radiotekhnika i Elektronika.* **30**, pp. 417-421

KOSHPARENOK, V.N., MELEZHIK, P.N., and SHESTOPALOV, V.P. (1979a): 'Electromagnetic oscillations in a circular cylinder with a wide slot', *Radiotekhnika i Elektronika.* **24**, pp. 228-235

KOSHPARENOK, V.N., MELEZHIK, P.N., and SHESTOPALOV, V.P. (1979b): 'On intertype interaction of oscillations in a cylinder with one and two longitudinal slots', *Radiotekhnika i Elektronika.* **24**, pp. 2350-2353

KOSHPARENOK, V.N., MELEZHIK, P.N., and SHESTOPALOV, V.P. (1980a): 'Helmholtz resonators in electrodynamics', *Doklady AN SSSR.* **250**, pp. 344-346

KOSHPARENOK, V.N., MELEZHIK, P.N., POYEDINCHUK, A.E., and SHESTOPALOV, V.P. (1980b): 'A rigorous method for the study of electromagnetic interaction of several resonant scatterers', *Doklady AN SSSR.* **252**, pp. 316-319

KOSHPARENOK, V.N., MELEZHIK, P.N., POYEDINCHUK, A.E., and SHESTOPALOV, V.P. (1983a): 'Wave interaction in open resonators', *Doklady AN USSR. Ser. A.* (1), pp. 46-49

KOSHPARENOK, V.N., MELEZHIK, P.N., POYEDINCHUK, A.E., and SHESTOPALOV, V.P. (1983b): 'Effect of a two-component splitting of the lowest spectral frequency in the two-surface Helmholtz resonator', *Zhurnal Tekhnicheskoi Fiziki.* **53**, pp. 2396-2397

KOSHPARENOK, V.N., MELEZHIK, P.N., POYEDINCHUK, A.E., and SHESTOPALOV, V.P. (1984a): 'Interaction of waves in open resonators', *Doklady AN SSSR.* **279**, pp. 1114-1118

KOSHPARENOK, V.N., MELEZHIK, P.N., POYEDINCHUK, A.E., and SHESTOPALOV, V.P. (1984b): 'Effect of a sharp variation of the Q-factor of open electrodynamic structures with internal inhomogeneities', *Pis'ma v Zhurnal Tekhnicheskoi Fiziki.* **10**, pp. 827-830

KOSHPARENOK, V.N., MELEZHIK, P.N., POYEDINCHUK, A.E., and SHESTOPALOV, V.P. (1985): 'Spectral theory of open two-dimensional resonators with dielectric inclusions', *Zhurnal Vychiclitel'noi Matematiki i Matematicheskoi Fiziki.* **25**, pp. 562-573

KRASNUSHKIN, P.E. (1947): 'A method of normal waves applied for plane-layered media', *Doklady AN SSSR.* **56**, pp. 687-690

KREIN, M.G. (1971): 'Linear equations in Banach spaces' (Nauka, Moscow)

KRUTIN, Y.I., and SIRENKO, YU.I. (1983): 'Solution to the problem of diffraction by sew-type absorbing surfaces having the form of an echelette' (AN USSR, Institut Radiofisiki i Elektroniki, preprint 235, Khar'kov)

KUPRADZE, V.D. (1967): 'On approximate solution to the problems of mathematical physics', *Uspekhi Matematicheskikh Nauk.* **22**, pp. 59-134

KUSAIKIN, A.A., and SIRENKO, YU.K. (1985): 'Effect of a quasi-complete non-mirror reflection with big telescopic coefficient', *Zhurnal Tekhnicheskoi Fiziki.* **55**, pp. 1241-1242

LADYZHENSKAYA, O.N., and URALTZEVA, N.A. (1964): 'Linear and quasilinear elliptic equations' (Nauka, Moscow)

LAX, P., and PHILLIPS, R.S. (1967): 'Scattering theory' (Academic Press)

LAZUTKIN, V.F. (1968): 'The formula for eigenfrequencies of a nonconfocal resonator with cylindrical mirrors taking into account aberration of mirrors', *Optica i Spectroscopia.* **24**, pp. 453-454

LAZUTKIN, V.F. (1981): 'Diffraction losses in open resonators. Geometro-optical approach', *Doklady AN SSSR.* **285**, pp. 1089-1091

LAZUTKIN, V.F.and TERMAK, D.Y. (1981): 'On a number of the "bouncing ball" quasi-modes'. In: Matematicheskie Voprosi Teorii Rasprostraneniya Voln, pp. 172-182 (Nauka).

LEVIN, L. (1951): 'Modern theory of waveguides' (Izdatelstvo Inostrannoi Literatury, Moscow)

LEVIN, L. (1958): 'A physical interpretation of impedance for rectangular waveguide', *Proc. Phys. Soc.* **1**, pp. 868-870

LIONS, J., and MAGENES, E. (1968): 'Problemes aux limites non homogenes et applications' (I. Dunod)

LISANOV, Y.P. (1958): 'Scattering by periodic rough surfaces', *Akusticheskii Zhurnal.* **4**, pp. 3-12

LORCH, E.R. (1962): 'Spectral theory' (Oxford University Press)

LOZHECHKO, V.V., and SHESTOPALOV, Y.V. (1994a): 'Properties of the spectrum of fundamental frequencies of a class of open cylindrical resonators', *Vestnik MGU, Seriya 3.* **35**, pp. 23-34

LOZHECHKO, V.V., and SHESTOPALOV, Y.V. (1994b): 'Methods of solution to the problems of direct and inverse scattering by a family of open domains', *Proceedings of the 1994 Journees Internationales de Nice sur les Antennes (JINA 94), France, November 7-10, 1994*, pp. 22-26

LOZHECHKO, V.V., and SHESTOPALOV, Y.V. (1995): 'On problems of excitation of open cylindrical resonators', *Zhurnal Vychiclitel'noi Matematiki i Matematichcskoi Fiziki.* **35**, pp. 71-83

LUBICH, Y.I. (1966): 'Classical and local Laplace's transform in an abstract Cauchy problem', *Uspekhi Matematicheskikh Nauk.* **21**, pp. 3–51

MARCHENKO, V.A., and KHRUSLOV, E.Y. (1972): 'Boundary-value problems in the domains with a fine-grained boundary' (Naukova Dumka, Kiev)

MARKUS, A.S., and MATZAEV, V.I. (1975): 'On the spectral theory of the operator pencils', *Funkcionalny Analis i Ego Prilozheniya.* **9**, pp. 76-77

MASALOV, S.A. (1977): 'Resonant scattering of light by an echelette', *Ukrainskii Fizicheskii Zhurnal.* **22**, pp. 1497-1501

MASALOV, S.A. (1981): 'The semi-inversion method and infinite systems of equations in some problems of waves diffraction', *Zhurnal Vychiclitel'noi Matematiki i Matematicheskoi Fiziki.* **21**, pp. 80-88

MASALOV, S.A., and SIRENKO, YU.K. (1980): 'Excitation of reflection gratings by a plane wave', *Izvestiya VUzaved.–Radiofizika.* **23**, pp. 479-487

MASALOV, S.A., SIRENKO, YU.K., and SHESTOPALOV, V.P. (1977): 'On applicability of the truncation method for some infinite systems of equations', *Doklady AN USSR. Ser. A.* (6), pp. 539-540

MASALOV, S.A., SIRENKO, YU.K., and SHESTOPALOV, V.P. (1978a): 'Solution to the problem of the plane wave diffraction by gratings with complicated structure of the period', *Radiotekhnika i Elektronika.* **23**, pp. 481-487

MASALOV, S.A., SIRENKO, YU.K., and SHESTOPALOV, V.P. (1978b): 'Anomalous scattering of waves caused by period's break in diffraction gratings', *Pis'ma v Zhurnal Tekhnicheskoi Fiziki.* **4**, pp. 286-290

MASALOV, S.A., SIRENKO, YU.K., and SHESTOPALOV, V.P. (1980): 'Conditions for the development of the Maluzhinec effect in gratings', *Pis'ma v Zhurnal Tekhnicheskoi Fiziki.* **6**, pp. 998-1001

MASALOV, S.A., and SOLOGUB, V.G. (1970): 'A rigorous solution to the problem of electromagnetic waves diffraction by a strip structure', *Zhurnal Vychiclitel'noi Matematiki i Matematicheskoi Fiziki.* **10**, pp. 639-715

MEIXNER, J. (1972): 'The behaviour of electromagnetic fields at edges', *IEEE Trans.* **AP-20**, pp. 442-446

MELEZHIK, P.N. (1980): 'Spectral characteristics of coupled open resonant objects', *Doklady AN USSR. Ser. A.* (2), pp. 65-69

MELEZHIK, P.N., KOSHPARENOK, V.N., POYEDINCHUK, A.E., and SHESTOPALOV, V.P. (1987): 'Spectral characteristics of two-mirror open resonators', *Doklady AN USSR. Ser. A.* (8), pp. 57-61

MELEZHIK, P.N., KOSHPARENOK, V.N., POYEDINCHUK, A.E., and SHESTOPALOV, V.P. (1988): 'On analytical nature of the phenomenon of intertype interaction of eigenoscillations', *Doklady AN SSSR.* **300**, pp. 1336-1339

MILNOR, J. (1971): 'Singular points of complex hypersurfaces' (Mir, Moscow)

MITTRA, R., and LEE, S.W. (1971): 'Analytical techniques in the theory of guided waves' (Macmillan)

MORSE, F.M., and FESCHBACH, G. (1958): 'Methods of theoretical physics' (Izdatelstvo Inostrannoi Literatury, Moscow)

MOYZHES, B.Y. (1955): 'Electrodynamic homogenized boundary-value conditions for metal gratings', *Zhurnal Tekhnicheskoi Fiziki.* **25**, pp. 158-166

MURAVEJ, L.A. (1978a): 'Analytic continuation of Green's functions of external boundary-value problems for two-dimensional Helmholtz equation', *Matematicheskii Sbornik.* **105**, pp. 63-108

MURAVEJ, L.A. (1978b): 'On asymptotic behaviour of solutions to external boundary-value problems for the wave equation with two space variables for big values of time', *Matematicheskii Sbornik.* **107**, pp. 84-103

MUSKHELISHVILI, N.I. (1962): 'Singular integral equations' (Nauka, Moscow)

NEFEDOV, E.I. (1979): 'Diffraction of electromagnetic waves by dielectric structures' (Nauka, Moscow)

NEFEDOV, E.I. (1982): 'Open coaxial resonance structures' (Nauka, Moscow)

NEFEDOV, E.I., and SIVOV, A.N. (1977): 'Electrodynamics of periodic structures' (Nauka, Moscow)

NIKOLSKY, V.V. (1967): 'Variational methods for the internal problems of electrodynamics' (Nauka, Moscow)

NIKOLSKY, V.V. (1977): 'Projectional methods in electrodynamics'. In: *Prikladnaya Elektrodinamika. Vipusk 1*, pp. 4-39 (Nauka).

NIKOLSKY, V.V. (1978): 'A decomposing approach for the problems of mathematical physics', *Doklady AN SSSR.* **243**, pp. 1426-1430

PETRUSHIN, A.A., and TRETYAKOVA, S.S. (1983): 'Dielectric plate in an open resonator'. In: *Radiotekhnika. Vipusk 31*, pp. 118-124 (KhGU).

POCHANINA, I.E., SHESTOPALOV, V.P., and YASHINA, N.P. (1989a): 'Intertype oscillations of open waveguide resonators', *Izvestiya VUzaved.-Radiofizika.* **32**, pp. 1000-1008

POCHANINA, I.E., TUCHKIN, YU.A., SHESTOPALOV, V.P., and YASHINA, N.P. (1989b): 'On intertaction of free oscillations in open waveguide resonators' (AN USSR, Institut Radiofisiki i Elektroniki, preprint 381, Khar'kov)

POCHANINA, I.E., SHESTOPALOV, V.P., and YASHINA, N.P. (1991): 'Interaction and degeneration of free oscillations of open waveguide resonators', *Doklady AN SSSR.* **320**, pp. 90-95

PODOLSKY, E.I. (1965): 'Justification of the short-wave asymptotics for the problem of a plane wave diffraction by a grating made of metal strips'. In: *Radiotekhnika. Vipusk 1*, pp. 58-68 (KhGU).

POPOV, M.M. (1974): 'On diffraction losses of open resonators', *Optica i Spectroscopia.* **36**, pp. 561-566

POPOV, M.M. (1977): 'Calculation of diffraction losses of open resonators by the Fox and Lee approximation'. In: Teoriya Difrakcii i Rasprostraneniya Voln. Tom 2, pp. 218-220 (Nauka).

POVZNER, A.Y. (1953): 'On expansion of arbitrary functions in series of eigenfunctions', *Matematicheskii Sbornik.* **32**, pp. 109-156

POYEDINCHUK, A.E. (1983a): 'On a spectral theory of open two-dimensional resonators with dielectric inclusions', *Doklady AN USSR. Ser. A.* (8), pp. 46-48

POYEDINCHUK, A.E. (1983b): 'Point spectrum of one class of open cylindrical structures', *Doklady AN USSR. Ser. A.* (8), pp. 48-52

POYEDINCHUK, A.E., SHESTOPALOV, V.P., and YASHINA, N.P. (1986): 'On a spectral theory of coaxial-waveguide resonator', *Zhurnal Vychiclitel'noi Matematiki i Matematicheskoi Fiziki.* **26**, pp. 552-562

POYEDINCHUK, A.E., TUCHKIN, YU.A., and SHESTOPALOV, V.P. (1987): 'On regularization of the spectral problems of wave scattering by unclosed screens', *Doklady AN SSSR.* **295**, pp. 1328-1342

POYEDINCHUK, A.E., TUCHKIN, YU.A., and SHESTOPALOV, V.P. (1993): 'Diffraction by curved strips', *Trans. Inst. Elect. Eng. of Japan.* **113-A**, pp. 139-196

RADZIEVSKI, G.I. (1973): 'Double completeness of eigenvectors and associated vectors of some classes of operator-valued functions', *Funkcionalny Analis i Ego Prilozheniya.* **7**, pp. 84-85

RADZIEVSKI, G.I. (1982): 'On the completeness of the system of eigenvectors and associated vectors of the operator pencils', *Uspekhi Matematicheskikh Nauk.* **37**, pp. 81-147

RAYLEIGH, O.M. (1907a): 'Note on the remarkable case of diffraction spectra described by Professor Wood', *Phil. Mag.* **6**, pp. 60-65

RAYLEIGH, O.M. (1907b): 'On the dynamical theory of grating', *Proc. Royal Soc. A.* **79**, pp. 399-403

RAJEVSKY, S.B. (1970): 'Free oscillations of an elliptic open resonator with a dielectric cylinder', *Izvestiya VUzaved.-Radioelektronika.* **13**, pp. 987-982

REED, M., and SIMON, B. (1978): 'Methods of modern mathematical physics Vol. 2' (Academic Press)

REICHARDT, H. (1960): 'Ausstralungsbedingungen für die Wellengle-ichung', *An. Math. Semin. Univ. Hamburg.* **24**, pp. 41-53

RIESZ, F., and NAGY, B. (1952): 'Lecons d'analyse fonctionnelle' (Gauthier Villars)

ROGER, A., and MAYSTRE, D. (1977): 'Quelques considerations numerique et theoretiques sur un probleme simple de jonction entre quides met-torigues infinment conducteures', *Rev. Phys. Appl.* **12**, pp. 1095-1103

RUD', L.A. (1984): 'Diffraction of waves by a T-junction of rectangular waveguides in the H-plane', *Radiotekhnika i Elektronika.* **29**, pp. 1711-1719

RUD', L.A. (1985): 'On a nature of resonance phenomena in a T-junction of rectangular waveguides', *Zhurnal Tekhnicheskoi Fiziki.* **55**, pp. 1213-1215

RUD', L.A., SIRENKO, Y.I., SHESTOPALOV, V.P., and YASHINA, N.P. (1986): 'Qualitative characteristics of spectra of open waveguide res-onators' (AN USSR, Institut Radiofisiki i Elektroniki, preprint 316, Khar'kov)

SAMARSKY, A.A. (1971): 'Introduction to the theory of difference schemes' (Nauka, Moscow)

SAMARSKY, A.A., and TIKHONOV, A.N. (1947): 'On representation of the field in waveguides as a sum of the TE and TM fields', *Zhurnal Tekhnicheskoi Fiziki.* **18**, pp. 959-963

SANCHEZ-PALENCIA, E. (1980): 'Non-homogeneous media and vibration theory' (Springer)

SHABAT, B.V. (1976): 'Introduction to complex analysis' (Nauka, Moscow)

SHESTOPALOV, V.P. (1971): 'Method of Riemann-Hilbert problem in the theory of diffraction and propagation of electromagnetic waves' (KhGU, Khar'kov)

SHESTOPALOV, V.P. (1976): 'Diffraction electronics' (KhUG, Khar'kov)

SHESTOPALOV, V.P. (1983): 'Summation-type equations in the modern theory of diffraction' (Naukova Dumka, Kiev)

SHESTOPALOV, V.P. (1985): 'Physical foundations of millimeter and sub-millimeter technology Vol. 1. Open structures' (Naukova Dumka, Kiev)

SHESTOPALOV, V.P. (1987): 'Spectral theory and excitation of open struc-tures' (Naukova Dumka, Kiev)

SHESTOPALOV, V.P. (1990): 'Resonance and intertype oscillations', *Doklady AN SSSR*. **305**, pp. 1131-1134

SHESTOPALOV, V.P. (1992): 'Morse critical points of dispersion equations' (Naukova Dumka, Kiev)

SHESTOPALOV, V.P. (1993): 'Morse critical points of dispersion equations of open resonators', *Electromagnetics*. **13**, pp. 239-255

SHESTOPALOV, V.P. (1994): 'Control of local evolution processes near the Morse points', *Doklady RAN*. **335**, pp. 440-442

SHESTOPALOV, V.P., KIRILENKO, A.A., and MASALOV, S.A. (1984): 'Matrix convolution-type equations in the diffraction theory' (Naukova Dumka, Kiev)

SHESTOPALOV, V.P., KIRILENKO, A.A., MASALOV, S.A., and SIRENKO, YU.K. (1987): 'Resonance scattering of waves Vol. 1. Diffraction gratings' (Naukova Dumka, Kiev)

SHESTOPALOV, V.P., LITVINENKO, L.N., MASALOV, S.A., and SOLOGUB, V.G. (1973): 'Diffraction of waves by gratings' (KhUG, Khar'kov)

SHESTOPALOV, V.P., and SHCHERBAK, V.V. (1968): 'Matrix operators in diffraction problems', *Izvestiya VUzaved.-Radiofizika*. **9**, pp. 185-206

SHESTOPALOV, V.P., and SIRENKO, YU.K. (1989): 'Dynamic theory of gratings' (Naukova Dumka, Kiev)

SHESTOPALOV, V.P., TUCHKIN, YU.A., POYEDINCHUK, A.E., and SIRENKO, YU.K. (1994): 'New methods of the diffraction theory' (Naukova Dumka, Kiev)

SHESTOPALOV, YU.V. (1980a): 'Justification of the spectral method for calculating normal waves of transmission lines', *Differentsialnye Uravneniya*. **16**, pp. 1504-1512

SHESTOPALOV, YU.V. (1980b): 'Properties of the spectrum of a class of non-self-adjoint boundary value problems for the systems of Helmholtz equations', *Doklady AN SSSR*. **252**, pp. 1108-1111

SHESTOPALOV, YU.V. (1981): 'On a spectrum of a family of non-self-adjoint boundary-value problems for the systems of Helmholtz equations', *Zhurnal Vychiclitel'noi Matematiki i Matematicheskoi Fiziki*. **21**, pp. 1459-1470

SHESTOPALOV, YU.V. (1982): 'Sufficient conditions for the existence of the point spectrum of normal waves of slot lines', *Doklady AN SSSR.* **264**, pp. 1131-1135

SHESTOPALOV, YU.V. (1983a: 'Existence of the point spectrum of normal waves of transmission lines with several dielectric layers', *Doklady AN SSSR.* **273**, pp. 594-596

SHESTOPALOV, YU.V. (1983b: 'Discrete spectrum of normal waves of an open slot line', *Zhurnal Vychiclitel'noi Matematiki i Matematicheskoi Fiziki.* **23**, pp. 1392-1401

SHESTOPALOV, YU.V. (1986): 'Normal waves of slot lines formed by domains with arbitrary cross-sections', *Doklady AN SSSR.* **284**, pp. 841-845

SHESTOPALOV, YU.V. (1990): 'Application of the method of generalized potentials in some problems of wave propagation and diffraction', *Zhurnal Vychiclitel'noi Matematiki i Matematicheskoi Fiziki.* **30**, pp. 1081-1092

SHESTOPALOV, YU.V. (1991a): 'On the theory of cylindrical resonators', *Math. Methods Appl. Sciences.* **14**, pp. 335-375

SHESTOPALOV, YU.V. (1991b): 'Fundamental frequencies of the parallel-plane resonator with a hole'. In: *Proceedings of the 3rd international school-seminar MMET-91*, pp. 113-126 (Test-Radio).

SHESTOPALOV, YU.V. (1992): 'Potential theory for Helmholtz operator and nonlinear eigenvalue problems'. In: KISHI, M. (Ed.): Potential theory, pp. 281-290 (Walter de Gruyter).

SHESTOPALOV, YU.V. (1993): 'Nonlinear eigenvalue problems in electrodynamics', *Electromagnetics.* **13**, pp. 5-18

SHESTOPALOV, YU.V. (1994): 'Calculation of open slotted structures with layered dielectric filling'. In: Digest of international conference on millimeter and submillimeter waves and applications, January, 1994, San Diego, California, USA, pp. 121-126.

SHTEINSHLEYGER, V.B. (1955): 'Mode interaction effects in electromagnetic resonators' (Oborongiz, Moscow)

SIRENKO, YU.K. (1978): 'Some mathematical questions in the problems of wave diffraction by waveguide-type gratings' (AN USSR, Institut Radiofisiki i Elektroniki, preprint 103, Khar'kov)

SIRENKO, YU.K. (1983): 'On the justification of the method of semi-inversion of matrix operators in the problems of waves diffraction', *Zhurnal Vychiclitel'noi Matematiki i Matematicheskoi Fiziki.* **23**, pp. 1381-1391

SIRENKO, YU.K. (1986): 'Analytical continuation of diffraction problems and threshold phenomena in electrodynamics', *Doklady AN USSR. Ser. A.* (8), pp. 65-68

SIRENKO, YU.K., and SHESTOPALOV, V.P. (1982): 'Rigorous theory of the scattering of waves by an echelette-grating with absorbing sides', *Doklady AN SSSR.* **263**, pp. 851-854

SIRENKO, YU.K., and SHESTOPALOV, V.P. (1985): 'On a uniqueness of solutions to spectral problems for gratings periodic in one dimension', *Doklady AN SSSR.* **285**, pp. 325-338

SIRENKO, YU.K., and SHESTOPALOV, V.P. (1986): 'Point spectrum of holographic gratings periodic in one dimension', *Doklady AN USSR. Ser. A.* (1), pp. 85-88

SIRENKO, YU.K., and SHESTOPALOV, V.P. (1987a): 'Free oscillations of electromagnetic field in gratings periodic in one dimension', *Zhurnal Vychiclitel'noi Matematiki i Matematicheskoi Fiziki.* **27**, pp. 262-271

SIRENKO, YU.K., and SHESTOPALOV, V.P. (1987b): 'On existence of free field oscillations in gratings periodic in one dimension', *Doklady AN SSSR.* **293**, pp. 87-90

SIRENKO, YU.K., and SHESTOPALOV, V.P. (1987c): 'On separation of physical solutions in the problems of the waves diffraction theory for gratings periodic in one dimension', *Doklady AN SSSR.* **297**, pp. 1346-1350

SIRENKO, YU.K., SHESTOPALOV, V.P., and YASHINA, N.P. (1987): 'Free oscillations in a coaxial waveguide resonator', *Radiotekhnika i Elektronika.* **32**, pp. 535-544

SIRENKO, YU.K., SHESTOPALOV, V.P., and YATSIK, V.V. (1985a): 'Elements of the spectral theory of gratings' (AN USSR, Institut Radiofisiki i Elektroniki, preprint 266, Khar'kov)

SIRENKO, YU.K., SHESTOPALOV, V.P., and YATSIK, V.V. (1985b): 'Algorithms for calculation of a quasi-stationary state of electromagnetic field in waveguide-type gratings', *Doklady AN USSR. Ser. A.* (9), pp. 60-64

SIRENKO, YU.K., SHESTOPALOV, V.P., and YATSIK, V.V. (1988): 'Free field oscillations and anomalous regimes of the scattering of waves by gratings periodic in one dimension', *Izvestiya VUzaved.–Radiofizika.* **31**, pp. 312-320

SIRENKO, YU.K., SHESTOPALOV, V.P., and YATSIK, V.V. (1990): 'Anomalous eigenregimes, intertype oscillations and waves of open periodic resonators and waveguides', *Radiotekhnika i Elektronika.* **35**, pp. 2507-2518

SIVOV, A.N. (1961): 'Excitation of a plane grating by a plane electromagnetic wave', *Radiotekhnika i Elektronika.* **6**, pp. 58-66

SLAVYANOV, S.YU. (1973): 'On the theory of open resonators', *Zhurnal Eksperimental'noi i Teoreticheskoi Fiziki.* **64**, pp. 785-796

SMIRNOV, V.I. (1981): 'Foundations of calculus Vol. 4' (Nauka, Moscow)

SMIRNOV, YU.G. (1987): 'On the completeness of the system of eigenwaves and associated waves of the waveguides with irregular boundary', *Doklady AN SSSR.* **297**, pp. 829-832

SMIRNOV, YU.G. (1990): 'Application of the method of operator pencils for the problem of normal waves in a partially filled waveguide', *Doklady AN SSSR.* **312**, pp. 597-599

SMIRNOV, YU.G. (1991): 'Method of operator pencils in the transmission boundary-value problems for a system of elliptic equations', *Differentsialnye Uravneniya.* **27**, pp. 140-147

SOBOLEV, S.L. (1966): 'Equations of mathematical physics' (Izdatelstvo Fiziko-Matematicheskoi Literatury, Moscow)

SOLOGUB, V.G. (1967): 'Oblique excitation of the periodic grating formed by rectangular bars by the H-polarized plane wave'. In: *Radiotekhnika. Vipusk 2*, pp. 61-68 (KhGU).

SOLOGUB, V.G. (1972): 'Diffraction of plane wave by strip grating in case of short wavelengths', *Zhurnal Vychiclitel'noi Matematiki i Matematicheskoi Fiziki.* **12**, pp. 974-989

SOLOGUB, V.G., and SHESTOPALOV, V.P. (1968): 'Resonance phenomena in the plane H-polarized wave diffraction on the grating formed by metallic bars', *Zhurnal Vychiclitel'noi Matematiki i Matematicheskoi Fiziki.* **38**, pp. 1505-1520

SOLOGUB, V.G., SHESTOPALOV, V.P., and POLOVNIKOV, G.G. (1967): 'Diffraction of electromagnetic waves by metal gratings with narrow strips', *Zhurnal Tekhnicheskoi Fiziki.* **37**, pp. 666-679

STREIFER, V., and GAMO, H. (1966): 'Application of Schmidt expansion for the analysis of oscillations in an optical resonator'. In: 'Quasioptica', pp. 236-244 (Mir).

SUKHININ, S.V. (1981): 'On a discreteness of fundamental frequencies of open acoustic resonators', *Dinamika Sploshnoi Sredi.* **49**, pp. 157-163

SUKHININ, S.V. (1984): 'Qualitative questions of the theory of scattering by periodic cylindrical obstacles', *Dinamika Sploshnoi Sredi.* **67**, pp. 118-134

SVESHNIKOV, A.G. (1950): 'On the radiation principle', *Doklady AN SSSR.* **75**, pp. 137-141

SVESHNIKOV, A.G. (1951): 'The limiting absorption principle for a waveguide', *Doklady AN SSSR.* **80**, pp. 345-347

SVESHNIKOV, A.G., and ILYINSKY, A.S. (1968): 'Methods of the study of irregular waveguides', *Zhurnal Vychiclitel'noi Matematiki i Matematicheskoi Fiziki.* **8**, pp. 363-373

SVISHCHEV, YU.V., TUCHKIN, YU.A., and SHESTOPALOV, V.P. (1990): 'On resonance readjustment of oscillations in open resonator with spherical mirrors', *Doklady AN SSSR.* **312**, pp. 1111-1114

TARAPOV, I.E. (1965): 'The problem of diffraction by a grating of arbitrary profiles', *Zhurnal Vychiclitel'noi Matematiki i Matematicheskoi Fiziki.* **5**, pp. 888-893

TIKHONOV, A.N., and SAMARSKY, A.A. (1978): 'Equations of mathematical physics' (Nauka, Moscow)

TIKHONOV, A.N., and SVESHNIKOV, A.G. (1966): 'Methods of mathematical physics' (Nauka, Moscow)

TITCHMARSH, E.C. (1946): 'Eigenfunction expansions associated with second-order differential equations Vol. 2' (Clarendon Press)

TUCHKIN, YU.A. (1985): 'Scattering of waves by arbitrary unclosed cylindrical screens in the case of the Dirichlet condition', *Doklady AH SSSR.* **285**, pp. 1370-1373

TUCHKIN, YU.A., and SHESTOPALOV, V.P. (1982): 'On a model class of the boundary-value problems of electrodynamics', *Differentsialnye Uravneniya.* **18**, pp. 663-673

TUCHKIN, YU.A., and SHESTOPALOV, V.P. (1985): 'Scattering of waves by a finite system of arbitrary unclosed cylindrical screens', *Doklady AN SSSR.* **284**, pp. 1008-1009

TWERSKY, V. (1956): 'On the scattering of waves by an infinite grating', *IEEE Trans.* **AP-4**, pp. 330-345

VAINIKKO, G.M., and KARMA, O.O. (974): 'On the rate of convergence of approximate methods in eigenvalue problems with nonlinear dependence on a parameter', *Zhurnal Vychiclitel'noi Matematiki i Matematicheskoi Fiziki.* **14**, pp. 1393-1408

VASILJEV, E.N., and MATERIKOVA, L.B. (1965): 'Excitation of a dielectric body of rotation', *Zhurnal Tekhnicheskoi Fiziki.* **35**, pp. 1817-1824

VASILJEVA, T.I., KIRILENKO, A.A., and RUD', L.A. (1986): 'Resonance phenomena in waveguide dielectric inclusions with oblique boundaries', *Radiotekhnika i Elektronika.* **31**, pp. 466-473

VEKUA, I.N. (1953): 'On the completeness of the system of metaharmonic functions', *Doklady AH SSSR.* **22**, pp. 715-718

VELIEV, E.I. (1976): 'Effect of total long-wave reflection of plane waves by a diffraction grating made of cylindrical strips', *Pis'ma v Zhurnal Tekhnicheskoi Fiziki.* **6**, pp. 1327-1330

VELIEV, E.I., VEREMEY, V.V., and SHESTOPALOV, V.P. (1981): 'Diffraction of waves by a finite system of unclosed cylindrical screens', *Doklady AN USSR. Ser. A.* (5), pp. 62-66

VERBITSKY, I.L. (1976): 'On existence of the angle of incidence of the total transmission of a plane wave through a bar grating', *Pis'ma v Zhurnal Tekhnicheskoi Fiziki.* **2**, pp. 73-75

VLASOV, A.A., and TALANOV, V.N. (1965): 'On a selection of axial oscillations in open resonators', *Radiotekhnika i Elektronika.* **10**, pp. 1115-1159

VINOGRADOV, S.S., TUCHKIN, YU.A., and SHESTOPALOV, V.P. (1981): 'On the theory of waves scattering by unclosed spherical mirrors', *Doklady AH SSSR.* **256**, pp. 1346-1350

VOSKRESENSKY, G.V., and ZHUKOV, S.M. (1976): 'Radiation of a plane waveguide with a flange', *Radiotekhnika i Elektronika.* **21**, pp. 1390-1395

WATSON, G.H. (1949): 'Theory of Bessel functions' (Izdatelstvo Inostrannoi Literatury, Moscow)

WEINSTEIN, L.A. (1955): 'Diffraction of electromagnetic waves by gratings made of parallel conducting strips', *Zhurnal Tekhnicheskoi Fiziki.* **25**, pp. 847-852

WEINSTEIN, L.A. (1969a): 'Open resonators and open waveguides' (Golem Press)

WEINSTEIN, L.A. (1969b): 'The theory of diffraction and the factorization method' (Golem Press)

WEINSTEIN, L.A. (1988): 'Electromagnetic waves' (Radio i Svyaz', Moscow)

WHITMER, R.M. (1948): 'Fields in non-metallic waveguides', *Proc. IRE.* **36**, pp. 876-879

WILKINSON, J.H. (1965): 'The algebraic eigenvalue problem' (Oxford University Press)

WINEBERG, B.R. (1972): 'On operator's eigenfunctions corresponding to the poles of analytical continuation of a resolvent through continuous spectrum', *Matematicheskii Sbornik.* **37**, pp. 293-308

WINEBERG, B.R. (1982): 'Asymptotic methods in the equations of mathematical physics' (MGU, Moscow)

WOOD, R.W. (1902): 'On a remarkable case of uneven distribution of light in diffraction grating problem', *Phil. Mag.* **4**, pp. 396-406

WOOD, R.W. (1915): 'Anomalous diffraction grating', *Phys. Rev.* **48**, pp. 928-936

YASHINA, N.P. (1978): 'Electrodynamic properties of sharp widenings of circular waveguides' (AN USSR, Institut Radiofisiki i Elektroniki, preprint 98, Khar'kov)

YASHINA, N.P. (1979): 'On some electrodynamic properties of a cross-sectional break of a circular waveguide', *Radiotekhnika i Elektronika.* **24**, pp. 165-167

YATSIK V.V. (1988): 'Complete transformation of wave packages by the structures periodic in one dimension'. In: Radiofizika i Elektronika Millimetrovikh i Submillimetrovikh Voln, pp. 16-28 (KhGU).

ZAKHAROV, E.I., and PIMENOV, YU.V. (1982): 'Numerical analysis of radio-waves diffraction' (Radio i Svyaz', Moscow)

ZAKHAROV, I.I., and TROITSKY, V.N. (1970): 'On a calculation of an open resonator with absorbing film', *Radiotekhnika i Elektronika.* **15**, pp. 2544-2546

Index